MAKING IT COUNT

HISTORIES OF ECONOMIC LIFE

Jeremy Adelman, Sunil Amrith, and Emma Rothschild,
Series Editors

Making It Count: Statistics and Statecraft in the Early People's Republic of China
   by Arunabh Ghosh

Empires of Vice: Opium and the Rise of Prohibition across Southeast Asia by Diana
   Kim

Pirates and Publishers: A Social History of Copyright in Modern China
   by Fei-Hsien Wang

Sorting Out the Mixed Economy: The Rise and Fall of Welfare and Developmental
   States in the Americas by Amy C. Offner

Red Meat Republic: A Hoof-to-Table History of How Beef Changed America
   by Joshua Specht

The Promise and Peril of Credit: What a Forgotten Legend about Jews and Finance
   Tells Us about the Making of European Commercial Society
   by Francesca Trivellato

A People's Constitution: The Everyday Life of Law in the Indian Republic
   by Rohit De

A Local History of Global Capital: Jute and Peasant Life in the Bengal Delta
   by Tariq Omar Ali

STUDIES OF THE WEATHERHEAD EAST
ASIAN INSTITUTE, COLUMBIA UNIVERSITY

The Studies of the Weatherhead East Asian Institute of Columbia
University were inaugurated in 1962 to bring to a wider public the results
of significant new research on modern and contemporary East Asia.

Making It Count

STATISTICS AND STATECRAFT IN THE
EARLY PEOPLE'S REPUBLIC OF CHINA

ARUNABH GHOSH

PRINCETON UNIVERSITY PRESS
PRINCETON & OXFORD

Published by Princeton University Press
41 William Street, Princeton, New Jersey 08540
6 Oxford Street, Woodstock, Oxfordshire OX20 1TR

press.princeton.edu

ISBN 9780691179476
ISBN (e-book) 9780691199214

British Library Cataloging-in-Publication Data is available

Editorial: Eric Crahan and Pamela Weidman
Production Editorial: Debbie Tegarden
Jacket/Cover Design: C. Alvarez-Gaffin
Production: Jacquie Poirier
Publicity: Alyssa Sanford
Copyeditor: Jay Boggis and Nancy Hearst

Jacket Credit: Graph depicting increase in industrial and agricultural production,
1949–1958, from *Weida de shinian* (Ten Great Years).

Publication of this book has been aided by the Weatherhead East Asian Institutute,
at Columbia University

Studies of the Weatherhead East Asian Institute of Columbia Uniersity were
inaugurated in 1962 to bring a wider public the results of significant new research on
modern and contemporary East Asia

This book has been composed in Arno and Adobe Kaiti Std R

Printed on acid-free paper. ∞

Printed in the United States of America

10 9 8 7 6 5 4 3 2 1

For my parents,
Indira and Partha

CONTENTS

ILLUSTRATIONS AND TABLES

Plates

Note: Plates follow page 166.

Figures

Tables

ABBREVIATIONS

AM	arithmetic mean
ANOVA	analysis of variance
BFES	Beijing Finance and Economics School
BMA	Beijing Municipal Archives
CAJ	China Academic Journals
CCP	Chinese Communist Party
CSO	Central Statistical Office (India)
DBTJGZ	*(Dongbei) tongji gongzuo* 统计工作 (Statistical Work; *1950–1954;* Journal of the Northeast Statistics Bureau)
DSJ	*Zhonghua renmin gongheguo tongji dashiji, 1949–2009* 中华人民共和国统计大事记 (Chronicle of Events in Statistics in China)
GLF	Great Leap Forward
GM	geometric mean
HM	harmonic mean
ISAAFE	International Statistical Association for Asia and the Far East
ISEC	International Statistical Education Centre
ISI	Indian Statistical Institute
JHTJ	*Jihua yu tongji* 计划与统计 (Planning and Statistics; *1959–1960)*
MPS	material product system
NAI	National Archives of India (New Delhi)
NGO	nongovernmental organization
NMML	Nehru Memorial Museum and Library (New Delhi)

NSB Northeast Statistics Bureau

NSS National Sample Survey (India)

PCMMMA P.C. Mahalanobis Memorial Museum and Archives (Kolkata)

PLA People's Liberation Army

PRC People's Republic of China

RMRB *Renmin ribao* 人民日报 (People's Daily)

RWTG *Re'ai women the tongji gongzuo* 热爱我们的统计工作 (Ardently Love Our Statistical Work)

SAC State Administrative Council

SIFE Shanghai Institute of Finance and Economics 上海财政经济学院

SPC State Planning Commission

SSB State Statistics Bureau

STS Science and Technology Studies

TJGZ *Tongji gongzuo* 统计工作 (Statistical Work; 1957–1958)

TJGZTX *Tongji gongzuo tongxun* 统计工作通讯 (Statistical Work Bulletin; 1952–1956)

WJB Waijiao bu 外交部 (Foreign Ministry Archives; Beijing)

ACKNOWLEDGMENTS

I FIRST ENCOUNTERED A truly large dataset as a research assistant at the Urban Institute in Washington, DC. The Medicaid Statistical Information System (MSIS) for 2001 contained data on each of Medicaid's 60 million beneficiaries, spread across 230 odd spending categories. By today's big data standards, a matrix with 60 million rows and 230 columns would perhaps constitute, at best, midsize data. Back in 2003, it was big enough to require several hours of batch processing. Often enough, I would set a program running in the evening, so that the results would be ready to analyze the following morning. I would like to think that it was in those moments that the seeds of a wider interest in the history of data were sown. The book's eventual focus on statistics, however, also has a more prosaic logic. My original intention had been to study China's 1953 census. It was the absence in the archives of detailed materials on the census, and the discovery instead of clues to fundamental debates about the nature of statistics itself, that gradually pointed me to a fascinating period in the histories of both statistics and data.

By the time I completed the dissertation on which this book is based, I felt I had already incurred more debts than I could repay. As the book now goes to print, that debt has only multiplied. Madeleine Zelin and Eugenia Lean have been sources of unstinting encouragement, necessary criticism, and intellectual guidance ever since I walked into their seminars at Columbia in 2005. Before that, Paul Smith at Haverford opened the door to the worlds of Chinese history. Graduate school at Columbia was a heady experience. Betsy Blackmar, Partha Chatterjee, Matt Connelly, Bob Hymes, Matt Jones, Lydia Liu, Bill McAllister, Ken Prewitt, and Carl Riskin, each in their own way and at different times, provided valuable feedback, guidance, and encouragement. Beyond Columbia, Jacob Eyferth has done more than I had any reason to expect. Today, it gives me tremendous pleasure to call them all my friends.

I was fortunate to receive feedback on the entire manuscript on two separate occasions. A postdoctoral fellowship at the Harvard Academy for

International and Area Studies gave me the opportunity to invite an eclectic mix of scholars for a book workshop. Loren Brandt, Matt Jones, Peter Perdue, Liz Perry, Ted Porter, Sigrid Schmalzer, and Asif Siddiqi made it an intensely rewarding and productive afternoon, pushing me to reframe the book in important and more ambitious ways. A couple of years later, my History Department colleagues, Bill Kirby, Charlie Maier, Terry Martin, and Emma Rothschild, read the revised and expanded manuscript, offering several crucial suggestions that substantially improved the book. The two anonymous readers for Princeton University Press provided trenchant comments that led to a much stronger final version. Jeremy Adelman, as the Academic Editor for "Histories of Economic Life," was the perfect shepherd, exercising a firm yet constructive hand.

Numerous colleagues at Harvard and beyond commented on parts of the manuscript or at talks and presentations that I delivered. Many others offered feedback or discussed points of common interest at conferences, workshops, and less formal occasions. Without their insights, probing questions, and invaluable suggestions, this book would be much poorer. They include Emily Baum, Sugata Bose, Felix Boecking, Andrea Bréard, Jeremy Brown, Timothy Cheek, Alex Cook, Henry Cowles, Rob Culp, Sasha Day, Dai Chaowu, Will Deringer, Neil Diamant, Jorge Dominguez, Prasenjit Duara, Manfred Elfstrom, Fa-ti Fan, Susan Greenhalgh, Emily Hill, Sheila Jasanoff, Jiang Lijing, Rebecca Karl, Elisabeth Köll, Daniel Koss, Liu Yanwen, David Luesink, Tong Lam, Casey Lurtz, Brian Lander, Fabio Lanza, Sean Lei, Elizabeth Lord, the late Rod MacFarquhar, Erez Manela, Yajun Mo, Manoranjan Mohanty, Thomas Mullaney, Rebecca Nedostup, Amy Offner, Jahnavi Phalkey, Ke Ren, Lukas Rieppel, Leon Rocha, Sigrid Schmalzer, Michael Schoenhals, Falguni Sen, Tansen Sen, Victor Seow, Amanda Shuman, Mindy Smith, Elena Songster, Hallam Stevens, Julia Strauss, Philip Thai, Madhavi Thampi, Malcolm Thompson, Karen Thornber, Glenn Tiffert, Eddy U, Nico Volland, Richard von Glahn, Rudolf Wagner, Zuoyue Wang, Jeff Wasserstrom, Benno Weiner, Jake Werner, Arne Westad, Yan Yunxiang, Yang Kuisong, Wen-hsin Yeh, Margherita Zanasi, Zhang Jishun, Zhang Jiucheng, and Zhang Ling. Although I have tried to attend to everyone's comments, any remaining flaws—of omission or commission—are mine alone.

In China and India, where most of the research for this book was conducted, I benefited from access to archives and libraries and from contact with helpful staff and local colleagues. In China, these include the Beijing Municipal Archives, the Foreign Ministry Archives, the Library and Archives of Tsinghua

University, the Resource Room of the Statistics Department at Renmin University, the library of the Institute of Economics at the Chinese Academy of Social Sciences, and the National Library of China. I am especially grateful to Professors Yuan Wei of Renmin University and Liu Beicheng of Tsinghua University for taking such an interest in my research and helping resolve what appeared at various times to be intractable problems. In India, I located important materials at the Nehru Memorial Museum and Library, the Indian Council of World Affairs, the Indian Statistical Institute, the Institute of Economic Growth, the National Archives of India, and the National Library. Oral history interviews provided an additional and important layer of substantiation for many of the arguments in this book. It is a matter of great sadness that many of the elderly gentlemen I interviewed in 2010 and 2011 are no longer with us. In the United States, where much of this book was written, I relied extensively upon resources at Columbia University Libraries, Columbia's Rare Book and Manuscript Library, and Harvard University Libraries.

This book would not have been possible without generous institutional and financial support. An International Traveling Fellowship from Columbia's Graduate School and an International Dissertation Research Fellowship from the Social Science Research Council made possible eighteen months of archival work in China and India. A Dissertation Completion Fellowship from the American Council of Learned Societies freed me from teaching during my last year at Columbia. The manuscript was revised and expanded over two wonderful years at the Harvard Academy for International and Area Studies (2014–2015; 2017–2018), where I was also able to begin work on new projects. The History Department, the Fairbank Center for Chinese Studies, and the larger China studies community at Harvard have provided a warm and intellectually stimulating home in which to complete the book. Special thanks to Bill Kirby, Emma Rothschild, Liz Perry, Mark Elliott, Sunil Amrith, Ian Miller, Dan Smail, Andy Gordon, and Michael Szonyi. The History Department also provided a generous publications subsidy. My editors at Princeton University Press, Brigitta van Rheinberg, Eric Crahan, Amanda Peery, Pamela Weidman, Debbie Tegarden, and Thalia Leaf have made the experience as pleasant and painless as I could have hoped for. I am also delighted that the book has been included as a Study of the Weatherhead East Asian Institute. My thanks to Ross Yelsey and Eugenia Lean. Hua Yang and Zhou Yun provided skillful research assistance. As copy-editors, Nancy Hearst and Jay Boggis cast their expert eyes over the manuscript, verifying elusive sources and wrangling with recalcitrant sentences.

Many friends have endured this project for nearly as long as I have. Daniel Asen, Sayaka Chatani, Anatoly Detwyler, Liza and Collin Lawrence, Peiting Li, Andy Liu, Meha Priyadarshini, Kristin Roebuck, Richard So, Brian Tsui, and Timothy Yang were present at the (somewhat nebulous) creation. Jennifer Altehenger, Hannah Barker, Kaustubh Chakraborty, Cyrus Chen, Divya Cherian, Rohit De, Manfred Elfstrom, Colm Fox, Sikha Ghosh, Gal Gvili, Toby Harper, Jeffrey Kahn, Abhishek Kaicker, Shoili Kanungo, Yumi Kim, Jenny Lah, Abhinav Madan, the late Jasbir Malik, Anjali Malik, Adhira Mangalagiri, Garrett McVaugh, Gaurav Pant, Surabhi Ranganathan, Aaron Scherb, Anand Taneja, Tal Unreich, Sören Urbansky, and Stacey van Vleet have been fast friends and important sources of support and sustenance. More locally, the warmth and camaraderie of friends in Boston has made these past several years an absolute joy. Joyita Bhaskar, Graham Chamness, Julia Chuang, Du Heng, Sanjay Krishnan, Brian Lander, Elizabeth Lord, Yajun Mo, David Mozina, Arijeet Pal, Steve Pieragastini, Teena Purohit, Aditya Sarkar, Victor Seow, Philip Thai, Paul Vierthaler, and Zhang Ling: thank you. A special thanks to Seema, who was with this book through its last stages. She bore with grace, wit, and patience, all my obsessions, even as she labored on her own book manuscript.

Passages in chapter 2 and chapter 3 appeared in "Lies, Damned Lies, and (Bourgeois) Statistics: Ascertaining Social Fact in Midcentury China and the Soviet Union," *Osiris* 33 (October 2018): 149–168. I thank the *Osiris* editors, Patrick McCray and Suman Seth, and the History of Science Society for permission to include them in the book. Chapter 7 is a modified version of "Accepting difference, seeking common ground: Sino-Indian statistical exchanges 1951–1959," *BJHS Themes* 1 (March 2016): 61–82. My thanks to the British Society for the History of Science (BSHS) for permission to include the revised version in the book.

I save family for last because in life they are first. My parents gave me the greatest gift of all—the confidence to pursue my interests and the equanimity to deal with the consequences. This book is dedicated to them. My sister Suparna has been a friend, counselor, and partner in all sorts of fun. That fun has only multiplied since Jensil, and now Ragini, entered her life and our family.

Frankfurt, Germany
8 November 2019

MAKING IT COUNT

1

INTRODUCTION

IN 1959 the State Statistics Bureau (SSB) of the People's Republic of China (PRC) compiled a volume entitled *Weida de shinian* (Ten great years). Part of nationwide celebrations to commemorate the tenth anniversary of the founding of the PRC, the volume declared that "an epic of world-shaking importance, forever worthy of being recalled," had been scripted. A smattering of text did little to distract from the substance of the book. Page after successive page, full of numbers, tables, and charts followed. A veritable barrage of statistical data, all corralled to provide indisputable proof that the Chinese people had indeed experienced "ten years of rebirth," "ten years of leaping progress in economy and culture."[1]

Statistics are rarely only about numbers and their truth claims. They exist at the crossroads where mathematical certainty encounters the messiness of quantifying and categorizing the inherently imprecise characteristics of human existence and activity. For many countries in the 1950s, and China is no exception, this encounter occurred against the backdrop of a postwar world of newly emerging postcolonial or postrevolutionary states and idealistic transnational institutions, all enamored of the positivistic promises of quantification. Imperatives to create accurate and scientific statistical systems as constituent parts of a technology of governance jostled with the political and ideological divides of capitalism and communism, even as relations between people and the state were being remolded, re-articulated, or fashioned anew.

---

1 SSB, *Weida de shinian*. In a 1983 article, Perkins ("Research on the Economy of the PRC," 347) characterized the 1950s, unlike the decades that followed, as a period when considerable data were available.

Typically, national statistical systems can be arranged along an axis whose extremes are populated by two idealized models: centralized or noncentralized. In a noncentralized system, a variety of agencies—central and local government organs, trade bodies, private institutions, research organizations, nongovernmental organizations (NGOs), and so forth—periodically collect and publish quantitative data on social and economic activities. The overall quality and comprehensiveness of the data rest on the number and diversity of the agencies collecting that data. When their density is high, the data they produce can represent a national whole. An example is the system that exists in the United States. At the other extreme is a centralized system in which a nationwide agency is responsible for standardization (of methods, concepts, and schedules), supervision and coordination (of public and private enterprises), and which has centralized control over the utilization and release of all national data. Centralized statistics are especially important in socialist states that rely on centrally planned economic growth. The former USSR is an obvious example. The case of the PRC after 1949 (at least during its first decade) is no different.[2]

The claims made in *Weida de shinian* are all the more remarkable when one considers the state of statistical activity in China in 1949. When Mao Zedong (1893–1976) strode up the ramparts of the Gate of Heavenly Peace in Beijing in October of that year and triumphantly declared the establishment of the PRC, the statistical apparatus of the country had largely been decimated. During the preceding four decades, starting with the collapse of the Qing empire in 1911, China had experienced warlordism, a Japanese invasion, a world war, and a debilitating civil war. Much to the chagrin of its director, Zhu Junyi (Jennings P. Chu, 1892–1963), the Nationalist government's central statistical agency commanded a mere 5,000 personnel on the eve of 1949 and, despite numerous attempts, had not been able to conduct a nationwide census.[3] What the Chinese Communist Party (CCP) won in 1949 was control over a fractured and withered state. For many CCP statisticians and economists, the long-term prospects of transforming the PRC into a true socialist utopia hinged, to a large degree, on being able to resolve this crisis of counting.

A Crisis of Counting

In its simplest form, the crisis of counting in the PRC was understood as a problem of building a centralized statistical system. In December 1950, Zhang

2 Li, *The Statistical System of Communist China.*
3 Chu, "The Independently Controlled Statistical System," 96.

Youuyu (1898–1992), vice mayor of Beijing, the first metropolitan area where the CCP formed a government, offered the following analysis:[4]

> Were there statistics in the past? No matter in liberated areas or in areas under the old regime, we cannot say there were no statistics, just that they were full of inadequacies. It is not that they did not value statistics; for example, in the liberated areas county and district committee bulletins did carry . . . reports and tables, but these materials in all likelihood were incomplete, inaccurate, and unsystematic, and therefore they could not serve as the basis [for anything]. As for the areas under the old regime, their numbers were even more unreliable since they are a product of formalism [形式主义; *xingshi zhuyi*].[5]

This was indeed a familiar criticism, in line with the basic imperatives of state-building, wherein expansion of state capacity is a central task for any government seeking to establish order after decades of strife and civil war. A decade later, the economist Li Choh-Ming's dismissal of the statistical infrastructure inherited by the CCP would largely echo Zhang Youyu's assessment:

> Since there was hardly any statistical system to speak of before 1949, did Peking manage to set one up that was actually workable? When did this happen and how did it develop? Where were official statistics produced and finalized? Were they used for planning purposes at different government levels? . . . What were the size and quality of the statistical work force?[6]

Within months of Zhang's analysis, however, a second, much more fundamental criticism of pre-1949 statistics was articulated by Li Fuchun (1900–1975), then a deputy head of the Central Finance and Economics Committee. This second critique did not waste time lamenting the lack of statistical data or institutions. After all, statistical infrastructure and activities could always be established where none or little existed. Instead, Li's critique called for a wholesale repudiation of existing statistical thought and practice:

---

4 As first vice mayor, Zhang Youyu was in charge of the city's day-to-day operations. Appointed shortly after the CCP took over Beijing, he was one of two vice mayors until 1957 (the other was Wu Han, head of the Beijing branch of the Democratic League). Peng Zhen was appointed mayor in February 1951, but Zhang claimed that for much of the 1950s it was he who really ran the city, not Peng Zhen. For more, see Mazur, "The United Front Redefined," 66–68.

5 BMA 002-020-000969: 2.

6 Li, "Statistics and Planning at the Hsien Level," 112.

In the past, China was a semi-colonial, semi-feudal country; strictly speaking, it did not possess any statistics [worth speaking of]. Statistics in old China was learned from the Anglo-American bourgeoisie. This kind of statistics cannot serve as our weapon; it is unsuitable for [the tasks of] managing and supervising the country . . . we need to build [a new] statistics for a New China. . . .[7]

According to this critique, the main problem with Anglo-American bourgeois statistics was that it served capitalists, whose sole purpose in turn was profit via the exploitation of labor. This argument would be developed and deployed during the rest of the decade by a range of interlocutors. An influential essay from the mid-1950s, for instance, made the case in the following way:

Bourgeois statistics exists in order to strengthen the exploitation of workers, in order to serve the interests of capitalists; it uses unscientific formalist mathematical doctrine to conceal the economic dangers of capitalism, whitewash class conflict, and deceive people. The viewpoints and methods of such statistical theory cannot meet the needs of national construction work and will directly endanger its progress.[8]

One year after Li's dismissal of pre-1949 statistics, Vice Premier Zhu De (1886–1976) noted that the establishment of a new comprehensive statistical system had already become an important task and anyone who lacked sufficient awareness of its significance was in error.[9] How this call to arms—to set up a new statistics for a New China—was answered is the principal subject of this book.

---

7 Li Fuchun, "Zhongyang renmin zhengfu zhengwuyuan caizheng jingji weiyuanhui Li Fuchun fuzhuren zai quanguo caijing tongji huiyi shang de zhishi" (Directive delivered by Li Fuchun, deputy head of the Central Finance and Economics Committee of the National Administrative Council, at the first National Finance and Statistics Meeting), in SSB, *Tongji gongzuo zhongyao wenjian huibian: Di yi juan*, 1–5.

8 Xu Qian and Liu Xin, "Guanyu zichan jieji tongji lilun (1955)," 28. The charge of formalism is used here to paint bourgeois mathematics as simply the manipulation of meaningless symbols, a focus on form over content. Such a definition appears broadly consistent with how formalism was deployed in arts and literature within the socialist bloc. To call something "formalist" was to label it elitist. It should be noted, however, that within the philosophy of mathematics, formalism is widely regarded as the investigation of systems of logic and it has its own specific genealogy.

9 Ibid., 28.

(Three) Modes of Counting

At the heart of the varied solutions attempted by Chinese statisticians was a contentious debate about the very nature of social reality and the place and efficacy of mathematical statistics—in particular, probability theory[10]—in ascertaining that reality. This debate played out against a backdrop populated by three divergent methodological approaches to statistics and statistical work. As a useful shorthand, let us label these approaches the Ethnographic, the Exhaustive, and the Stochastic. Each approach answered differently the question of how best to count and had implications for the types of data that were collected as well as for the methods used to collect and analyze that data. The resolution of the debate meant that for much of the decade it was the Exhaustive approach that dominated, but the Ethnographic and Stochastic approaches also enjoyed moments of contrasting prominence, especially toward the end of the 1950s.

From the perspective of PRC statisticians, the most indigenous among these approaches, because it could be traced to Mao's 1927 *Report on an Investigation of the Peasant Movement in Hunan* as well as to his later essays, such as "On Book Worship" and "On Practice," was the Ethnographic approach. As its label suggests, it relied on a method that placed the researcher in the middle of the people and the phenomena he was surveying. His personal presence on the ground, interacting in-depth with people, observing and recording phenomena first-hand, were deemed indispensable to his ability to understand the objective reality of a place and a situation. Such a "typical" or "paradigmatic" understanding could then be extrapolated to produce wider, more comprehensive knowledge of social, economic, or cultural trends. Direct experience was necessary because it alone was the source of the surveyor's authority. Readers will recognize this as a form of qualitative sampling, an important methodology that continues to undergird vast domains of social science and historical research today. It has a long history of use within statistical work as well.[11] The Maoist version will be introduced at the end of chapter 2, but we will encounter it again in greater detail in chapter 8, when it became the basis for the reformulation of statistical work during the Great Leap Forward (GLF) (1958–1962).

---

10 The branch of mathematics focusing on the study of random phenomena.

11 See, for instance, ch. 7 ("The Part for the Whole: Monographs or Representative Samplings") in Desrosières, *The Politics of Large Numbers.*

The most pervasive among the three approaches was the Exhaustive, because it was both the *de jure* and the *de facto* approach to statistics during much of the 1950s. Less dominant in subsequent decades, it nonetheless continued to serve as the basis of statistical theory and practice in China into the early 1980s. The Exhaustive approach was based on defining statistics as a social science, as opposed to a natural science. Most significant to this definitional distinction was the rejection of mathematical statistics, in particular probability theory and its attention to questions of randomness and chance. Instead, drawing direct inspiration from Soviet statistics, the resultant approach—socialist statistics—favored exhaustive enumeration through periodic complete counts. Although qualitative sampling was acknowledged as an ancillary method, its use was restricted to those instances where a complete count was inconvenient or impracticable. The dominance of exhaustive enumeration was, as we shall see, instrumental in the shaping of new bureaus, the designing of regimes of statistical work, and the training of personnel. It also generated tremendous incapacities—a country as large and as diverse as China was not easy to enumerate.

One of the consequences of the growing frustration with the Exhaustive approach was an openness, especially by late 1956, to the youngest of the three approaches—the Stochastic. Unlike qualitative sampling or the census method, which had been around in some form for millennia, the Stochastic approach was only a few decades old. It relied explicitly on recent advances in mathematical statistics and probability theory to promote what was in the 1950s a contentious but exciting new technology—large-scale random sampling. Compared to exhaustive enumeration, large-scale random sampling carried the promise of not only generating more accurate data but also of being both cheaper and faster. In their desire to learn more about its possible applications, the Chinese turned to a group of Indian statisticians who were at the forefront of international efforts to convince practitioners of the efficacy of this method.

Each of these approaches offered specific advantages, but each also had its limitations: the Ethnographic was easily biased; the Exhaustive was frequently inefficient and, in certain sectors (such as agriculture), impracticable; and the Stochastic was technically demanding and, given its novelty, still mired in theoretical and methodological controversy. No single method was a panacea, a fact that is as true today as it was in the 1950s. The uneven prominence the various methods enjoyed over the course of the decade also does not lend itself to neat temporal phases. Instead, such unevenness highlights the importance of the interplay between technical considerations and broader shifts in domestic and international politics. A more capacious approach, employing a judicious mix of all three, would quite possibly have allowed the Chinese state to have a better sense of its activities and achievements. But for most Chinese statisticians such a capacious approach remained elusive or downright theoretically unacceptable through much of the 1950s.

The Significance of Statistics

Abstract ideas about the nature of the world, whether defined by chance or certainty, have real world consequences.[12] Chinese deliberations over such questions and their engagement with the Ethnographic, Exhaustive, and Stochastic approaches during the 1950s exemplify some of those consequences. Unpacking these choices and tracing how statistics in its various forms—as a (social) science, as a profession, and as an activity—came to be formulated and practiced sheds light on fundamental questions germane to the histories of the People's Republic, statistics and data, and mid-century science.

My approach to these questions is directly shaped by the sources I was able to consult. These include unpublished documents, letters, institutional archives, memoirs, oral histories, and newspaper reports. They were, for the most part, produced by statisticians or statistical bureaus, and they focus on statistical activities. Such an internal perspective allows me to tell the story primarily from the inside out; that is, from the perspective of statisticians and statistics itself and not of political leaders, planners, or others with an interest in statistics, broadly construed. Nevertheless, the benefits of this perspective—insights into how social facts were understood and conceptualized—come with costs. I am less able, for instance, to delve into detail about how statistics were consumed, how they shaped the regime,[13] or about the nature

12 Scott, *Seeing Like a State.*

13 Take the case of statistics and planning. In the 1950s institutional hierarchy of China, the State Planning Commission (SPC) took precedence over the SSB. This hierarchy was replicated from Beijing—where the SPC quite literally operated in the same building, but on the floor directly above the SSB—down to the provincial and district committees. The SSB's task was to provide data and analysis to the SPC, but it was not expected to participate in the planning process. At the highest levels of the leadership, however, these tasks did indeed converge; SSB director Xue Muqiao (1904–2005) also served on the SPC for much of the 1950s. For the most part, however, the materials consulted here suggest that statisticians operated under and outside of the planning process. Accordingly, with the exception of a brief discussion in chapter 5, I do not engage in a sustained discussion of the economic planning process.

of the relationship between statistics and accounting.[14] To do justice to such questions would require a different book project, one that would entail perhaps a dozen or more detailed case studies. But such a book would still require the conceptual and substantive foundation provided in the pages that follow.

Histories of the People's Republic

As the first historical study of the development of statistics in Mao-era China, this book is a part of a recent renaissance of PRC history.[15] In the China field, 1949 long marked a boundary that historians rarely transgressed. The post-1949 years were almost exclusively the domain of political scientists, sociologists, economists, and anthropologists. But during the last fifteen years, no longer hostage to Cold War geo-politics and disciplinary or temporal boundaries, and encouraged by the increasing openness of archives, historians have offered new perspectives on the early PRC.[16] While some have facilitated a reassessment of 1949 as a rupture,[17] others have investigated aspects of the transition to Communist rule, exploring subjects such as marriage, gender relations, skill

14 Accounting (会计; *kuaiji*) is not discussed here independently but rather through the prism of statistics and statistical work. At the broadest level, the two can be differentiated according to both scope and scale. Accounting typically focuses on financial information, whereas the purview of statistics ranges across a much wider set of quantifiable activities and objects. Also, the scale of accounting is relatively limited, often operating at the level of the factory or collective farm. In Chinese discussions, each of these distinctions seems to be at work. Statistics was identified as a tool to investigate plan completion, the relationship between various productive elements of the economy or its various bureaus, and so on. Accounting, in contrast, focused on specific units to ascertain information, such as capital stocks, profits, yields, costs, and so on. These tasks naturally overlapped at the lowest levels of data collection, and it was common for the village or the factory accountant to double as the statistician. But as one proceeded higher up the chain of collection and collation, these tasks became more distinct. For representative discussions, see Anon., "Tongji, kuaiji he yewu"; Jiang Xinming, "Tongji yu kuaiji de guanxi."

15 The only extant book-length study on Chinese statistics in the 1950s is the contemporary institutional analysis in Li, *The Statistical System of Communist China*, which focuses primarily on questions of accuracy and reliability. Perkins, in Appendix A of *Market Control and Planning* also addresses these questions.

16 Strauss, "The History of the People's Republic," in particular the introduction. Indications are that the period of archival openness is coming to an end.

17 Among the earliest arguments against understanding 1949 as a moment of rupture is the work by Kirby, "Continuity and Change," who makes the case for continuities in economic planning in both Taiwan and the PRC. For more recent work that stresses continuity, see Bian, *The Making of the State Enterprise System*; and Mullaney, *Coming to Terms with the Nation*.

and rural industries, urban transformation, film, urban outcasts, the urban-rural divide, and much else.[18] Notable in this new scholarship is a focus on science, where historians are taking seriously the claims of China's socialist scientists to understand the era's scientific and state-building activities on their own terms.[19] Much of this work on PRC history is interesting because it asks new questions or approaches old questions with fresh materials, thereby offering a more finely grained sense of the period. This has also spurred the writing of PRC history from a transnational perspective, exploiting not only the newly available archival materials within the PRC but also archives and repositories the world over.[20]

Among the questions on which this book offers fresh perspectives is the nature of the early PRC state. For too long, our understanding of this question has been dominated by a focus on the campaign-style governance that was characteristic of the Mao era (1949–1976) as a whole. Mention of the 1950s thus evokes images of campaigns and movements, such as the Three and Five Antis (1951–1952); the purge of hidden counter-revolutionaries (1955); the Hundred Flowers (1956); the Anti-Rightist (1957); and many others. Exceptions to such campaign chronologies consist of two periods defined primarily by economic activity: the three years of economic recovery (1949–1952) and the First Five-year Plan (1953–1957). For certain topics, these campaign chronologies obscure more than they reveal, most obviously when it comes to issues about everyday life, but also to some extent about institution-building and knowledge-generation, which often have their own temporality.[21] The

18 Representative works include: Altehenger, *Legal Lessons*; Brown, *City Versus Countryside*; Brown and Johnson, *Maoism at the Grassroots*; Brown and Pickowicz, *Dilemmas of Victory*; De-Mare, *Mao's Cultural Army*; Diamant, *Revolutionizing the Family*; Eyferth, *Eating Rice from Bamboo Roots*; Gao, *The Communist Takeover of Hangzhou*; Hershatter, *The Gender of Memory*; Ho, *Curating Revolution*; Lü and Perry, *Danwei*; Smith, *Thought Reform*; and Strauss, "The History of the People's Republic."

19 "Focus: Science and Modern China"; Schmalzer, *The People's Peking Man*; Schmalzer, "Self-Reliant Science"; Schmalzer, *Red Revolution*; Fan, "Collective Monitoring"; Wang, "The Cold War and the Reshaping of Transnational Science"; Gross, *Farewell to the God of Plague*; and Hu, "Science, Technology, and Medicine."

20 For a recent dissertation that places PRC history, albeit of a slightly later period, in a transnational context, see Scarlett, "China After the Sino-Soviet Split." Also see, Ghosh and Urbansky, "China from Without."

21 Based on discussion of "campaign time" and how such received chronologies have a tenuous connection with women's memories of the decade, Hershatter, *The Gender of Memory*, points to the inadequacy of canonical periodization. In similar fashion, Eyferth, *Eating Rice from*

result is an emphasis on the informal and the ad hoc at the expense of the formal, the planned, and the personal. The new PRC scholarship, despite the numerous new horizons it has charted, retains many elements of the imbalance between the informal and the planned. For the various fresh perspectives that have been generated, we remain in the dark about aspects of the state's formal structure and the institutional ambitions of its functionaries. This book encourages us to acknowledge their significance.

A key aspect of that significance relates to issues about state ideology and state capacity.[22] Why did the Exhaustive approach dominate statistical work during the 1950s? What kinds of capacities and incapacities did such a choice generate? How did it affect the Chinese state's ability to collect and analyze data? Adapting James Scott, then, we may ask, what does it mean to "see like a socialist state?" As the chapters in Parts II and III show, the adoption of socialist statistics led to two distinct kinds of state incapacity: infrastructural and technoscientific. The first draws upon Michael Mann's ideas about the infrastructural power of the state and focuses on issues of personnel and training.[23] The second, inspired by Donald MacKenzie's work on financial markets, helps us recognize that throughout the 1950s the selection or rejection of specific statistical methods imposed limitations on both how and how fast data could be collected, reported, and analyzed.[24]

Attention to the vicissitudes of statistical debate and activity is especially relevant in considering the singular event that animates most people's imaginations when we juxtapose China, statistics, and the 1950s. One of the twentieth century's worst tragedies, the famine of 1959–1961 and the GLF (1958–1962), which largely caused it, form a teleological end-point in early PRC history, often constraining our ability to study the 1950s on its own terms. Rejecting this teleology makes it possible to place changes in statistical practices during the GLF within a longer trajectory of choices and deliberations. Such a perspective rejects the reductive idea that the GLF disaster was caused by the collapse of the statistical system. Instead, I show that the shifts in practice

---

Bamboo Roots, points to longer trends, such as the de-skilling of rural industries, which occurred during the first three decades of the PRC.

22 Among works that have approached related questions for the earlier Republican era, exemplary are Kirby, *Germany and Republican China*; Strauss, *Strong Institutions in Weak Polities*; and Lam, *A Passion for Facts*.

23 Mann, "The Autonomous Power of the State."

24 MacKenzie, *An Engine Not a Camera*; also see, Morgan, *The World in the Model*.

during those tumultuous years must be understood in the context not only of the immediate politics of the GLF but also as an ongoing and decade-long engagement with and critique of statistical theory and methods.

Taking the theory and practice of statistics seriously also helps to disentangle the ways in which data might appear to be manipulated or biased. There is a common perception today that China "jukes the stats." Most analyses of this phenomenon, in the popular press or in academic scholarship, focus on what I label "post-hoc manipulation," that is, on the possibility and the degree to which a statistical datum—such as GDP today, GVIAO in 1950s China—was manipulated after it was generated in order to conform to political compulsions.[25] Such analyses are undoubtedly crucial, and scholars have also explored contemporary institutional and structural issues in China's statistical work that might produce inaccurate data.[26] This book highlights a different process that can also result in data being skewed in specific ways; a process that is about first principles and not post-hoc manipulation. Chinese statisticians' initial assumptions about the nature of social reality generated path-dependencies that constrained the types of methods they could use and, in turn, affected the data they collected and the analyses they performed.[27]

Histories of Statistics and Data

Although historical writing on statistics and quantification has focused primarily on the early-modern and early-twentieth-century West, this book brings that history into the twentieth century, when states, multinational institutions, and private actors, regardless of their ideological hue, mobilized statistics on behalf of positivist social science, economic planning, and statecraft. In so doing, it challenges a central assumption in the field: the universal rise of probabilistic thinking and the attendant spread of probabilistic methods during the early-modern and modern eras. Central to this process has been what Ian Hacking has identified as "the taming of chance" and what Theodore

25 Holz, "The Quality of China's GDP Statistics"; Wallace, "Juking the Stats"; Clark, Pinkovskiy, and Sala-i-Martin, "China's GDP Growth." Obviously, this is not exclusively a Chinese problem. See, for instance, Coyle, *GDP*.

26 Holz, "China's Statistical System."

27 Travers, "Bias in China's Economic Statistics," makes a similar point in a discussion of sampling practices in China during the early 1980s. For both broader and narrower definitions of path dependency, see Pierson, "Increasing Returns."

Porter has described as "chance subdued by science."[28] To know something through numbers remains one of the most powerful ways of knowing in the modern world. Powerful not because such knowing is necessarily or always nearer the truth (were we to grant the singularity of such a thing), but powerful because numbers offer a tool of persuasion and a basis for rational, methodical, calibrated, and repeatable actions that remain unmatched. These characteristics make statistics (and quantification more broadly) an indispensable tool to adjudicate between competing political, administrative, and ideological agendas.[29] Such power has become all the more desirable as we have come to realize that common-sense understandings of the world are often erroneous.[30] It is for these reasons that statistics and quantification have gained such traction over the past several centuries.

Our current all-pervasive zeal for Big Data is symptomatic of this general impetus to quantify, but it has come at a time when the relationship between statistics and data appears to be at a crossroads. In an influential paper published nearly two decades ago, the statistician Leo Breiman spoke of two cultures within statistics, inference (which he called stochastic data modeling) and prediction (which he called algorithmic modeling), pointing out that theoretical statisticians work primarily on the former and data scientists are principally concerned with the latter.[31] Breiman called for statisticians to overcome their traditional reticence and to embrace algorithmic modeling as well. In this, he was probably anticipating statistics' possible future marginalization.

28 See, for instance, Daston, *Classical Probability*; Desrosières, *The Politics of Large Numbers*; Gigerenzer et al., *The Empire of Chance*; Hacking, "Biopower and the Avalanche"; Hacking, *The Taming of Chance*; Hacking, "Making Up People"; Krüger, Daston, and Heidelberger, *The Probabilistic Revolution*; Patriarca, *Numbers and Nationhood*; Poovey, *A History of the Modern Fact*; Porter, *The Rise of Statistical Thinking*; Porter, "Chance Subdued by Science"; Stigler, *The History of Statistics*; and Tooze, *Statistics and the German State*.

29 Porter, *Trust in Numbers*, inter alia 19, 123.

30 Watts, *Everything Is Obvious*.

31 Breiman, "Statistical Modeling." Put differently, this is the distinction between explaining "why" things happen (inference) and ascertaining "if" they will happen (prediction). The first is about establishing a causal mechanism, i.e., understanding how things work; the second is about greater knowledge regarding what will happen. The methods and tools for each end up being substantially different. As I have learned from discussions with colleagues in Science and Technology Studies, this distinction has interesting moral and philosophical implications, since rules in society are based on normative ideas. If a judge or a jury cannot sufficiently determine the motive for a particular action, how can they make a suitable judgment? My thanks to Martha Poon for this last insight. See also Grimmer, "We are All Social Scientists Now."

Indeed, in 2013 Andrew Gelman provocatively claimed that statistics was the least important part of data science.[32] But in a talk delivered the previous year, Gelman had noted that no quantitative analysis was possible without a strong grasp of two foundational statistical concepts: statistical significance and random sampling.[33] That statisticians are now fully engaged in responding to a disciplinary crisis has been recognized by David Donoho, whose influential paper at the Tukey Centennial Workshop in 2015 offered reflections on the recent histories of statistics and data science, and their possible futures.[34] Even more recently, in 2017, the science journal *Nature* carried short contributions by several eminent statisticians on how to "fix" statistics.[35]

Much of this hand-wringing is informed by recent leaps in data storage and computational capacity and leaves open the question of whether this is something fundamentally new. How do we understand and assess the impact of quantum leaps in capabilities? As the case of China in the 1950s demonstrates, enthusiasm for the transformative power of quantification is hardly new. Since the nineteenth century we have arguably experienced at least three major waves of quantitative positivism. The first was during the late nineteenth and early twentieth century, when the use of numbers to produce actionable knowledge in society received a major boost through the activities of figures such as Francis Galton, Karl Pearson, and Émile Durkheim.[36] The ethos—confidence in quantitative analysis—that drove Galton, Pearson, and their contemporaries gave rise to disciplines such as statistics, demography, and sociology.

The second major wave of quantitative positivism took place in the 1950s, the period of time that is the focus of this book. In chapter 3 I provide a more systematic treatment of the promise of postwar statistics. For now, I would like to stress that the belief that any problem could be diagnosed and remedied as long as enough data were collected pervaded not only the worlds of science,

32 Gelman, "Statistics is the *Least* Important Part."

33 Gelman, "Little Data."

34 Donoho, "50 Years of Data Science." John Tukey (1915–2000) was an influential statistician who made contributions across a vast range of areas. Tukey was also a philosopher of statistical practice. He distinguished exploratory data analysis from confirmatory data analysis, pointing out that statisticians devote too much attention to the latter, a view that put him in a small minority. For more on Tukey, see McCullagh, "John Wilder Tukey."

35 Leek et al., "Five Ways to Fix Statistics."

36 On Galton, see Kevles, *In the Name of Eugenics*; on Pearson, see Porter, *Karl Pearson*; and on Durkheim, see Pickering and Walford, *Durkheim's Suicide*.

social science, and governance but was a part of the *zeitgeist* itself. This faith, for that is what it was, is perhaps most evocatively captured in Isaac Asimov's Foundation Series of science-fiction novels.[37] Asimov wrote the stories in the 1940s, a time when new statistical techniques, such as operations research, large-scale random sampling, and decision theory, were fundamentally altering our ability to ascertain (social) fact and to engineer change. In the Foundation Series, Asimov introduces us to the "psychohistorian" Hari Seldon, who calculates that the Galactic Empire is in terminal decline and because of this the galaxy will enter an extended period of chaos. Seldon performs this analysis by using the science of psychohistory, a field that he developed from "a set of vague axioms" to "a profound statistical science."[38] A neologism coined by Asimov, psychohistory combined history, psychology, sociology, and mathematical statistics to make general predictions about the future of large masses of people: "The individual human being is unpredictable, but the reactions of human mobs, Seldon found, could be treated statistically. The larger the mob, the greater the accuracy that could be achieved."[39] Asimov was clearly applying the law of large numbers to prospective large-scale human action.[40] Elsewhere in the series, Asimov wrote: "The laws of Psychohistory are statistical in nature and are rendered invalid if the actions of individual men are not random in nature.... In other words, they would no longer be perfectly predictable."[41] Enthusiasm for quantitative positivism was clearly not confined only to the domains of science or social science.[42]

37 Asimov, *Foundation, Foundation and Empire, Second Foundation*. The nine stories that comprise *The Foundation Trilogy* were originally published in serial form in the science-fiction magazine *Astounding* over an eight-year period in the 1940s. They were compiled in three volumes in the early 1950s.

38 *Asimov, Foundation, Foundation and Empire, Second Foundation*, 7.

39 *Asimov, Foundation, Foundation and Empire, Second Foundation*, 411.

40 "A 'law of large numbers' is one of several theorems expressing the idea that as the number of trials of a random process increases, the difference in percentage between the expected and actual values declines to zero (accessed at: http://mathworld.wolfram.com /LawofLargeNumbers.html). So, for example, the larger the number of coin tosses, the greater the probability that the number of tails will equal the number of heads. For a more mathematically robust exposition, see Dodge, *The Oxford Dictionary of Statistical Terms*, 229.

41 Asimov, *Foundation, Foundation and Empire, Second Foundation*, 500.

42 On how the series inspired one of the pre-eminent economists of the last several decades to take up a career in the social sciences, see Krugman, "Asimov's Foundation Novels." Another prominent economist inspired by the Foundation Series is Chicago Business School professor and former governor of the Reserve Bank of India, Raghuram Rajan ("Professor Raghuram Rajan talks about his return to Chicago Booth").

And yet, even if the 1950s were an era of pervasive data enthusiasm, such enthusiasm manifested itself in more than one way. As the chapters that follow demonstrate, the scientific community in China and the Soviet Union remained divided over the relationship between probability theory and statistics. Resolving this issue involved not only epistemological and theoretical debates on the unity or disunity of statistical science but also practical considerations regarding state capacity building. In reformulating statistics explicitly as a social science, they eschewed probabilistic methods and instead chose to valorize exhaustive enumeration and the seductive idea of total or comprehensive information that it promised. Socialist statistics was, for them, the perfect anticapitalist antidote to the problem of accurate and correct knowledge production in the social world.

In a somewhat ironic twist, our current wave of Big Data positivism, the third in my reckoning, has witnessed the return of the dream of total information, though unsurprisingly, it is also accompanied by legitimate fears about the growth of the all-powerful and all-seeing state and corporation. Indeed, it is in this context that the case of 1950s China continues to remain relevant. Socialist statistics' idealization of exhaustive enumeration in the 1950s is echoed in our own contemporary moment of Big Data enthusiasm and its attendant disdain for "traditional" statistical theory; the past indeed is prologue.

Cold War and Postcolonial Science

In *Trust in Numbers*, Ted Porter offered a basic declaration of faith, noting that human actors make science, but they cannot make it however they choose— they are constrained by what can be seen in nature, created in a lab, and by social processes.[43] More recent work in Science and Technology Studies (STS) has "adopted as its foundational concern the investigation of knowledge societies in all their complexities."[44] Sheila Jasanoff, who has been at the forefront of such efforts, has argued that "in broad areas of both present and past human activity; we gain explanatory power by thinking of natural and social orders as being produced together."[45] For Jasanoff, co-production helps us understand that science is as much about positive understandings of the world

43 Porter, *Trust in Numbers*, 12.
44 Jasanoff, *States of Knowledge*, 2.
45 Ibid.

as it is about normative formulations of how the world ought to be.[46] To understand scientific activity, therefore, we have to pay attention to the "constant intertwining of the cognitive, the material, the social and the normative."[47] My approach to understanding statistics and statistical work in China in the 1950s is informed by these insights. Ideological commitments, political imperatives, and material constraints—from the macro level of international politics to the micro level of individual relationships—all influenced and were influenced by the articulation of statistics as a scientific pursuit in China in the 1950s.

While historians of science of twentieth-century China have persuasively demonstrated modern China's active participation in globally evolving technologies, their focus has tended to be either on the Republican years (1912–1949) or on the reform era (1978–).[48] In taking seriously the claims of China's socialist scientists and understanding the era's scientific and state-building activities on their own terms, this book joins other works in the now vibrant field of Mao-era science, including those on agricultural sciences, Sino-American scientists, earthquake prediction, ethnic classification, and public health.[49] It shows that statistical activity functioned under the twin pressures of the need for Marxist fidelity and the search for postimperial/postcolonial autonomy.[50] This focus on socialist science in China is important because it helps us understand science in contexts that are non-Western or nonliberal, or both.[51]

This story of statistics in China in the 1950s also refuses to isolate the West from the non-West (or the North from the South) and seeks "multi-directional

46 Jasanoff, *Designs on Nature*, 19.

47 Jasanoff, *States of Knowledge*, 6.

48 Exemplary among these are Asen, *Death in Beijing*; Bréard, "Reform, Bureaucratic Expansion and Production of Numbers"; Chiang, *Social Engineering and the Social Sciences in China*; Greenhalgh, *Just One Child*; Hu, *China and Albert Einstein*; Lam, *A Passion for Facts*; Rogaski, *Hygienic Modernity*; Schmalzer, *The People's Peking Man*; Seow, "Carbon Technocracy"; Shen, *Unearthing the Nation*; and Trescott, *Jingji Xue*.

49 Representative works include Fan, "Collective Monitoring"; Gross, *Farewell to the God of Plague*; Mullaney, *Coming to Terms with the Nation*; Schmalzer, *Red Revolution*; and Wang, "The Cold War and the Reshaping of Transnational Science." See also "Focus: Science and Modern China."

50 In certain contexts, such as agriculture, the need for autonomy was articulated by an emphasis on self-reliance and native/local knowledge. See, for instance, Schmalzer, "Self-Reliant Science; and Schmalzer, *Red Revolution*.

51 See, for instance, Jasanoff, *States of Knowledge*, 32, for a discussion of Polanyi and Yaron Ezrahi's claims that "modern science provides the template for a particular form of politics: liberal democracy." See also the collection of essays in Phalkey and Lam, "Science of Giants."

influences and channels simultaneously."[52] The book's focus on the global mobilization of new technologies in the service of new governance agendas also goes hand in hand with the wider turn away from earlier Cold War paradigms that foreclosed the possibility of meaningful comparisons outside of the geopolitical blocs.[53] From the establishment of the PRC on 1 October 1949 to the Sino-Soviet split in 1960, the Chinese had hewn close to the Soviet Union as a role model. That the People's Republic was in the Soviet camp fits a well understood Cold War paradigm. The world had two centers, with their own zones of influence, and they each vied for control and influence over the vast regions that lay beyond. This center-periphery framework has come to dominate recent studies of the period, but it does not always account for periphery-periphery links and what they can tell us.[54] The Sino-Indian exchanges outlined here remind us that experimentation and innovation took place in many contexts after 1945. They also allow us to better appreciate the frustrations as well as the achievements of statistics and the agency of statisticians in the early People's Republic.

Statistics and *Tongji* 统计—An Etymological Excursus

The Chinese word for statistics is a compound of two characters: *tong* (统), which means "all" or "together," and *ji* (计), which means to "count" or "calculate."[55] Together, *tongji* (统计) is thus defined as the "collection, sorting, calculation, and analysis of numerical data associated with a given phenomenon," or more simply as "summary calculation" (总括地计算; *zongkuo de*

---

52 Abraham, "The Contradictory Spaces"; see also Anderson, "Postcolonial Specters."

53 There is a growing literature on the history of development and aid in the post–World War II era, much of it through the prism of U.S.- or Soviet-centered networks of aid and influence. Exemplary among these are: Cullather, *The Hungry World*; Ekbladh, *The Great American Mission*; Engerman, *The Price of Aid*; the materials in the Cold War International History Project at the Wilson Center; Bernstein and Li, *China Learns from the Soviet Union*; Immerwahr, *Thinking Small*; Krige and Rausch, *American Foundations*; and Westad, *Brothers in Arms*. Even more recent works have shifted the lens somewhat to look at Sino-Soviet or Sino-American competition: Friedman, *Shadow Cold War*; and Brazinsky, *Winning the Third World*. Finally, De Grief and Olarte, "What We Still Do Not Know About North-South Technoscientific Exchange," offer an insightful critique of scholarship on the history of North-South scientific exchanges in the post–World War II years.

54 For the importance of transnational flows of people and expertise, see, for instance, Connelly, *Fatal Misconception*; and Iriye, "Internationalizing International History."

55 Yao Naiqiang, *Hanying shuangjie Xinhua zidian*, 287, 652.

jisuan).[56] This is, of course, a relatively recent definition. But the *Hanyu dacidian* (Unabridged Dictionary of Chinese) lists instances of its use in this context as early as the Ming dynasty (1368–1644). Explicit links between *tongji* and statecraft, however, do not appear to have been made at that time. Instead, it is the word *kuaiji* (会计; accounting) that was the standard word for talking about numbers and statecraft through much of Chinese history. Some scholars have contended that in ancient and late imperial China the relationship between numbers and statecraft was captured in the statement, "[if] accounting is proper [correct], [then] it is already enough [for governance to be effective]."[57] The phrase is from *The Mencius* and was purportedly uttered by Confucius when he was a minor official in charge of warehousing (委吏; *weili*). At least one Chinese statistician has made the case that the invocation of the phrase during succeeding dynasties should be considered evidence of the importance that was accorded to proper statistical work in Chinese history.[58]

The rise of statistics as a modern discipline in China, however, is frequently traced to the work of Robert Hart (1835–1911) at the Imperial Maritime Customs Service (1854–1950) during the late nineteenth century.[59] Hart's understanding of statistics was informed by its evolution within a European context. In the English language, the word statistics can be traced to French (*statistique*) and German (*Statistik*) antecedents dating from the mid-eighteenth century. In both the French and German cases, the word originally referred to "the study of the state, of statecraft, or of the conditions, circumstances, and politics of a state, the study of numerical data concerning society."[60] Prior to the adoption of statistics, the term that incorporated a similar meaning in English was "political arithmetic," which is credited to the seventeenth-century English economist William Petty.[61] An understanding that also incorporated "techniques of mathematical interpretation applied to phenomena for which an

56 Luo Zhufeng, *Hanyu dacidian*, 846.

57 See Mencius, *Wan Zhang—II*. Accessed at: Chinese Text Project, http://ctext.org/mengzi/wan-zhang-ii.

58 Yu Yue, "Kuaiji dang eryi yi."

59 Eberhard-Bréard, "Robert Hart and China's Statistical Revolution." See also Boecking, *No Great Wall*.

60 See, for instance, entries for Statistic and Statistics in the *Oxford English Dictionary* (Online edition).

61 For more on Petty and political arithmetic, see McCormick, *William Petty and the Ambitions of Political Arithmetic*.

exhaustive study of all data is impractical," that is, probability theory, took nearly another century to make its appearance.[62] In China, this modern understanding of statistics was denoted by *tongji*, which reappeared as a "return graphic loan"; a "Kanji" term derived from Classical Chinese and used in Japanese to translate modern European words that were subsequently, in the nineteenth century, re-imported into modern Chinese.[63] Both these definitions of *tongji*/statistics—one that links statistics to statecraft and one that is primarily mathematical in nature—and the different ways in which they were perceived in the early People's Republic are central to the arguments in this book.

Structure of the Book

The book is divided into three parts comprising seven chapters, which are followed by a conclusion. Part I, "A Statistical Revolution," consists of three chapters that explore what was new about statistics in the People's Republic after 1949. Chapter 2, "A New Type of Standardized Statistical Work," explores early statistical work in the PRC's Northeast, arguing that this work and the practical experience so gained was the foundation upon which the rest of the country's statistical apparatus was based. The establishment of socialist statistical work thus preceded its theoretical and ideological justification, which is the subject of chapter 3: "Ascertaining Social Fact." The chapter also provides an assessment of Soviet technical aid and introduces the Soviet statistical experts who were instrumental in helping organize statistical activity in the PRC. Chapter 4, "No 'Mean' Solution: Reformulating Statistics, Disciplining Scientists," explores how this new understanding of statistics became dominant in the 1950s and how it affected the valuation of key concepts and methods.

Taken together, the three chapters in Part I bring us back to an important issue in early PRC history: understanding the extent to which the shift to Communist rule after 1949 was a rupture and tracing the continuities that persisted nevertheless. Statistics exhibits elements of both rupture and continuity, depending on the timeframe as well as the geographic focus. A closer look at the statistical activities in the Northeast produces a strong case for a substantial rupture—new statistical methods and practices that were introduced under the direction of Soviet experts quickly became extensive. The

---

62 *Oxford English Dictionary* (*Online Edition*). The latter mathematical definition was first recorded in English in 1843.

63 Liu, *Translingual Practice*, 302 (fn), 338.

picture for the rest of the country until 1952 is much less clear. Thus, in the early years the rupture appears to have been regionally determined. In contrasting fashion, statistical education continued unchanged during the first two to three years of the regime. Many academic statisticians maintained their positions after 1949 and continued to be involved in the training of statistical workers. It was only in 1951, and more properly after 1952, that a concerted effort was mounted to redefine statistics, both by changing the academic curriculum and through targeted criticisms of well-established statisticians and their textbooks. Changes in practice thus preceded changes in education and academic discourse. Collectively, such findings help us rethink 1949 not as a singular moment of rupture but instead call for a separation of the rhetoric of rupture (as promoted by the CCP) from actual changes on the ground, which were both temporally and regionally variegated.

The two chapters in Part II, "Seeing Like a Socialist State," focus on statistics in practice, tracing some of the implications of the theoretical, scientific, and administrative decisions investigated in Part I. Chapter 5, "The Nature of Statistical Work," draws upon statistical reports generated from all levels of the statistical system—internal work bulletins, and materials from conferences at the local, provincial, and national levels—to uncover the messiness of actual statistical work and its relationship to planning. The chapter captures not only the centralizing impetus of the expansion but also the varieties of challenges that were encountered in putting into practice the methods that were at the heart of socialist statistics: (1) the periodic reporting system, and (2) the various forms of typical sampling. Even though acknowledgment of the problem of the excess issuance of reports and the chaos such reports generated were present before 1949, the problem became increasingly ominous throughout the 1950s, fueled in part by a table-as-product rationale, which is explored in chapter 6, "To 'Ardently Love Statistical Work': State (In) Capacity, Professionalization, and Their Discontents." This chapter analyzes the variety of stratagems—training, supplementary training, self-study, motivation, and rewards—that were employed to professionalize and maintain a cadre of statistical workers that by 1956 numbered as many as 200,000. The chapter argues that by the mid-1950s the state found itself incapable of training adequate numbers of personnel to meet the demands of the periodic reporting system.

Part III, "Alternatives," consists of two chapters, demonstrating the new paths that opened up in the late 1950s as Chinese statisticians sought solutions to the challenges generated by socialist statistics. Chapter 7, "Seeking Common Ground Amidst Differences: The Turn to India," explores the first of these. Based on a study of key figures, such as the deputy director of China's State Statistics Bureau, Wang Sihua, and the Indian statistician P. C. Mahalanobis, the chapter unearths a series of heretofore largely forgotten exchanges between Chinese and Indian statisticians. Focusing on Chinese interest in the emerging technology of large-scale random sampling, in which Mahalanobis and the Indian Statistical Institute were global innovators, the exchanges point to alternative frameworks for Cold War scientific exchanges while also placing in stark relief the extent to which Chinese statisticians and leaders clearly understood both the strengths and shortcomings of their own statistical system.

In spite of the optimism generated by the possible adoption of random sampling, the exchanges with India were stymied by early 1959. Instead, as is discussed in chapter 8, "A 'Great Leap' in Statistics," an "on-the-spot meeting" in the northern city of Baoding in the summer of 1958 launched statistics in China down an altogether different path. During the ensuing months, the tussle between socialist statistics and its probabilistic alternatives was largely overwhelmed by Maoist mass science. In statistics this meant a rebadging and valorization of typical sampling, which was now explained as Mao Zedong's synthesis of Marxist–Leninist theory with the practice of revolution in China. Mao's 1927 *Report on an Investigation of the Peasant Movement in Hunan* became the foundational text for this "revolutionary" method. It is this notion of mass science, with its antiexpert and antiprofessional credos, that has come to dominate our understanding of much of the Mao era.

The Conclusion returns to the main themes of the book before ending with a brief overture to developments in 1979. In that year, the statistician Dai Shiguang published two influential articles calling for a complete overhaul of the system of socialist statistics. Dai Shiguang's articles had an electrifying impact in China's statistical world. Within a few years, socialist statistics was cast aside, much like an old cloak, and mathematical statistics was formally reintegrated into the discipline and practice of statistics. An entire way of knowing society came to an end. Facts that the state had both seen and cherished disappeared, along with the periodic reporting systems that supplied them. Other facts, which did not exist before the 1990s grew to become central.

PART I

A Statistical Revolution

2

A New Type of Standardized
Statistical Work

IN JULY 1950, Wang Sihua (1904–1978), recently designated head of the Northeast Statistical Bureau (NSB), proudly declared to his audience that not only were they participating in the first meeting of statistical workers in the history of the Northeast but their meeting was also the largest in the nation's history.[1] This was clear evidence, according to Wang, that New China was already on the path toward achieving a planned economy, because no planned economy was possible without the presence of scientific statistical work. In his address, Wang observed that since promulgation of the Decision Regarding Strengthening Statistical Work on 10 April 1950, statistical work in New China had entered a "new phase."[2] As a result, "statistical work in the Northeast began to move from a state of decentralization to integration, from chaos to the correct path, from the non-scientific to the scientific." Wang explained that the help offered by Soviet experts was critical to this change. The Soviet Union possessed over thirty years of experience and it was its expertise and aid that led Wang to declare: "We have set up a statistical system, we possess method, [and] we possess confidence."[3]

A few years later, in December 1952, when addressing the third annual meeting of the Northeast Statistics Bureau, Wang was even more emphatic.

1 Wang Sihua, "Tongji gongzuo de renwu," 7. A slightly edited version of this address is included in a collection of Wang's works published posthumously, *Wang Sihua tongji lunwenji*, 1–14.

2 Wang Sihua, "Tongji gongzuo de renwu," 7. For the text of the decision, see "Guanyu jiaqiang tongji gongzuo de jueding" in Northeast Statistics Bureau, *Tongji gongzuo: Di yi xuanji*.

3 Wang Sihua, "Tongji gongzuo de renwu," 8.

During the previous three years, he claimed, the bureau had successfully established "a new type of standardized statistical work" (新型正规化的统计工作; *xinxing zhengguihua de tongji gongzuo*).[4] For him, such standardized statistical work was the hallmark of socialist countries and was fundamentally different from statistics in capitalist countries. It was also different from what he dismissively characterized as the older small-scale, handicraft-like practices of conducting the odd statistical survey that was common in the pre-1949 years.[5]

By the time that Wang delivered the latter speech, a national bureau of statistics (State Statistics Bureau; SSB) had been set up in Beijing and a rapid expansion of a national network of statistical offices and personnel at provincial, county, and village levels was underway. As the first subnational statistical agency in the PRC to fully incorporate Soviet/socialist approaches to statistics, the Northeast Statistics Bureau became a model for this rapid expansion.[6] Under Wang's leadership, the Northeast Bureau had not only carried out statistical work since early 1950 but had also undertaken the first major attempt at translating and publishing Soviet treatises on statistical theory and methods.[7]

---

4 Wang Sihua, "Sannianlai dongbei tongjiju." An edited version is included in Wang's collected works, *Wang Sihua tongji lunwenji*, 31–49.

5 Wang Sihua, "Sannianlai dongbei tongjiju," 3.

6 Oral history interview with Wu Cangping (1922–), Beijing, 11 March 2011. Wu, a demographer and statistician, was trained in the West before being invited back to China by the Communist government in the early 1950s. Offered a position at Renmin (People's) University, he was active in demographic research and was a member of the four-person team that visited India in late 1956. Since the 1980s he has been among the pioneers in the field of gerontology in China. For the Northeast's distinctiveness and prototypical qualities in the early PRC more generally, see MacFarquhar and Fairbank, eds., *The Cambridge History of China*, 82–83: "The outstanding case of regional particularity was the Northeast . . . it reflected the fact that the Northeast was the most advanced region and served as a bellwether for the rest of the country for a number of reasons. First, having benefited from industrialization under Japanese rule, the Northeast had the most developed economic base. It provided 34 percent of China's industrial output in 1949 and 52 percent in 1952. Second, by virtue of being the first region totally liberated, the Northeast could move more quickly toward comprehensive policies and was able to begin regional planning by 1950. And finally, proximity to the Soviet Union and Soviet holdings in the regional railroads and the port of Luta (Port Arthur–Dairen) combined to provide easy access to Soviet aid and influence. Thus the Northeast instituted Soviet methods of economic management, albeit with difficulty due to shortages of skilled personnel, and these methods were generally endorsed by the central leadership in Peking for extension to China as a whole."

7 For more, see Liu Chang and Zhang Yun, *Ershi shiji Zhongguo*, 118.

The NSB's activities had a seminal impact not only on the ways in which statistical work was organized and carried out but also on the very definition of statistics. By valorizing the Soviet Union's "advanced experience," debates within China, beginning as early as 1950, came to reformulate statistics as a social science in contradistinction to a natural or universal science. This social science's function lay in helping to build a socialist society. As a result, the new statistics stood bifurcated from what was labeled the tainted, bourgeois, and socially unproductive pursuit of mathematical statistics. Statistical work, most notably first in the Northeast under Wang Sihua and then progressively elsewhere, followed suit. By 1952, these changes began to be implemented within statistical education as part of a general reorganization of the system of higher education.

And yet, in extant studies of PRC statistics, this early period garners almost no attention. Li Choh-Ming, one of the first scholars to focus attention on the issue of statistics in the PRC, was dismissive of these initial years. As he wrote in 1961, "from 1949 to 1952 no effort was made to develop a statistical system national in scope."[8] This emphasis on a national system led Li to begin his longer book-length study on PRC statistics not with the founding of the PRC in 1949, nor with statistical activities in the Northeast, but with the founding of the SSB in Beijing on 7 August 1952. To be fair, Li did recognize the Northeast as the only region capable of producing reliable statistical data prior to 1952, and he also acknowledged the early presence of Soviet statisticians in the Ministries of Railways (1950) and Health (1951–1952).[9] But aside from this somewhat perfunctory acknowledgment, Li's approach only took seriously statistical activities national in scale, which were only made possible after the formation of the SSB in 1952.[10]

The early period from 1949 to 1952, however, is crucial to understanding statistics and statistical work throughout the decade that followed. Unpacking the constituent elements of Wang's "new type of standardized statistical work" permits an articulation of the ways in which this new type of statistics was regarded as more standardized, correct, and scientific than what preceded it in the years before 1949. At the same time, the neat binary—of socialist and bourgeois statistics, of preliberation and postliberation statistics—idealized

8 Li, "Communist China's Statistical System," 500.

9 Li, *The Statistical System of Communist China*, 7–8.

10 Writing a little more than ten years later, Orleans, "Chinese Statistics," 50, echoed this assessment: "The responsibility of standardizing and centralizing all statistical work in the country was finally vested in the State Statistical Bureau in August of 1952."

by Chinese (and Soviet) statisticians is too simplistic and incomplete. The 1950s were shot through with the influence, rarely pedagogical or technical in nature but rather all-pervasive and methodological, of what by 1959 would come to be termed the Maoist model of ascertaining fact. This Maoist model, with its emphases on direct experience over technical knowledge and on ideological fervor as a source of expertise, rendered the earlier binary and the clean solutions it offered much more conflicted. As a result, statisticians and statistical workers throughout the 1950s found themselves not only transitioning from a "nonsocialist" to a "socialist" system of data collection and analysis but also operating in a milieu that particularly valued personal, individual, practical experience.

Statistics in China before 1949

In September 1947, Washington, DC, played host to the twenty-fifth meeting of the International Statistical Institute's biannual conference. China was well represented at the nearly two-week-long event.[11] Director of Statistics of the Chinese National Government, Jennings P. Chu (Zhu Junyi), led a six-member delegation representing a variety of fields and specializations from within and outside the government. Of the six members, two were already stationed in Washington, DC: D. K. Lieu (Liu Dajun, 1891–1962) was commercial counselor at the Chinese Embassy as well as Chinese representative on the United Nations Statistical Commission, and Bangnee Alfred Liu (Liu Pengnian, 1901–1988) was director of the Chinese News Service. These three men were joined by the well-known sociologist, demographer, and Tsinghua University professor Ta Ch'en (Chen Da, 1892–1975), who had traveled from Beijing; by the Shanghai-based director of the Accounting Department of the Central Bank of China, Kuo-Pao King (Jin Guobao, 1893–1963); and by the head of the Department of Mathematics at National Central University in Nanjing, Pei-Ching Tang (Tang Peijing, 1903–1988).[12] Born in the late Qing, some of the members of the delegation had benefited from Boxer Indemnity fellow-

11 The names of the participants are listed in International Statistical Institute, *Introduction: International Statistical Conferences*, Vol. I: 42; see also Liu Chang and Zhang Yun, *Ershi shiji Zhongguo*, 70.

12 Among them, three delivered papers: Jennings Chu delivered a paper titled "The Independently Controlled Statistical System of the Chinese Government"; Kuo-Pao King delivered a paper titled "A Census of Nine Hsien in Szechwan, China," and Ta Ch'en delivered a paper titled "Factors of Urban Growth in China." For additional details, see International Statistical

ships and been educated in the United States, whereas others had been trained closer to home in Japan. Some had conducted and published research using statistical data and methods, whereas others had written textbooks and had been involved in the administration of statistical work in China. They were each exposed to contemporary social and scientific theories, and to a large extent their approaches reflected the dynamic and varied characteristics of the world of social science research and statistical activity during the first half of the twentieth century in China. It was this world, and its intellectual and administrative traditions, that Wang Sihua, Li Fuchun, and Zhu De forcefully rejected in the early 1950s.

The dynamism and variety in the world of Chinese social science research during this period was the outcome, at least partially, of an intense search for the causes of, and remedies for, China's relative political, economic, and military weaknesses on the world stage. As a once-powerful empire transitioned into a weak republic, elites (intellectuals, administrators, and the like) had to contend with what they saw as the long shadow of enduring imperial practices and traditions. Enfolded within this long shadow were the twin practices of statecraft and data collection.

Imperial Chinese states routinely gathered information on prices, rainfall, salt, grain holdings, famine, and population.[13] In 1793, five years before Malthus published his famous tract, *An Essay on the Principle of Population*, the Qing official Hong Liangji (1746–1809) had already raised the specter of unchecked population growth, thereby challenging the deeply rooted theory that population increase reflected good governance.[14] Toward the end of the Qing,

---

Institute, *Introduction: International Statistical Conferences*, Vol. I: 75–76, 78–79, and 94; Vol. II: 92–97, and 167–185.

13 For wide-ranging data collected by the Qing state, see, for instance, Li, *Fighting Famine*. Mention should also be made of the decimal-based *lijia* and *baojia* systems (with their origins in the Song dynasty during the life of Wang Anshi (1021–1086), though Qing methods, population censuses, fish-scale registers, and so forth, probably cannot be traced farther back than the early Ming (1368–1644). In general, barring central projects such as the Grand Canal and famine relief, most public services and local security services were handled at the local or regional levels. For the kinds of tensions this generated in the fiscal structure during the High Qing, and the attendant reforms that were attempted, see Zelin, *The Magistrate's Tael*.

14 "China's Population Problem," in deBary and Lufrano, *Sources of Chinese Tradition*, Vol. 2, 174–76. In similar fashion, Rowe, "Bao Shichen," demonstrates that the nineteenth-century statecraft thinker and agrarian reformer Bao Shichen (1775–1855), made national estimates of crop yields and population and dabbled in statistical ideas about state output, future economic and fiscal policies, and concepts of state income.

increasing incapacity forced the state to abandon attempts to quantify and collect information. During this same period, heightened interest in Western forms of knowledge set off an enduring debate among Chinese elites on the "usefulness" of Western knowledge in contrast to the "essence" embodied in Chinese learning.[15] As a result, institutions such as the Jiangnan Arsenal, a shipyard established near Shanghai in 1865 to manufacture modern arms and to study Western technology and languages, devoted considerable resources to the translation of Western tracts, including treatises on statecraft that included some statistical analyses. By the turn of the century, the Qing civil-service examination had begun to incorporate questions drawn from modern social sciences and mathematics.[16]

Within these broader stories of translation and knowledge acquisition, the advent of "modern" statistics in China has at least two origin stories. D. K. Lieu provided the first of them as early as 1930, explaining in his paper delivered at the nineteenth session of the International Statistical Institute in Tokyo that it was the publication of the first maritime customs report in 1859 that was the harbinger of modern statistics in China.[17] This 1859 report was published by the Imperial Maritime Customs Service, an agency that had been established in collaboration with European powers in 1854 to address the Qing state's inability to collect customs duties in the face of the unrest caused by the rampaging Taiping armies of Hong Xiuquan. The Maritime Customs Service thus took the place of the defunct Canton System of Trade (1757–1842), which had ceased to exist with the signing of the Treaty of Nanjing at the conclusion of the First Opium War (1839–1842). Formally a part of Qing imperial adminis-

---

15 Formulated in Chinese as *ti-yong* (体用; Western learning for its usefulness, Chinese learning for its essence), this instrumental use of Western knowledge to strengthen an increasingly weak China polarized Chinese intellectuals. For the translator and editor Yan Fu (1854–1921), without a broader engagement with Western civilization and its political and scientific forms, such an attempt was bound to end in failure. Others, such as the scholar Feng Guifen (1809–1874), held that the only things that China needed to compete on level terms were the material products of Western civilization.

16 Elman, *On Their Own Terms*, 331–334. Among the examples that Elman cites is the Chinese translation of Henry Fawcett's *A Manual of Political Economy* (富国策; *Fuoguo ce*). For a digital copy of the book, see http://www.sino.uni-heidelberg.de/encyc/Fuguo_ce/index.htm. On the development of statistics and the translation into Chinese of mathematical and statistical works during the nineteenth century, see Wang Youjun and Andrea Bréard, "Tongjixue zai jindai Zhongguo."

17 Lieu, "Statistical Work in China," 3. This was likely a report on "Annual Returns of Trade." For a list of reports published by the Maritime Customs Service through 1940, see Chinese Maritime Customs Project, *List of Chinese Customs Publications*.

tration, the Maritime Customs Service was staffed at senior levels by Europeans and Americans. Its most influential inspector-general was Robert Hart (1835–1911), under whose leadership a Customs Statistical Department was set up in Shanghai to publish regular, and increasingly detailed, statistical compendia related to China's foreign trade. Even so, well into the last decade of the Qing, the Customs Service remained isolated from the rest of the imperial administration and employed only a handful of Chinese staff.[18]

The second origin story takes as its point of departure China's loss in the Sino-Japanese war of 1894–1895, an outcome that shocked the Chinese and forced them to acknowledge both the superiority of Japan and the failure of China's Self-Strengthening Movement (ca. 1861–1895; a period of institutional reforms that was initiated in the wake of military defeats and concessions to the Western powers). The result was greater attention to Japanese statecraft practices, including statistical work. It was not until the late Qing New Policy reforms (新政; *Xinzheng*) were well underway after 1901, however, that new forms of statistical work were promoted.[19] Japan-educated students who had returned to China introduced Japanese statistical ideas and practices, and translated Japanese manuals in hopes that China too would adopt methods that were themselves originally influenced by German social statistics (*gessellschaftslehre*) and early Dutch methods.[20] It is against this background that statistics was introduced into the curricula of the New Schools of Law and Administration (法政学堂; *fazheng xuetang*) after 1905.[21] In 1907, a Central Statistical Bureau was established, but it operated entirely independently of the Imperial Maritime Customs Service, which by then boasted of several decades of statistical experience and expertise.[22]

18 Hall, *The Chinese Maritime Customs.* On statistics at the Maritime Customs Service and the role of Robert Hart, see Eberhard-Bréard, "Robert Hart and China's Statistical Revolution." For a general history of the Maritime Customs Service, see Van de Ven, *Breaking with the Past.*

19 Eberhard-Bréard, "Robert Hart and China's Statistical Revolution," 622. Subsequent reforms announced in 1905 included plans for representative constitutional government and a modern census. For more on the 1908 census, see Lam, *A Passion for Facts,* ch. 3. The New Policy reforms were a series of wide-ranging programs implemented after the Boxer Rebellion (1898–1900) and aimed to prolong Qing rule.

20 Bréard, "Reform, Bureaucratic Expansion and Production of Numbers," 3–5, 31–32.

21 In Japan, the first specialized schools dedicated to statistics were established in 1881. See Bréard, "Reform, Bureaucratic Expansion and Production of Numbers," 11–12.

22 Ibid., xx, 2, 22, 36, 55. Elsewhere, Bréard notes that "several reasons explain why the Qing implemented the model pioneered by the Customs Service on a broader scale without referring to it. A lack of fiscal pressure, the fear of the consequences of upsetting local arrangements, and

Translation of ideas, terms, and practices were central to the emergence of this new statistical culture.[23] According to Andrea Bréard, this new statistics posed a conceptual problem for provincial and metropolitan bureaucrats, "requiring a new understanding of the very object of statistics; [and] concerned with social dynamics and not with static entities."[24] Much of this shift remained within the domain of descriptive statistics and, following Japanese practice, continued to be limited to the social realm and devoid of any engagement with statistical theory. It was only in 1910 that Gu Cheng (1882–1947?) published what was likely the first translation of a statistics manual that combined both administrative and mathematical practices.[25]

The fall of the Qing in 1911, while aborting many of the Xinzheng reforms, helped accelerate the spread of nascent Western social sciences and witnessed attempts at new forms of governance and state-building. A number of Chinese received training as demographers, sociologists, anthropologists, and ethnographers, usually in the West or Japan, and began to work for the Chinese state and to collaborate on data collection with Western institutions and scholars. No longer understood purely through an Impact-Response framework, recent research into their activities has stressed the level of hybridity—drawing upon their Western educations, but also deeply cognizant of local realities and traditions—that these social scientists achieved in their research.[26] Through the

---

resistance by powerful local officials and landholders were probably important. The Customs Service, on the other hand, sought to safeguard its statistical monopoly and was slow to transmit statistical techniques." See Eberhard-Bréard, "Robert Hart and China's Statistical Revolution," 622.

23 Bréard, "Reform, Bureaucratic Expansion and Production of Numbers," 6, 146.

24 Ibid., 31.

25 Ibid., 107–112. At the time Gu Cheng was professor of Statistics at the Beijing School of Law and Politics, director of the Statistical Department of the Ministry of Education, and director-general of the Central Statistical Association. For a discussion on combinatorics and probability during the nineteenth century, see Bréard, "Reform, Bureaucratic Expansion and Production of Numbers," 115–142. It appears that among China's early reformers, only Liang Qichao (1873–1929) showed some interest in translating works on probability theory (135). Liang also wrote enthusiastically about the possibilities of historical statistics. See, for instance, Liang Qichao, "Lishi tongjixue."

26 There has been much recent work on the rise of several modern (social) science disciplines, which highlight the frequently transnational intellectual milieu from which they emerged, the critical importance of European and American influence through training, the role of institutions such as the Rockefeller Foundation, and collaboration among international and Chinese scholars. See, for instance, Asen, *Death in Beijing* (Forensics); Buck, *American Science and Modern China* (American Science); Chiang, *Social Engineering and the Social Sciences*

ensuing decades of the 1920s and 1930s, there emerged what Tong Lam has identified as a "Social Survey Movement," in which the desire to document, count, tabulate, and survey the nation permeated not only the elite, trained practitioners of the social sciences but also the layman. As a result, between 1927 and 1935, as many as 9,000 surveys were carried out in China.[27] For Tong Lam, these surveys operated on two levels. "The idea of social science and social survey research presumed an external social reality with internal mechanisms waiting to be discovered and dissected. On the other hand, Chinese social thinkers themselves also vehemently argued that China did not have a real organic and functioning society and that their goal was to create one."[28] By 1949, social scientists, such as the demographer and sociologist who headed the Institute of Census Research in the 1940s, Chen Da, the population geographer Hu Huanyong (1901–1998), the sociologist Chen Hansheng (1897–2004), and the economist Ma Yinchu (1882–1982), all participated in a well-established, sophisticated, and methodologically and ideologically diverse tradition of social science research and statistical analysis.

In the field of statistics and statistical activity, three broad trends are discernible. The first is the expansion and diversification of educational materials relating to statistics and a concomitant engagement with various approaches to statistical theory and practice. In the years immediately following the establishment of the Republic of China in 1912, translated texts reflected an awareness and appreciation of the co-existing strands of mathematical statistics (with its focus on probability theory), Anglo-American statistics, as well as the more descriptive statistical practices prevalent in Germany.[29] From the 1920s to the 1940s, numerous original works on statistical theory and application were published.[30] Other volumes offered comparisons of the differing statistical traditions and their historical trajectories. For instance, in the 1935 publication *Tongjixue gangyao* (An outline of statistics), Liu Hongwan observed:

---

in China (Sociology); Hu, *China and Albert Einstein* (Physics); Mullaney, *Coming to Terms with the Nation* (Ethno-linguistic Classification); Schmalzer, *The People's Peking Man* (Evolutionary Science); Schneider, *Biology and Revolution* (Biology and Genetics); and Trescott, *Jingji Xue* (Economics).

27 Lam, *A Passion for Facts*, 4.

28 Ibid., 9–10.

29 Indeed, Wang Youjun and Andrew Bréard, in *"Tongjixue zai jindai Zhongguo,"* see the long-term trajectory from the late nineteenth into the first half of the twentieth century to be a transition from social statistics to mathematical statistics.

30 Liu Chang and Zhang Yun, *Ershi shiji Zhongguo*, 66, list at least six texts on statistical theory and ten texts on applied statistics.

The view of the School of Social Statistics is that the objects of statistical research are social phenomena, and the method of analysis is large-scale observation [大量观察法; *daliang guanchafa*]; the view of the School of Mathematical Statistics is that it is a mathematical method that helps arrive at statistical solutions [统计解析; *tongji jiexi*].[31]

In contrast, Wu Zaoxi, in his *Jingji tongjixue* (Economic statistics), contended that statistics was not an independent social science. Instead, it was "a research method that aided the pursuit of other forms of knowledge."[32] Indeed, during much of the first two decades after 1911, statistics was not taught as an independent subject. In 1927, with the establishment of the Central Politics School, a statistics division was set up within the Economics Department. It was another ten years before a dedicated Department of Statistics was established at National Chongqing University.[33] Even so, general practice in most places dictated that statistical courses were offered in other departments, such as Economics, Law, and Business.[34]

Concurrent with the gradual expansion and diversification of statistical education was a gradual expansion of official statistical work. During the first few years of the Republic, offices and bureaus charged with carrying out statistical work were set up at various levels. Built partly on the foundations of pre-existing local institutional mechanisms, Bréard maintains that the reforms were nonetheless distinct enough to signal a major rupture with Qing-era practices.[35] By 1916, a State Council/Cabinet Office was re-established and contained within it a Bureau of Statistics. Therefore, the general picture up to 1927 is one of various government bureaus, some dedicated to statistical work, but most involved in a variety of spheres that collected and disseminated data germane to their briefs.[36]

31 Liu Hongwan, *Tongjixue gangyao*, 1935, cited in Liu Chang and Zhang Yun, *Ershi shiji Zhongguo*, 67.

32 Wu Zaoxi, *Jingji tongjixue* (Economic statistics), 1943, cited in Liu Chang and Zhang Yun, *Ershi shiji Zhongguo*, 68.

33 Liu Chang and Zhang Yun, *Ershi shiji Zhongguo*, 89–93. Several schools offering short-term courses in statistical methods had existed since the mid-1920s. It is also worth noting that globally the establishment of statistics departments was a late development.

34 Statistics was introduced as a subject of study within several disciplines as early as 1913, and by the 1920s several universities were offering courses on statistics.

35 Bréard, "Reform, Bureaucratic Expansion and Production of Numbers."

36 For a detailed description, see Lieu, "Statistical Work in China," 7–12. Also see Liu Chang and Zhang Yun, *Ershi shiji Zhongguo*, 79–85.

As described by D. K. Lieu, the years following 1927 witnessed renewed attempts to centralize and expand statistical activity. But Lieu considers much of this "mushroom growth" unsystematic and, in the final analysis, unhealthy, since it meant duplication of work and a significant waste of limited resources. Lieu was among those who convened a conference in February 1930 that sought to address these issues. The conference recommended that an independent Central Statistical Bureau be established. It also recommended the concurrent establishment of a Central Statistical Commission, which would coordinate the statistical activities of the various ministries. During the time it took to set up these institutions, the conference suggested forming a Central Statistical Association that could immediately coordinate statistical activities that were already underway. It also suggested carrying out population and agricultural censuses in 1931. A final recommendation focused on the creation of a Chinese Statistical Society, which would provide a private forum for academic discussions on statistics.[37] Pursuant to these guidelines, the Directorate-General of Budgets, Accounts, and Statistics was established on 1 April 1931.

D. K. Lieu's discussion is indicative of two broad features of statistical activity during the latter half of the Republican period. The first is that state statistical activities, though they ranged across subjects such as agriculture, population, industry, labor, communications, finance, and so on, were fragmentary and had an essentially limited scope in terms of the types of data collected, geographical reach, and representativeness. As the demographer Leo Orleans noted in a 1974 essay, "the statistical system [of Republican China] never developed beyond the rudimentary stage and almost never reached down to the *hsien* [county] levels."[38] The point is driven home forcefully by the number of nonstate, in many cases non-Chinese, actors whose work has become an important source of data for the period. Lieu himself noted approvingly the statistical surveys conducted by the likes of Sydney Gamble, John Lossing Buck, and Chen Da, among many others. In so doing, he points us to the second broad feature of the period—the lively and fairly diverse private engagement with statistics and statistical work throughout China during the 1930s and 1940s.[39] Indeed, outside of government activity, a lively community of scholars who participated in research on social issues also developed, generating copious amounts of statistical data. Such collaborative social research, frequently

37 Lieu, "Statistical Work in China," 12.
38 Orleans, "Chinese Statistics," 48.
39 Lieu, "Statistical Work in China," 14–17.

heavily statistical in nature, involved Western social scientists, Chinese social scientists trained in the West, and locally trained social scientists.[40] International organizations, such as the Institute of Pacific Relations and the Rockefeller Foundation, also played a significant role, frequently serving as the principal funding or organizing agencies.[41]

By 1947, Jennings P. Chu reported to participants at the International Statistical Institute's biannual conference that the Republic of China possessed a Statistical Directorate that had proven to "be of immense value in government planning, execution and investigation."[42] He identified five main characteristics of the system: independence (from political interference), liaisonship (i.e., a combination of centralized and decentralized administration), uniformity (across bureaus and types of data), specialization (of personnel), and finally interdependence (among planning, execution, and investigation). The duties of the Statistical Directorate included:

1. Direction and supervision of statistical personnel and statistical functions
2. Design and issuance of statistical forms and enforcement of uniform procedures
3. Assignment of statistical work across various departments
4. Compilation of national census statistics
5. Compilation of national statistical report
6. Investigation into administrative efficiency and compilation of statistics on administrative results

The Statistical Directorate's activities were summarized in a variety of publications, such as *The National Statistical Report and Abstract, The Statistical Analysis of National Problems Series, The Statistical Monthly*, and the first edition of *The China Statistical Yearbook*, released in 1947.[43]

40 See, for instance, Kulp, *Country Life in South China*; Fei Xiaotong, *Peasant Life in China*; Fei Xiaotong, *Earthbound China*; Gamble, *Ting Hsien*; Gamble, *North China Villages*; Lin, *The Golden Wing*; and Buck, *Land Utilization in China*.

41 For more on the activities of the Rockefeller Foundation, see Chiang, *Social Engineering and the Social Sciences in China*, ch. 9.

42 Chu, "The Independently Controlled Statistical System," 93. In 1963, Zhu drafted a book on statistical work during the Republican years. From what I can tell, it was not published until 1988 (Zhu Junyi, *Minguo shiqi*). For a description of the types of statistical work carried out between 1931 and 1949, see also Ma Min and Lu Hanwen, "Minguo shiqi zhengfu tongji gongzuo."

43 Chu, "The Independently Controlled Statistical System," 92–95.

These activities are suggestive of a large statistical establishment in the post-war years, but the actual number of bureaus and personnel that Zhu enumerated paint a very different picture. According to Zhu, at the end of June 1947 there were eight Bureaus of Statistics and 660 Offices of Statistics in the principal and subordinate organs of the central government. An additional thirty-two bureaus and 1,196 offices were located in the provinces, municipalities, and counties. The total number of statistical units in the country was 1,905, and the total number of personnel was 5,066, of which 120 were employed in the Directorate of Statistics.[44] In spite of such a seemingly small number of personnel for such a large country, Zhu was optimistic about the future of statistical work in China. The Japanese had been defeated, and the Communists seemed to be on the run. A stable future beckoned, and Zhu looked forward to the establishment of a complete statistical network, more trained personnel, and the "attainment of comprehensive, accurate, and timely results."[45]

The optimism that fueled Zhu's grand statistical aspirations, we now know, was premature. By 1948, the tide of the civil war that had raged since 1946 had begun to turn in the CCP's favor. As such, perhaps Leo Orleans summed the situation up best when he noted that "despite some well-meaning efforts by the Nationalist government, when the Communists took over the mainland in 1949 China continued to be 'a land where the statistician may perish for want of a few figures, where records are more romantic than mathematical.'"[46]

Wang Sihua and the Northeast Statistics Bureau, 1949–1954

Well before the People's Republic was officially established, the CCP had begun to organize statistical work and economic activity in those regions that were under its control. After the end of World War II, the party moved into the northeastern provinces that had been vacated by the Japanese. Wang Sihua was assigned to aid in the governance of these areas. Working under leaders

44 Ibid., 96. Individual ministries (much like the Maritime Customs Service) likely had their own staff of dedicated statisticians, but I have been unable to estimate their numbers with any degree of reliability.

45 Ibid., 96. Zhu concluded: "The world is united now more than ever, and international cooperation is aiming at worldwide stabilization and prosperity. All this leads to an extensive demand for accurate and timely statistical information for internal as well as external use, and toward this goal the Chinese government is determined to keep abreast of the advanced countries of the world."

46 Orleans, "Chinese Statistics," 48. Orleans quoted Martin, *Strange Vigour*, 3.

such as Li Fuchun and Gao Gang (1905–1954), Wang's first assignment was to take over administration of West and North Liaoning. In early 1947, he was also appointed deputy secretary-general of Heilongjiang province and director of both the Northeast Statistics Office and the Standing Finance Committee.[47]

Wang Sihua was well suited to these tasks. Born in 1904 in coastal Laoting county in the northern province of Hebei, Wang attended Peking University during the 1920s. As a student there, he enjoyed frequent contact with a fellow Laoting county native—the firebrand intellectual and founding father of the CCP, Li Dazhao (1888–1927). Under Li's influence, Wang soon became involved in revolutionary work. In 1926, he traveled to Europe and spent four years studying in France and England. It was during those four years that he began translating *Das Kapital* into Chinese. Wang returned to China in the winter of 1930 and was appointed professor of political economy at Beiping University and at the Sino-French University.[48] At the same time, he also became part of an underground network of left-leaning professors. With the support of the CCP and help from Hou Wailu (1903–1987), he continued his translation of *Das Kapital*. In June 1936, the results of their labors were published by the International Society as *Zibenlun, diyi juan* (资本论, 第一卷, *Capital*, Vol. 1). An accompanying commentary was also published.[49]

A turn toward more active involvement in revolutionary work took place in 1934 after Wang was arrested and then released on bail following charges of participating in intelligence work. Starting that year, Wang not only collaborated with the Anti-Imperialist League and the Anti-Japanese Society, he was also selected to be a member of the Beiping Second War Zone Mobilization Committee. After the fall of Beiping in early August 1937, Wang made his way

47 Wang's biographical details are drawn from the following sources: (1) Wang Sihua, *Wang Sihua tongji lunwenji*, 224–226; (2) Editorial Department of *Tongji*, "Jinian Wang Sihua," 5–6, 14; (3) Li Deng, "Jinian Wang Sihua tongzhi"; (4) Trescott, *Jingji Xue*, 241.

48 Beijing was renamed Beiping (Northern Peace) in late June 1928 in the wake of the Guomindang's decision to relocate the national capital to Nanjing. The Sino-French University (*Université Franco-Chinoise de Pékin*) was set up in 1920 to train students in French and a range of other disciplines. For more, see Duanmu Mei, "Cong qingong jianxue dao Zhong-Fa daxue."

49 Wang Sihua, *Wang Sihua tongji lunwenji*, 224. Wang's biography notes that he wrote under the pseudonym Wang Shenming. The companion volume was titled *Zibenlun jieshuo*. Hou Wailu was a prominent historian and philosopher, and also one of the major experts on Marxism in the PRC. After 1949 he taught at Northwest University. I was unable to locate information about the 1936 publication. The earliest extant (possibly abridged) edition appears to be from 1950 (Wang Sihua, *Zibenlun jieshuo*).

to Yan'an, arriving there in September. He entered the party rolls the following year and was assigned to teach political economy at the Central Party School. During the rectification campaign that began in 1942, Wang, who by then was directing research on the Chinese economy within the Central Research Institute, conducted an agricultural survey in northern Shaanxi. His primary research method was a form of typical sampling (典型调查; *dianxing diaocha*). The following year, Wang became deputy director of the Northwest Financial Planning Council, which was under the leadership of Chen Yun (1905–1995).[50]

By the time World War II drew to a close in Asia, Wang had garnered significant experience in economic administration and was an ideal candidate to contribute to administration in the Northeast. In October 1948, he was appointed to the newly established Statistical Survey Unit of the Northeast Finance and Economics Committee. Among his early tasks was the assessment of economic conditions in the Northeast. Wang organized surveys of industrial and commercial activity in Harbin, led investigations of the recent economic development in the region, and compiled existing statistical data on the preliberation economy.[51]

It was organizing statistical activity for the newly liberated Northeast, however, that was to become Wang's principal mission. In April 1950, the Northeast Statistics Bureau (NSB) was officially set up in Shenyang (formerly Mukden, the largest city in northeast China), and Wang was appointed director.[52] Under Wang, the NSB established a foundation for a comprehensive statistical apparatus, began a systematic translation of Soviet statistics treatises, and disseminated information through its monthly journal *Tongji gongzuo* (Statistical Work) (cited hereafter as *DBTJGZ*), about the bureau's work and experience.

50 Today, Chen Yun (1905–1995) is regarded in China as one of the "eight immortals," especially for his role in promoting economic reforms during the post-Mao era. During the 1940s and 1950s he was among the senior leaders responsible for economic and financial affairs. For more on Chen's post-1978 role, see Lardy and Lieberthal, *Chen Yun's Strategy for China's Development*.

51 These activities were coordinated through an institute set up to research Japanese rule and the activities of the Manchukuo puppet regime from 1932 to 1945. The institute's work was aided by a staff of thirteen Japanese, who helped edit historical materials and also helped calculate fixed prices for 1943. Upon completion of these activities, the staff returned to Japan. See National Bureau of Statistics, "Huiwang."

52 The offices were originally located near Zhongshan Square, but later they were moved to Northwest Horse Street in Heping District. The deputy director was Lin Lifu (1909–2001) and the secretary of the bureau was Yang Jianbai (1911–2004).

Over the course of its four years of existence, the NSB would register many firsts in PRC history: the first region to set up a statistics bureau, the first to publish a specialized statistics journal, the first to publish an annual statistical report, and the first to pass a regional ordinance (地区条例; *diqu tiaolie*) regarding statistical work.[53]

Under Wang's leadership, the NSB experienced a more than fivefold expansion of its rank and file. In a July 1950 address, Wang reported that there were 5,860 cadres involved in statistical work (this estimate also included part-time workers). Of these, most were tasked to track industrial statistics (4,778), with the remainder divided unevenly across trade, railways, and agriculture.[54] Two years later, in a December 1952 address, Wang noted that by then the NSB had 21,000 full-time statistical cadres who were aided by about 9,000 part-time workers.[55] He also noted with approval that various statistical forms had been unified and rationalized and their overall number brought under control. He explained that during this nearly three-year period, the bureau managed to reduce the number of forms from 3,318 to the much more manageable 396. This was a first step in controlling duplication and excess issuance of forms.[56]

The monthly journal *DBTJGZ*, which began publication in May 1950, was the primary venue through which the bureau discussed a wide variety of statistical issues. A total of thirty-two issues were produced through the end of 1952 before the journal was absorbed into its national-level namesake, which began publication in January 1953.[57] On occasion, important articles were also compiled and published as anthologies.[58] The subjects covered in these an-

53 In August 1954, along with the dissolution of the six regions, the NSB was dismantled and most of its personnel were transferred to provincial statistics bureaus; some fifty personnel were sent to Beijing. See National Bureau of Statistics, "Huiwang," 22.

54 Wang Sihua, "Tongji de renwu," 8.

55 Wang Sihua, "Sannianlai dongbei tongjiju," 1. About 125 statistical personnel worked in the head office in Shenyang.

56 Wang Sihua, "Tongji de renwu," 8; Wang Sihua, "Sannianlai dongbei tongjiju," 1. As early as July 1950, nearly 600 forms were earmarked for deletion. Other strategies involved merging two or more types of forms, correcting the forms, and, on rare occasions, issuing new forms.

57 Six monthly issues constituted a volume (卷; *juan*). In 1952, the system of volumes was replaced by issue numbers. Up until vol. 2, no. 2, it was an internal journal; thereafter it could be purchased in bookstores. Beginning with vol. 2, no. 5, it was published by Xinhua Press. See National Bureau of Statistics, "Huiwang," 23.

58 The library of the Institute of Economics of the Chinese Academy of Social Sciences in Beijing possesses the first, third, fourth, and fifth volumes of such anthologies; the Tsinghua University Library possesses the fourth and sixth volumes. In all, I was able to access four volumes, though at least six were produced. These volumes are not included in the digitized China

thologies can broadly be divided into four categories. The first category was instructional and included essays on how to complete statistical tables and forms for industry, agriculture, and other areas; how to assess and address problems with calculation, tables, original records, and accounting; the nature of training; how to conduct analyses; and finally, how to provide explanatory textual material to accompany numerical data. A second category consisted of problems that the bureau was encountering and wanted to address. Examples include the excess issuance of forms; problems of leadership; and dealing with conservatism in work. The third category included the publication of important statistical laws and regulations. The fourth and final category dealt with the various kinds of statistical activities carried out by the bureau, and also included summaries of work and responsibilities; descriptions of surveys of (state and private) industry, crop production, as well as some early essays on typical sampling. The picture that emerges from such categorization is one of a clear emphasis on establishing procedures, unifying systems, and training people to engage in statistical work. Surprisingly, there was little discussion of actual statistical results.

The Primacy of Exhaustive Enumeration

In terms of method, the system built under Wang Sihua's leadership consisted of two principal means of collecting statistical data. The first, and by far the more significant, was exhaustive enumeration via a complete enumeration periodic reporting system.[59] Its bedrock was the institutionalization and rationalization of original records. The idea was that data would be collected primarily through a vast network of comprehensive and periodic statistical reports spanning all sectors of the economy. As Wang noted, each and every table had clear instructions regarding how to fill in the data and on the timing and sequence of filing forms. Three important laws were passed in 1950 to facilitate the process:[60]

Academic Journals (CAJ) database and are not available in various libraries in Beijing. It is possible that individual issues can still to be found in libraries or archives in Harbin, Shenyang, or Changchun.

59 "Complete Enumeration Periodical Report System" is how the Chinese themselves had translated 全面定期统计报表制度 (*quanmian dingqi tongji baobiao zhidu*). See WJB 105-00530-05: 11–12.

60 For the texts of these laws, see Northeast Statistics Bureau, *Tongji gongzuo: Diyi xuanji*, 127–132.

1. The 10 April law, Regarding the Strengthening of Statistical Work (关于加强统计工作的指示; *Guanyu jiaqiang tongji gongzuo de zhishi*), which established the NSB, called for the setting-up of a centralized system of reports, and required that statutory forms (法定表格; *fading biaoge*) be used by all levels. All pre-existing forms were to be discarded.

2. The 22 August law, Regarding the Unified Issuance of Statistical Numbers (关于统一公布统计数字的决定; *Guanyu tongyi gongbu tongji shuzi de jueding*), required that all reported data be checked by the corresponding statistical office before being sent to a higher level within the statistical organization.

3. The 13 October law, Temporary Regulations for Statistical Reports in the Northeast Region (东北区统计报告暂行规程; *Dongbeiqu tongji baogao zanxing guicheng*), was the first regional statistical ordinance in post-1949 China. It clarified the relationship among different levels in the statistical apparatus and mandated the use of a common set of forms and tables.

In other parts of the country, where statistics bureaus had yet to be set up, similar directives were issued several months later by the city or provincial governments based on the authority of the Statistics Office of the Finance and Economics Committee of the Central People's Government. For instance, a directive to establish and strengthen statistical work was issued on 15 December 1950 in Beijing. It designated the city's Research Office, to which all statistical tables were to be sent, to serve as coordinator and distributor.[61]

In March 1951, the NSB issued the Decision Regarding Trials of a Two-Track Statistical Reporting System (关于试行统计报告双规制的决定; *Guanyu shixing tongji baogao shuangguizhi de jueding*), which set up a two-track system for reporting and collecting data. In the first track, grassroots units reported data to the enterprise (business, company, or organization) with which they were affiliated, and the enterprise then prepared a consolidated statement to be forwarded to the NSB. However, these grassroots reporting units were also required to report their data using the statistical reporting system. Thus, in theory, data about the same unit arrived at the statistical offices in Shenyang via two parallel routes: one via the enterprise and the other via

---

61 BMA 136-001-00014: 1–6; BMA 002-002-00223: 1–6. Along with this directive, two sets of instructions became available, one explaining the reporting methods and the other detailing the methods on how to review the forms.

the statistical system. The frequency of these reports could be as short as a ten-day cycle or as long as a year, and they also included several possibilities in between: monthly, seasonal, semiannual, and so on. These reports would be sent up through each successive level until they reached the headquarters of the NSB in Shenyang. Starting in 1951, some of the seasonal and annual reports were also published, such as the Northeast Report on Implementation of the National Economic Plan (东北区国民经济计划执行情况的报告; *Dongbeiqu guomin jingji jihua zhixing qingkuang de baogao*).[62] The first annual report was published on 1 April 1952 and provided data for all of 1951. By 1954, this system had expanded to include data about industry, capital construction, agriculture, trade, transportation and communications, private enterprises, and numerous other fields.

When data were required irregularly, one-time censuses were also employed. A national census on industry was completed at the end of 1950, for the first time giving planners in Beijing an idea of the basic production situation in each factory. On the factory side, the census was used to understand production, discover unharnessed potential (后备力量; *houbei liliang*), correct management shortcomings, and carry out economic calculations. The census also helped raise awareness at the factory level about the importance of statistical work, partly through the presence of the many statistical cadres who had been trained for the census and were now familiar with production processes and related technical issues. The NSB was at the forefront of the national census on industry, carrying it out from April through December of 1950. A census of private industry began on 19 July 1950, via the Decision Regarding Conducting a Census of Private Industry in the Northeast (关于进行全东北私营工业普查的决定; *Guanyu jinxing quan dongbei siying gongye pucha de jueding*). This census, covering 6 provinces and 5 cities, enumerated 39,539 private industrial units, and took 8 months to complete.[63]

Unlike in industry, statistical work in agriculture did not easily permit exhaustive enumeration, whether via a periodic reporting system or via a onetime census. Accordingly, the second method of data collection relied on sample surveys. The NSB designed surveys to produce statistical estimates of agricultural production. These surveys invariably were based on some form of typical sampling. In designing the surveys, the Chinese drew upon both Soviet

---

62 In the third season its title was changed to Report on Implementation of the People's Economic Plan (*Renmin jingji jihua zhixing qingkuang de baogao*) and subsequently, in 1952, it became a biannual report (National Bureau of Statistics, "Huiwang," 22–23).

63 Such units either used machines or employed three or more workers.

expertise and Chinese experiences in the 1930s. They discovered that Soviet experts were generally not overly enthusiastic about the use of such surveys. In the materials produced by the NSB, the discounting of surveys by Soviet statistical experts is perhaps best expressed by the absence of the word for representativeness (代表性).

The CCP's experience conducting rural surveys during the 1930s was thus a critical factor in determining the design of surveys in the early 1950s. In his memoirs, Xue Muqiao (1904–2005), an economist and the first director of the SSB, observed that the purpose of the surveys in the 1930s was to estimate production and to organize the levying of taxes. Every province and city had established production survey committees or production survey offices, which were under the leadership of the local planning committee. These committees and offices used meetings (会议的方式; huiyi de fangshi) to disseminate what they were doing at the county levels. Based on the maxim "divide by type, select [a] typical [sample], really cut [grain] to really measure," 8,000 survey-ors investigated 1,175 archetypical (典型; dianxing) villages.[64] First, sown area was surveyed; then areas were divided based on the nature of the harvest (good or bad); typical cases were selected and surveyed from within these divisions, after which overall estimates were generated based on the relative weights of good and bad harvest areas.[65] Such typical sampling formed the basis of several other surveys carried out during this period.[66]

Other Activities

In addition to the collection and collation of data, other statistical activities included the calculation of price indices, the periodic rectification of original records, and the cleaning up of statistical tables. In 1950, Soviet methods were used to compile a price index. A common program for calculation was subse-quently assigned to all provinces and cities so as to calculate local price indices. For instance, a survey of 2,000 products was carried out in Shenyang in August 1950, and in the following month the NSB released the Shenyang Wholesale

---

64 National Bureau of Statistics, "Huiwang," 22–23. Also see Yang Jianbai, "Sulian zhuanjia bangzhu women."

65 As Xue Muqiao, in Xue Muqiao: Huiyilu, 174, notes, this was basically a rather simple form of survey sampling.

66 These included investigations of rural surpluses, production surveys, farm censuses, rural ancillary activities and livestock surveys, and surveys of private industry with more than ten workers, and so forth. See National Bureau of Statistics, "Huiwang," 22–23.

Price Index. In October 1950, the NSB released a State-Owned Wholesale Price Index. Beginning in December, such indices were published monthly in *Dongbei ribao* (Northeast Daily). In the following year, the NSB released a price index for the ten big cities in the Northeast. Thereafter, provincial- and city-level indices for the Northeast were calculated on a ten-day, monthly, or some other predetermined cycle.[67]

The rectification of original records, another practice that would continue throughout the remainder of the decade, initially began in the Northeast in 1951 and covered statistical data on industry, capital construction, and trade. During the following two years, most industrial units underwent two rounds of rectification, and as a result their data were widely considered to be reliable. Local state-owned industrial units also underwent at least one round of rectification. In capital construction, rectification was modeled after that in the industrial cities of Angang and Fushun in northeast China. These cities published the newspaper *Zuzhang ribao* (Section Leader Daily), which shared information on a daily basis about work conditions: work completion, labor time, worker development, use of materials, and so on. By collecting all such data in one report, these daily papers became the basis for unified, regular, and accurate statistical data about the various industries. This experience was subsequently shared with the rest of the country, and this method spread throughout the country.[68]

In addition, statistical reports were put in order on two occasions. The first instance was in 1950, by order of the Northeast People's Government. The second instance was in 1953, following a directive from the recently established SSB. The clean-up involved discarding illegal and unnecessary reports in order to reduce the burden on grassroots units.

"A New Type of Standardized Statistical Work"

Taken together, these activities begin to provide a picture of the basic contours of statistical work in the Northeast. Wang Sihui's "new type of standardized statistical work" consisted of three basic features. The first of these was its "extensiveness" (广泛化; *guangfanhua*). For Wang, this meant that socialist statistics, unlike their bourgeois counterpart, had a mass character (大众化; *dazhonghua*) that ran through every sector of the economy and society:

67 Ibid.
68 Ibid., 23.

Not only did it include all branches of the national economy but also culture, education, and health care; its content was extremely extensive, and the procurement of materials and the collation of numbers depended on a comprehensive and unified reporting system.[69]

A unified reporting system also meant that each person and each element within the system was absolutely critical. If even one person's work suffered from a fault, then the entirety of the statistical work was at fault.

Accordingly, a second characteristic was the system's "completeness" (整体性; *zhengtixing*). This completeness was a corollary to the system's extensiveness in that it sought to conceive of the statistical work of every unit as an integral part of a unified statistical apparatus. The key to achieving such completeness required standardization of forms and types of data so that comprehensive and comparable data were readily available. Equally important was the creation of norms for entering the data on forms and for their accompanying textual explanations.[70] Since the integrity of such an extensive and complete system could easily be threatened by a kink at any one level, it was imperative to create a unified structure of reporting that permitted data (i.e., forms and tables) to be sent up and down different levels of the statistical and administrative systems.[71]

Such extensiveness and completeness were the basis for the third feature of socialist statistics, namely its "objectivity" (客观性; *keguanxing*). Unlike bourgeois statistics, which were predicated on profit and thus meant that statistical data were frequently distorted (歪曲; *waiqu*) to serve the interests of the few and to protect their business secrets, Wang reassured his audience that socialist statistics was unafraid of exposing weakness. Quite to the contrary, they welcomed criticism and even took part in self-criticism in the pursuit of objectivity and truth (真实性; *zhenshixing*).[72]

69 Wang Sihua, "Tongji gongzuo de renwu," 13–14.

70 Ibid., 14.

71 Ibid.

72 Ibid. It is worth noting that Wang Sihua's concerns about objectivity appear to focus on the possible manipulation of data and not on the objectivity of measurement and representation themselves, a form of mechanical objectivity that is the subject of the seminal work by Daston and Galison, *Objectivity*. In that sense, his concerns resonate with the formulation in Porter, *The Rise of Statistical Thinking*, 229, that states that "objectivity means knowledge that does not depend too much on the particular individuals who author it." In later articulations, by Wang and other Chinese statisticians, a definition of objectivity emerges that valorizes the ability of exhaustive enumeration to provide a like-for-like (i.e., one-to-one) representation of reality. This

None of the above three ideals could be achieved without a disciplined cadre of statistical workers who were able to calculate and analyze data. Wang therefore noted that it was necessary that discipline undergird the entire edifice of statistical work. Discipline meant that numbers were diligently recorded, that reporting deadlines were strictly observed, and, somewhat oddly, since Wang had just accused capitalists of the same failing, that all statistical numbers were guarded as secrets.[73]

These features—extensiveness, completeness, objectivity—exemplified what was deemed scientific by Wang, because only by counting everything could an objective sense of social reality be achieved.[74] Such exhaustive counting in turn was the basis for the socialist corrective to bourgeois statistics. Indeed, the implication was that a system that was socialist was by definition also scientific and, therefore, also correct. The three terms—*socialist, scientific,* and *correct*—were mutually interchangeable. Several years later, in 1957, the Indian statistician P.C. Mahalanobis would be impressed by these very qualities of the by-then national statistical system, noting that it was "appreciably better than that in India in respect of coverage, availability, and accuracy of data required for purposes of planning and current policy decisions."[75] Wang's description of the system he had overseen in the Northeast was without a doubt an idealization. On the ground, a variety of challenges existed: some on account of local variations across China, some because of problems of calculation, and others because of incentives and disincentives that were systemic in nature. Many of these will be explored in later chapters. For now, let us turn to an alternate method of ascertaining social fact, which drew its authority directly from Mao Zedong.

Maoist Social Investigation as Method

Mao is known to have personally conducted at least two investigations into social conditions in the countryside, the first in 1927 and the second in 1930.[76]

---

contains within it elements of what Megill, "Introduction: Four Senses of Objectivity," characterizes as absolute objectivity, that is, representing things as they really are.

73 Of course, unlike private secrets, these were the public's secrets, i.e., the nation's secrets, to be kept safe from meddling by bourgeois nations.

74 For a brief but insightful discussion on the "reality" of statistics, especially through an analysis of its practitioners, see Desrosières, "How Real Are Statistics."

75 Mahalanobis, "Some Impressions," 12–13.

76 In 1927 Mao spent thirty-two days, from 4 January to 5 February, in five counties in

The reports on these investigations, along with other writings, such as his essays "Oppose Book Worship" (1930) and "On Practice" (1937),[77] articulated a form of social investigation that was at odds not only with the bourgeois "social survey movement" of the 1930s but also with the established Marxist methods of social science research, in both China and elsewhere.[78] For Mao, "the basic method for knowing conditions is to concentrate on a few cities and villages . . . use the fundamental viewpoint of Marxism, that is, the method of class analysis, and make a number of thorough investigations. Only thus can we acquire even the most rudimentary knowledge of China's social problems."[79]

Mao's principal method was to carry out "investigation meetings." This meant gathering a small select group of informants from whom he would gather not only knowledge about the situation on the ground but also information about local group dynamics, solidarities, and interpersonal tensions, all with a view to deploying such information to achieve specific political ends. This method valorized the personal experience of conducting such interviews. It discounted methodological issues such as representativeness and comprehensiveness in favor of the ability to effectively create change and bring about consensus.[80] The method's importance and its impact before and after 1949 have been the subject of earlier works, most notably those by Thomas Mullaney, Yung-chen Chiang, and Roger Thompson.[81] While Chiang

---

Hunan. His report was published as Mao Tse-tung, *Report on an Investigation of the Peasant Movement in Hunan*. Some years later, in May 1930, Mao conducted a survey in the town of Xunwu (寻乌), which is located in southern China where the provinces of Jiangxi, Fujian, and Guangdong meet. See Mao Tse-tung, *Selected Works of Mao Tse-tung*, Vol. 1; and Thompson, *Report from Xunwu*.

77 On the controversy over when exactly "On Practice" and the essay on contradiction were composed, see Knight, "Mao Zedong's *On Contradiction*."

78 On the former, see Lam, *A Passion for Facts*; and on the latter, see Chiang, *Social Engineering and the Social Sciences in China*.

79 Mao Tse-tung, "Preface to *Rural Surveys*," 17 March 1941, 11.

80 For Mao, the stakes were admittedly different: an interest in equity, self-reliance, decentralization, antagonism to bureaucratism and elitism, collective aims and discipline, and the continuing desire for mass struggle. For a discussion of these themes, see Mao Tse-Tung, *A Critique of Soviet Economics*, 13, 14, and 17–18. Also see Xue Muqiao, "Di yige wunian jihua qijian," 13, where this method is listed as one among a set of supplementary methods used by statisticians.

81 For detailed treatments of Mao's social-ethnographic method that explore some of these themes, see Chiang, *Social Engineering and the Social Sciences in China*, 150–158; Mullaney, *Coming to Terms with the Nation*, 97–101; and Thompson, *Report from Xunwu*, 23–27.

and Mullaney disagree on the degree to which Mao was self-reflexive about the methodological strengths or weaknesses of his method, both insist that the method retained its primacy in the post-1949 period.[82] While that may indeed have been true in some spheres, in the realm of statistical work it occupied at best an ancillary space during much of the 1950s. As a putative method of social investigation, it was also very much a source of tensions for its practitioners and theoreticians.

Two things within Mao's method were particularly at odds with the socialist statistics espoused by Wang Sihua. The first was that Mao placed tremendous value on individual experience—drawing authority from one's presence in the field to personally carry out a survey.[83] Personal subjectivity was the basis for the pursuit of objective reality. The theoretical basis of this position was best articulated in Mao's essay "On Practice." In it, he noted that "the peoples with real personal knowledge are those engaged in practice the wide world over."[84]

If you want to know a certain thing or a certain class of things directly, you must personally participate in the practical struggle to change reality, to change that thing or class of things. . . . Only through personal participation

82 As Chiang notes: "The power and prestige of Mao were such that his vision and approach dictated the way social investigations were conducted in the People's Republic until the 1970s" adding, "his group fact-finding sessions became the ubiquitous format according to which all social investigations have been conducted in China." Mullaney is in general agreement: "Following 1949, this mode of field research—one in which the boundary between information and transformation was porous by design—continued to be employed." See Chiang, *Social Engineering and the Social Sciences in China*, 156, 157; Mullaney, *Coming to Terms with the Nation*, 99. The impact of the *Report on Xunwu*, "the earliest surviving example of a formal local investigation by Mao," is harder to assess because it remained largely unknown until its eventual publication in the early 1980s (Thompson, *Report from Xunwu*, 27–36).

83 Some Republican-era social scientists also made similar claims of authoritative knowledge based on their personal experience of social conditions, often focusing in particular on their ability to endure personal hardship in pursuit of social facts. For more, see Lam, *A Passion for Facts*; and Mullaney, *Coming to Terms with the Nation*, 96–97. Such an approach also has antecedents in other parts of the world. Porter, *Trust in Numbers*, 99–100, notes that "inquiries on social questions" in late nineteenth- and early twentieth-century Britain often favored depending upon those with direct experience of accounting and economic activity over those with academic (expert) knowledge.

84 "Leaving aside their genius, the reason why Marx, Engels, Lenin and Stalin could work out their theories was mainly that they personally took part in the practice of the class struggle and the scientific experimentation of their time; lacking this condition, no genius could have succeeded." Mao Tse-tung, "On Practice," in *Selected Works of Mao Tse-Tung*, Vol. 1, 299.

in the practical struggle to change reality can you uncover the essence of that thing or class of things and comprehend them. . . . All genuine knowledge originates in direct experience . . . [and] there can be no knowledge apart from practice.[85]

In his Hunan report, Mao exemplified this position in practice. From the very first sentence of the report, Mao placed himself squarely in the middle of the investigation:

> During my recent visit to Hunan I made a *first-hand* investigation of conditions in the five counties. . . . I called together *fact-finding conferences* in villages and county towns, which were attended by experienced peasants and by comrades working in the peasant movement, and I *listened attentively* to their reports and collected a great deal of material (emphasis added).[86]

Mao was present "first-hand," "fact-finding," and "listened attentively." The local informants he preferred to meet were those who had a good understanding of local conditions.[87] But he added that picking up information less systematically was just as important: "I met all sorts of people and picked up a good deal of *gossip*."[88] Later in the report he noted, for instance, "When I was in the countryside, I heard the gentry say. . . ."[89] Mao's authority to speak also drew from his subjective position of being personally present in the field, which allowed him to not only interview informants but also witness events as they unfolded: "I myself saw two [evil gentry] captured in Changsha. . . ."[90] Personal testimony was augmented by direct quotes from the people whom he met and interviewed, thereby adding the kind of local flavor that statistical data collected through a periodic reporting system could never hope to attain: "Whenever their conversation turns to the heads of the *tu* and *tuan*, the peasants say angrily, 'That bunch! They are finished!' "[91] It was such experience that

85 Ibid. This ideological position would lead to the famous dictum: "no investigation, no right to speak."

86 Mao, *Report on an Investigation of the Peasant Movement in Hunan*, 1.

87 For somewhat contrasting descriptions of Mao's group-interview process and the selection of interview candidates, see Mullaney, *Coming to Terms with the Nation*, 97; Chiang, *Social Engineering and the Social Sciences in China*, 155, 156.

88 Mao, *Report on an Investigation of the Peasant Movement in Hunan*, 7.

89 Ibid., 13.

90 Ibid., 29.

91 Ibid., 35; for other examples see ibid., 16–17, 34, 41. In most cases these are representative quotes and not specific statements by any one person.

Mao would extol but also claim was imperative because it allowed one to gather information and to assess the situation.[92] And the process was not without excitement of impending revolutionary change: "If your revolutionary viewpoint is firmly established and if you have been to the villages and looked around, you will undoubtedly feel thrilled as never before."[93]

Mao's enthusiasm for such investigative strategies was accompanied by a lack of precision where actual numbers were concerned. Throughout the Hunan report he frequently dealt in whole numbers or expressed ranges that could, at best, only be taken as possessing a token representative value. Thus, in discussing the number of spears among the spear corps in the five counties that were the focus of the report, he would offer numbers such as 100,000 in one county (Xiangxiang), or ranges such as "70,000–80,000 or 50,000–60,000 or 30,000–40,000" in the remaining counties.[94] The number of rifles among the landlord class was discussed with somewhat greater precision, but still with a propensity for neat, whole figures: "An average of 600 rifles for each county would make a total of 45,000 rifles for all seventy-five counties [in central Hunan]. . . ."[95] Elsewhere, he noted that "a crowd of fifteen thousand peasants," demonstrated at the houses of six evil gentry.[96] Finally, it remains unclear how Mao arrived at the results of his surveys of Changsha county (where he claimed that poor peasants comprised 70 percent, middle peasants 20 percent, and landlord and rich peasants the remaining 10 percent of the population) and Hengshan (85 percent of peasant leaders are capable and hard-working, whereas 15 percent retain some bad habits).[97]

92 On Mao's understanding of cognition and knowledge, it is again useful to consult "On Practice," where Mao notes: ". . . the dependence of rational knowledge upon perceptual knowledge. Anyone who thinks that rational knowledge need not be derived from perceptual knowledge is an idealist. In the history of philosophy there is the "rationalist" school that admits the reality only of reason and not of experience, believing that reason alone is reliable while perceptual experience is not; this school errs by turning things upside down. The rational is reliable precisely because it has its source in sense perceptions, otherwise it would be like water without a source, a tree without roots, subjective, self-engendered and unreliable. As to the sequence in the process of cognition, perceptual experience comes first. . . ." See Mao Tse-tung. "On Practice," in *Selected Works of Mao Tse-tung*, Vol. 1, 302–303.

93 Mao, *Report on an Investigation of the Peasant Movement in Hunan*, 8.

94 Ibid., 37.

95 Ibid., 36.

96 Ibid., 27.

97 Ibid., 18–19, 21. On the impossibility of determining what kind of data Mao actually collected during his investigations because the field data and notebooks are all likely lost, see Chiang, *Social Engineering and the Social Sciences in China*, 151.

The second feature of Maoist social research at odds with socialist statistics was the invocation of a telling example as a representation of a larger social reality.[98] As Mao would state in his Hunan report, "quite a few such cases have occurred. There was a case recently at Machiaho, Hsiangtan County, where . . .," or "once in Hsiangtan when a district peasant association and a township peasant association could not see eye to eye."[99] In similar fashion, when discussing the effect that the execution of one landlord could have, Mao cited the example of his native county of Hsiangtan. Socialist statistics was intrinsically uncomfortable with such a strategy of using a part as a representative of the whole. The use of a carefully selected example was acceptable for illustration, but it could not be the basis of one's knowledge about the totality of social reality.

If Mao's ethnographic method was a way of ascertaining social fact, so indeed was socialist statistics.[100] For PRC-era socialist statisticians, Mao's method was essentially a form of typical sampling (i.e., much like a case-study method). In his July 1951 speech, Li Fuchun spoke of methods of data collection that were supplementary to the periodic reporting system, pointing specifically to Mao's use of investigation meetings and noting that they had been in use for well over a decade.[101] But within socialist statistics, such typical sampling was considered a secondary method, only to be used when periodic reports and exhaustive enumeration were impossible.[102] Wang's "new type of standardized statistical work" was predicated on abstraction and an ability to collate vast amounts of data. An individual statistician or a statistical cadre drew his authority and knowledge neither from his personal experience in the field nor from having participated in data collection at the village and county levels, but primarily from the power of abstraction. It was not the observation of habit, the application of common sense, or any other individual personal experience that was critical, but rather it was specialized training that allowed

98 The rhetoric of example as discussed by Alain Desrosières has a long history and remains a powerful form of social analysis to this day. Before the rise of randomized survey sampling, examples provided the basic ideas behind systematized forms of purposive or typical sampling. See Desrosières, *The Politics of Large Numbers*, 210–235 ("The Part for the Whole: Monographs or Representative Sampling").

99 Mao, *Report on an Investigation of the Peasant Movement in Hunan*, 27.

100 Yang Jianbai, "Sulian zhuanjia bangzhu women."

101 *DSJ*, 5.

102 For a discussion of an early instance of the use of socialist statistics to gather information about local residence permits (*hukou*) in Beijing and Tianjin during 1949–1950, see Graminius, "Building New China."

him to correctly record data, process it, and analyze it. The knowledge and practice that Mao spoke of were not the knowledge and practice that statisticians practicing socialist statistics specialized in.[103] To further complicate matters, Mao's method was also not the kind of typical sampling that had actually been conducted in the Northeast under Wang Sihua.

Socialist statistics remained in uncomfortable tension with the Maoist approach for much of the 1950s. In fields that had putatively ceased to exist, such as sociology, the methodological vacuum was filled by the Maoist method. But in the case of statistics, the approach went against Soviet experience as well as against the very framing of socialist statistics. Given the latter's dominance during the 1950s, the more emotional appeal of Mao's ethnographic style served to make it difficult to convince cadres, in particular grassroots cadres, of the efficacy and need for something as expansive and alienating as the periodic reporting system and exhaustive enumeration. It was only after 1958, when socialist statistics was forsaken during the GLF, that the Maoist method came to occupy center stage as the principal statistical method of social investigation.

Conclusion

In marking a strong break from Republican-era statistical activity, the NSB under the directorship of Wang Sihua provided a blueprint for emulation by the rest of the country. Much as China was to follow the Soviet Union's advanced experience, so would each of China's five other regions (and eventually each of China's provinces) follow the advanced experience of the Northeast. For Wang Sihua, three features exemplified the strengths and scientific character of the Northeast's statistical system: extensiveness across economy and society, completeness of the statistical system, and objectivity of social facts. With periodic reports at the core of this system, Wang articulated a preference for exhaustive enumeration. Only in this way, that is, by counting everything, did he and other Chinese statisticians claim to achieve an objective sense of social reality. In contradistinction to their system of high calculation was the personal, anecdotal, and ethnographic model of Mao's early social analyses. For much of the 1950s, the abstraction in the periodic report was the normative ideal in the world of PRC statistics. That is not to say that its alienating

---

103   "There can be no knowledge apart from practice." See "On Practice," in *Selected Works of Mao Tse-tung*, Vol. 1, 301.

tendencies did not generate tensions (a subject we shall turn to in later chapters), but such tensions were largely kept in abeyance. It was only with the launch of the GLF in the summer of 1958 that Mao's ethnographic approach was rebadged as a bonafide method with its own theoretical underpinnings and elevated above other methods of statistical data gathering.

3

Ascertaining Social Fact

WHERE DID THE INSPIRATION FOR, and confidence in, exhaustive enumeration via a periodic reporting system come from? In the most concrete
terms, it came from the desire to emulate the success of the Soviet Union. As
Maurice Meisner noted, to the Chinese (and many other socialist states after
1945), the Soviet Union "provided the only historical model for industrializing
an economically backward country under socialist political auspices."[1] Mao
Zedong had articulated much the same point in his 1937 essay "On Practice,"
where he noted:

> The struggle of the proletariat and the revolutionary people to change the
> world comprises the fulfillment of the following tasks: to change the objec-
> tive world and, at the same time, their own subjective world—to change
> their cognitive ability and change the relations between the subjective and
> the objective world. Such a change has already come about in one part of
> the globe, in the Soviet Union. There the people are pushing forward this
> process of change. The people of China and the rest of the world either are
> going through, or will go through, such a process."[2]

By 1949, Mao was even more direct: "The Communist Party of the Soviet
Union is our best teacher and we must learn from it."[3] The Soviet Union responded with great enthusiasm. A principal element in Soviet aid was the

---

1 Meisner, *Mao's China and After*, 109.

2 Mao, Tse-tung, *Selected Works of Mao Tse-tung*, Vol. 1, 308.

3 Mao made this claim on 30 June 1949 in his speech on "The People's Democratic Dictatorship" (*Selected Works of Mao Tse-tung*, Vol. 4, 411–423). This was also the speech in which Mao
used the phrase "leaning to one side" to describe Chinese policy. On the speech and this policy,
see Chen and Yang, "Chinese Politics and the Collapse of the Sino-Soviet Alliance," 246–94,

sharing of technical knowledge. By 1960, when the Sino-Soviet split became final, as many as 18,000 Soviet and Eastern European experts had spent time in China sharing their expertise.[4] A handful of the experts were statistical experts—consultants, administrators, and teachers—whose advice was instrumental in the organization of the NSB and, later, the SSB.

Confidence in exhaustive enumeration was not, however, a mere artifact of emulation; rather, embedded in this choice was a deep-seated and sophisticated theoretical criticism of statistics as it had evolved and was evolving at that time. Put differently, the confidence in exhaustive enumeration came from a particular understanding of how to count and ascertain social reality, and the place of mathematics in that process. Within this theoretical framework, exhaustive enumeration came to serve as the basis for the socialist corrective to what came to be labeled "bourgeois" statistics. What socialist statistics and its preference for exhaustive enumeration did was to deny the universal applicability of mathematics in general and mathematical statistics in particular.

The purpose of this chapter is twofold: first, to uncover and understand the socialist critique of statistics and, second, to understand the role of the Soviet statistical experts who spent time in China and who were instrumental in the rise of socialist statistics to a position of epistemological and administrative dominance. I begin with a discussion of the 1950s (or, more accurately, the years after 1945) as a period when the imperative to ascertain social fact took on added urgency throughout the world. There existed, however, competing approaches to ascertaining social fact. The following sections focus on the rise of socialist statistics, in particular its rise in the USSR, and contrast it with other approaches to statistics. I then explore the Soviet experts who spent extended periods of time in the PRC, examining the variety of ways—teaching, translation of textbooks, and consultation—by which their expertise was mobilized by the Chinese SSB as it sought to disseminate a correct understanding and implementation of socialist statistics.

Ascertaining Social Fact in the 1950s—A Global Imperative

During the last two to three decades, the history of quantification, probability, counting, and facticity in the early modern and pre-twentieth-century world

---

esp. 247–50. For similar statements extolling the virtues of the Soviet Union, see also Bernstein and Li, *China Learns from the Soviet Union*, 1.

4 Shen Zhihua, *Sulian zhuanjia zai Zhongguo*.

has garnered increasing attention. The rich studies that have emerged have established quantification and statistics as independent subjects of inquiry, situating them at the very heart of the modern and postmodern experience.[5] Among the many insights that are offered by this rich literature is the significant role played by the state and by the study of the social world (in addition to the natural world), particularly through the expansion of bureaucracies and administrative duties, to realize what Porter has called the "statistical perspective."[6]

If these were essentially nineteenth-century developments, largely limited within the boundaries of nation-states, or in certain cases imperial formations, the twentieth century presented challenges of a new nature. The experiences of World War I led to the desire for new international/global institutions of arbitration and peace. Exemplary perhaps is the founding of the League of Nations in 1919. One of the activities of the League was involvement in the design and organization of commensurable economic and financial data.[7] At the same time, the duties of the state also expanded far beyond their nineteenth-century norms. Fueled in part by the rise of communism and the experience of the Great Depression, which, in turn, facilitated the rise of Keynesian economics, the state began to take on an ever-expanding responsibility to regulate the economy and provide services to its citizens.[8] The decade witnessed the earliest articulations of the idea of an "economy" as a bounded object that could be tracked via new inventions, such as National Income

5 See, for instance, Porter, *The Rise of Statistical Thinking*; Porter, *Trust in Numbers*; Porter, *Karl Pearson*; Hacking, *The Taming of Chance*; Poovey, *A History of the Modern Fact*; Krüger, Daston, and Heidelberger, *The Probabilistic Revolution*; Patriarca, *Numbers and Nationhood*; and Desrosières, *The Politics of Large Numbers*. Exemplary, for instance, is Porter's discussion of the life and work of Karl Pearson, in Porter, *Trust in Numbers*, 20–21; and Porter, *Karl Pearson: The Scientific Life*.

6 "[W]hat we might call the statistical perspective owed more to the work of census bureaus and statistical offices concerned with health, crime, trade, and education than to the pure theory of mathematical probability." Porter, "Chance Subdued by Science," 474. In a review of two works on Russian and Soviet statistics, Gordin, "Statistique et revolution en Russie," 803–810, writes of "the importance of statistics as a lens for examining the rise of Soviet governance," a claim that applies more generally to the rise of governance.

7 As early as August 1919 the League convened meetings to discuss international statistical cooperation. It mooted the establishment of an International Statistical Commission, which met for the first time in Paris in October 1920. For more, see Nixon, *A History of the International Statistical Institute*, 28–29, 42.

8 For more, see, for instance, Desrosières, *The Politics of Large Numbers*; and Porter, *Trust in Numbers*.

Accounts and Input-Output Matrices, and quantified through made-up statistical indicators, such as Gross Domestic Product.[9] From a technological standpoint, each of the world wars also played a role, as they provided opportunities for the design, experimentation, and deployment of advanced techniques for data collection and analysis. The field of operations research, for instance, grew substantially on the basis of wartime exigencies.[10] After World War II, the earlier efforts by the League of Nations were renewed through the United Nations, which established various institutions for global order and communications, including a Statistical Commission in 1947.[11]

Finally, the collapse of empires in the wake of World War II heralded the arrival of several new nation-states, each with its own modernizing mission. A principal way to be modern, and to demonstrate one's modernity to others, was to know oneself and at the same time exhibit such self-knowledge to others. A statistical system that gathered and disseminated such knowledge using numbers, and employed the latest methods, became the hallmark of a modern nation-state.[12] For a long time, a standard imperial argument had been that locals did not even possess an accurate and scientific means of ascertaining (social) fact. Thus, empire's civilizing mission, included this "gift" of fact and numerical representation. In the Chinese case, the discourse about the lack of facts was keenly felt. Almost on par with the nineteenth-century Western imperial assertion that China lacked a rational legal system, the lack of factual self-knowledge came to inform a narrative of national humiliation. As Tong

9 Especially crucial in this regard were the contributions by Simon Kuznets (1901–1985), who drafted U.S. Bureau of Foreign and Domestic Commerce, *National Income, 1929–32*, which provided the basis for a system of national accounts; and Wassily Leontief (1905–1999), who developed input-output accounts. For more, see Landefeld, "GDP: One of the Great Inventions"; Coyle, *GDP: A Brief but Affectionate History*; and Karabell, *The Leading Indicators*.

10 On the wartime histories of operations research in the United Kingdom and the United States, see Kirby, *Operational Research in War and Peace*; Wakelam, *The Science of Bombing*; and Schrader, *History of Operations Research*.

11 The Statistical Commission was established through Resolution 8(I) of the Economic and Social Council of 16 February 1946. The council took supplementary actions on 18 February 1946, and the resolution was subsequently amended by Resolution 8(II) of 21 June 1946. The Statistical Commission's latest terms of reference are based on Resolution 1566 (L), which was adopted on 3 May 1971. For more, see https://unstats.un.org/unsd/statcom/. China's nominee to the commission was D. K. Lieu.

12 In his discussion of planning in postcolonial India, for instance, Chatterjee, "Development Planning and the Indian State," 82–103, has shown how planning for the new nation was formulated based on the logic of numbers. Statistical indicators were precisely the metrics through which progress could be measured and contrasted with colonial underdevelopment.

Lam has shown, the response was the rise of a "social survey movement," whose principal goal was to assess what China was, so as to make her legible, knowable, and thereby eventually changeable.[13]

By the end of World War II, these developments, and the desire for a new world order, contributed in part to what might be termed a more globalized epoch of statistics-driven numerical positivism.[14] New attempts at global organization coincided with the postcolonial wave of newly independent countries. Statistics came to be viewed as a key tool for national and international development and stability. Newly formed nation-states sought to establish statistical bureaus, collect data, and address issues of governance and growth. At the global level, newly formed international organizations, such as the United Nations and the World Bank, sought to render such data commensurable by establishing norms for their collection, collation, and analysis.[15] Each of these developments depended on and demanded new levels of data collection and analysis. This desire for data, or for facts, was summed up by UN Secretary-General Trygve Lie (1896–1968) in 1947. Addressing the inaugural session of the International Statistical Institute's twenty-fifth meeting in Washington, DC, Lie observed:

> The free exchange of information on economic and social affairs among all countries in the world is absolutely necessary to economic and social advancement. We cannot cure our troubles unless we know in the first place what those troubles are. Likewise, we cannot achieve international understanding, which is the basis of advancement, unless the peoples of the world are given the facts about each other. . . . *There is no substitute for facts, for clear and systematically organized facts. They alone can be relied upon to measure resources and potentialities for progress and to direct policies and actions designed to achieve the objectives of all civilized people* (emphasis added).[16]

13 Lam, *A Passion for Facts*.

14 On how some of these trends played out in the articulation of a particular "Cold-War rationality" in the United States, see Erickson et al., *How Reason Almost Lost its Mind*; and Lemov, *Database of Dreams*.

15 For a history of the UN and its role in this process, see Ward, *Quantifying the World*.

16 Lie, "Opening Address," 151. At the same meeting, discussions were also held to set up an International Association for Research in Income and Wealth. It was felt that "the solution to many important problems in the field could be sought most effectively by the close cooperation of those most familiar with the income and wealth statistics in the various countries . . . [and that] such an organization would serve to bring the scholars in the field together; to develop greater familiarity on the part of members with work in other countries; and to provide

Speaking, as he was, at the pre-eminent gathering of international statisticians, the desire for "clear and systematically organized facts" was predicated on a form of numerical positivism—numbers alone could help understand basic realities, design solutions to problems, and plan for the future. As a result, the postwar years were a time when several statistical innovations achieved tremendous legitimacy.[17] These innovations included large-scale random sampling, operations research, development theory (most influentially, W. W. Rostow's theory on the stages of economic growth), demographic transition theory, random sampling methods such as Monte Carlo, and the spread of economic planning.[18]

A Universal Science

Most of these innovations were possible because of a specific shift in the understanding of statistics, quantification, and probability that occurred during the nineteenth century. Until the late nineteenth century, statistical activities broadly construed progressed along two nonintersecting tracks, one descriptive and the other analytic. The descriptive track traced its origins to the seventeenth-century "political arithmetic" of William Petty and, among others, the German school of descriptive statistics (*Statistik*, from which the mod-

---

a continuous stimulus to the solution of ever present problems of definition and analysis and to the formulation of new approaches." See International Statistical Institute, *Introduction: International Statistical Conferences, September 6–18, 1947*, 308–311.

17 Porter, in *The Rise of Statistical Thinking*, 3, has observed that statistical tools used in modern sciences had been worked out during the previous some 100 years. The phase of invention took place mainly during the nineteenth century and the phase of development during the late nineteenth and early twentieth centuries. Extending his periodization, we can label the remainder of the twentieth century, starting with 1930s, as a period of application and refinement.

18 An interesting case-in-point is the career of the Kennedy and Johnson era secretary of defense Robert McNamara (1916–2009), whose obsession with statistics is well-documented. During World War II, he worked in the Air Force's Office of Statistical Control, applying statistical and management theory (the "Harvard Method") to enhance the efficiency of B-29 bombers. Following the war, McNamara put his skills to good use by helping resuscitate the ailing Ford Motor Company. During the Vietnam War his obsession with "statistical control" led to an overt reliance on numbers and models over information coming from other, in retrospect, more reliable sources, thereby contributing to the prolonging of the war. From 1968 to 1981 McNamara also served as president of the World Bank. See Rosensweig, "Robert S. McNamara and the Evolution of Modern Management"; and Halberstam, *The Best and the Brightest*.

ern word originates), and through them to a much more fundamental and age-old global phenomenon—the desire of the state to collect information about its constituents, its land, its produce, and so on. The analytic, which can be traced at least as far back as the sixteenth and seventeenth centuries, was principally concerned with concepts of chance, probability, and estimating and controlling error, primarily in an abstract, mathematical sense.[19] The shift that occurred in the nineteenth century involved a new understanding of error not as a reflection of chaos but rather as variations in a given natural order. As Porter notes, "the acceptance of indeterminism constitutes one of the most striking changes of modern scientific thought. With few exceptions, scientists and philosophers previous to the late nineteenth century would have agreed with Augustus De Morgan that to say an event occurs by chance is to say that it occurs for no reason at all."[20] Similarly, Hacking has shown how the erosion of determinism made possible "autonomous laws of chance."

> The idea of human nature was displaced by a model of normal people with laws of dispersion. These two transformations were parallel and fed into each other. Chance made the world seem less capricious: it was legitimated because it brought order out of chaos. The greater the level of indeterminism in our conception of the world and of people, the higher the expected level of control.[21]

19 This is a gross simplification of an era during which ideas about probability were in tremendous flux. As Daston, *Classical Probability in the Enlightenment,* has shown, a key feature of this period was that the probabilists did not differentiate between objective and subjective aspects of probability, the former an observable feature of the world and the latter based on a degree of personal judgment about the outcome of specific events. In a recent article, Gelman and Hennig ("Beyond Subjective and Objective in Statistics") propose eliminating the subjective/objective dyad and replacing it with a broader set of attributes: "objectivity replaced by *transparency, consensus, impartiality* and *correspondence to observable reality,* and subjectivity replaced by awareness of *multiple perspectives* and *context dependence.*" See also Poovey, *A History of the Modern Fact.*

20 Porter, *The Rise of Statistical Thinking,* 149–150. Porter continues: "Still, this story is not simply one of a new appreciation of the empire of chance. Randomness first attained real standing in scientific thought not as a source of massive uncertainty, but as a small-scale component of an overarching order. The recognition of chance stemmed not from the weakness of science, but from its strength—or rather, its aggressive imperialism, the drive to extend scientific determinism into a domain that had previously been seen by most as the realm of inscrutable whimsy."

21 Hacking, *The Taming of Chance,* vii.

For Hacking and Porter, these developments owed as much to the study of natural phenomena as they did to social phenomena.[22] It was not error analysis per se, but rather the use of probability as a modeling tool to analyze variation in nature *and* society, a mutual cross-pollination across disciplines and intellectual pursuits that contributed to the evolution of statistics as a distinctly mathematical field of inquiry. This, in brief, is the liberal story in which statistics emerges as a universal science, capable of studying and discerning patterns in both society and nature, and employing a catholic approach to methodology—incorporating probabilistic and nonprobabilistic methods.[23] But statistics also has a parallel history, which, at least for a significant portion of the twentieth century, enjoyed great legitimacy and widespread use.

A Critique of Universality

Lenin is famously said to have observed that "socialism is first of all accounting."[24] Within the USSR he enjoyed fame as a skilled statistician who used numbers to expose the oppression and injustices of imperialist tsarist rule and, after 1918, as someone who vigorously promoted the development of state statistical institutions.[25] Stalin, too, is known to have noted the fundamental importance of statistical activities and data.[26] In accepting the centrality of statistics as a tool for understanding and shaping economic phenomena, Marxist and Soviet thinkers and practitioners shared much in common with thinkers in other parts of the world.[27] But in the selection of theory and meth-

---

22 As Porter, *The Rise of Statistical Thinking*, 91, noted: "Quetelet's insight concerning distribution of traits in human populations . . . confirms the more general argument of this book concerning the role of social science in the development of statistical thinking."

23 See for instance, the discussion in Desrosières, "How Real Are Statistics?" It is perhaps instructive to recall that Porter's thesis, the basis for *The Rise of Statistical Thinking*, was entitled "The Calculus of Liberalism: The Development of Statistical Thinking in the Social and Natural Sciences of the Nineteenth Century." My thanks to William Deringer for pointing this out.

24 As Haber, "Socialist Realist Science," 87, notes, this was a popular paraphrasing of a sentiment Lenin articulated in December 1917. See also Lenin, "How to Organise Competition?" 404–415.

25 On Lenin as a statistician, see Kotz and Seneta, "Lenin as a Statistician."

26 "Comrades, no constructive work, no state activity, no planning is conceivable without proper accounting. And accounting is inconceivable without statistics. Without statistics, accounting cannot advance one inch." Stalin, "Organisational Report of the Central Committee," 24 May 1924, 225.

27 "In the U.S.S.R., statistics cannot be separated from economics either in the field of

ods to pursue these goals, these thinkers drew a sharp line between themselves and those in a liberal tradition. Indeed, while Marx was likely cognizant and appreciative of developments in mathematics and statistics made possible by the works of such notable figures as Quetelet, Poisson, and Laplace, and by the numerical turn within German statistics, Soviet and Chinese statisticians were quick to point out that he restricted his own discussions on the efficacy of statistics solely to the realm of political economy.[28]

In Soviet and Chinese statisticians' reading of Marx and Engels, and subsequently of Lenin and Stalin, there were no universal laws of economics and statistics.[29] Rather, different historical epochs and different social systems generated different statistical and economic laws. As Yugow noted in a 1947 review of Soviet statistics, it was felt that:

> The statistical methods applied in the accounting of the results of private economic activity, and the patterns established for economies in which the chief economic regulating forces are the laws of supply and demand, have proven inapplicable to statistical accounting in a planned state economy.[30]

The nature of the economy and social organization in a particular country, which, in turn, reflected a particular stage in its history, required its own sets of statistical and economic laws. The distinction between the laws that govern society, and thus appropriately belonged to social science, and those that govern nature was one that could not be surmounted. Accordingly, Chinese and

---

theory, in the establishment of the methodology of study of new economic phenomena, or in practical work in the accounting and analysis of primary statistical data." See Yugow, "Economic Statistics in the U.S.S.R.," 243 (Yugow is identified as a "writer on Russian economics, New York City," in the paratext to the journal issue).

28 Xu Qian, "Tongji xueshuo fazhan bianhua de lunkuo," 24.

29 That Marx's own views on the matter were more complex perhaps requires little belaboring. Even so, his writings provided enough space for such a selective interpretation to proceed. Consider, for instance, Marx's assertion in *Capital*, Vol. 1, 693, that "every special historical mode of production has its own special laws of population, historically valid within its limits alone. An abstract law of population exists for plants and animals only, and only in so far as man has not interfered with them." But some twenty years earlier, Marx had also stated in *Economic and Philosophic Manuscripts of 1844*, 111, that "History itself is a *real* part of *natural history*—of nature's coming to be man. Natural science will in time subsume under itself the science of man, just as the science of man will subsume under itself natural science: there will be *one* science (emphasis in original)."

30 Yugow, "Economic Statistics in the U.S.S.R.," 243.

Soviet statisticians latched onto the numerous occasions when Marx identified statistics as a weapon that could be used in service to the proletariat classes, in particular by providing trenchant critiques of existing political-economic conditions and as a tool for the realization of workers' statistics. In such a formulation, statistics was fundamentally political arithmetic and Marx's regard for William Petty, the originator of the phrase, as the founder of both statistics and political economy was seen as further evidence of the validity of the natural/ social dyad.[31]

Although Marx did not enjoy the political power necessary to put many of these ideas about the distinction between the natural and social world into action, his self-proclaimed intellectual heirs were in a position to do so. They were able to apply statistics to statecraft—to use it as a tool, not just for analysis and critique, but also for administering change. The manner in which they chose to do this had implications for the evolution of statistics. For instance, in picking up Marx's mantle, Lenin also gave primacy to statistics as a practical tool, even to the extent of de-emphasizing academically oriented activities among statisticians.[32] Lenin was especially keen to make the Soviet Central Statistical Board the primary source for all data related to society and economy. In a letter to the board, Lenin noted that "the Central Statistical Board, which *lags behind* an unofficial group of writers, is a model bureaucratic institution. In about two years' time it may provide a heap of data for research, but that is not what we want." He proceeded to demand that "for our practical work we *must* have figures and the Central Statistical Board *must* have them *before anybody else*" (emphasis in original).[33] The result, initially in the Soviet Union and subsequently in at least several of the places where its influence

31 "William Petty, the father of Political Economy, and to some extent the founder of Statistics. . . ." Marx, *Capital*, Vol. 1, 299. For Marx, William Petty symbolized a leap in quality as far as statistical activities were concerned (Xu Qian, "Tongji xueshuo fazhan bianhua de lunkuo," 18–21). On Marx and his views on William Petty, see also McCormick, *William Petty and the Ambitions of Political Arithmetic*, where the epilogue, in particular, explores what Mc-Cormick calls Marx's resurrection of Petty as the founder of modern political economy.

32 In a 26 May 1921 letter to his friend, the economist, planner, and then chief of GOSPLAN, G. M. Krzhizhanovsky (1872–1959), Lenin observed: "The Central Statistical Board should be made into an organisation that does analysis for us, current, not 'scientific' analysis. . . . Statisticians must be our *practical assistants*, not engage in scholastics (emphasis in original)." Lenin, "To G. M. Krzhizhanovsky" (dated 26 May 1921), 498. The latter half of the citation above, about practical assistants, is also quoted in Kotz and Seneta, "Lenin as a Statistician," 87.

33 Lenin, "Letters to the Central Statistical Board" (dated 16 August 1921), 31.

extended, was to create a distinction in statistical practice and, eventually, in statistics as a discipline. Fueled by such an understanding of human activity and how to account for it, an understanding of statistics that was specific, context-based, and nonuniversal began to attain legitimacy by the 1930s.

> Statistics [thus] became an instrument for planning the national economy. Consequently, its basis is the Marx-Lenin political economy; it represents a social science or, in other words, a class science. The law of large numbers, the idea of random deviations, and everything else belonging to the mathematical theory of statistics were swept away as the constituent elements of the false universal theory of statistical science.[34]

Prior to the 1930s, Russian and Soviet statisticians enjoyed a fair degree of freedom and saw themselves as participants and contributors to the evolving universal discipline of statistics, the bedrock of which was mathematics itself. Keenly aware of global developments in the field, they also actively sought engagement with methods devised outside the Soviet Union. The application of theory to practical problems was a hallmark of this period.[35] But this was also a period during which numbers as measures of performance became absolutely critical. As Michael Gordin has observed in a review of two studies on Russian and Stalinist statistics: "For the Soviet state, given its self-professed character as a scientific state, these numbers become crucial for measures of how much socialism was an improvement over tsarist capitalism...."[36] As a result, statisticians faced tremendous pressures, both in terms of how they understood their discipline and in terms of the actual numbers they produced and analyzed.[37]

34 Zarkovic, "Note on the History of Sampling Methods in Russia," 338.

35 On Russian and Soviet contributions to sampling theory, see, in particular, Seneta, "A Sketch of the History of Survey Sampling in Russia." For recent histories of statistics during Tsarist Russia and the early Soviet Union, see Mespoulet, *Statistique et révolution en Russie*; and Blum and Mespoulet, *L'anarchie bureaucratique*. A useful summary and review of both volumes can be found in Gordin, "Statistique et revolution en Russie." For contemporary accounts of early Soviet statistical work and organization, see Ezhov, *Soviet Statistics*; and Ezhov, *Organisation of Statistics in the U.S.S.R.*

36 Gordin, "Statistique et revolution en Russie," 807.

37 For instance, Pavel Il'ich Popov (1872–1950), the first director of the Soviet Central Statistical Administration, argued in a letter to Stalin on 22 December 1925 that there was no difference between socialist and bourgeois statistics. In the following year, he was removed from his post. Gordin, "Statistique et revolution en Russie," 809, citing Blum and Mespoulet, *L'anarchie bureaucratique*.

By the late 1930s, *Vestnik Statistiki* (Messenger of Statistics), the leading statistics journal in the Soviet Union, began to reject submissions that used mathematical approaches to deal with statistical problems.[38] As a result, over the course of the 1930s and 1940s, many statisticians withdrew from statistical work, finding a relative safe haven in universities or research institutes where they could continue their statistical research without identifying it as such.[39] The practitioners of statistics, in turn, grew increasingly ignorant of the latest statistical methods and their potential applications to statistical work.[40] Indeed, by the 1930s the "most persistent difficulty for these statisticians stemmed from their dual task: they were supposed to provide accurate

38 This claim is based on Zarkovic, "Note on the History of Sampling Methods in Russia," 338. However, Seneta, "A Sketch of the History of Survey Sampling in Russia," 122–123, observes that *Vestnik Statistiki*, which began publication in 1919 as an organ of the Central Statistical Office (*Tsentralnoe Statisticheskoe Upravlenie*) of the USSR, endured a near-twenty-year hiatus from 1930 until 1949. During that period, many of its functions, especially those pertaining to official statistics, were taken over by the journal *Planovoe Khozaistvo* (Planned Economy). It is likely this combined record that Zarkovic referenced in 1956. Indeed, a bibliographic record of the journal published in 1971 (Onoprienko, *Bibliograficheskii Ukazatel Statei i Materialov po Statistike i Uchetu. Zhurnal Vestnik Statistiki za 50 let [1919–1968]*) covers the fifty years from 1919 to 1968, clearly combining the two journals under the banner of *Vestnik Statistiki*.

39 Symptomatic is the case of Andrey Kolmogorov (1903–1987), who was recognized as among the leading probabalists of his time, but was rarely of interest to statisticians within the USSR. It is indeed possible that a general disciplinary demarcation may have also had the benefit of freeing up the harder sciences, most notably nuclear research, which could then use advanced mathematical statistics without falling prey to ideological battles. Such an analysis was offered by the commentator (identified as J. M.) to an essay on the 1954 Soviet meeting on statistics: "One very practical reason for this distinction [the bifurcation of statistics and mathematical statistics] is the need to legitimize statistical work in technology and the natural sciences, which was questioned by some attitudes of Marxist orthodoxy expressed in the published discussion. Now that such work is labelled mathematical statistics and thus distinguished from socio-economic statistics, it can be pursued with freedom from doubts induced by high theory." On Kolmogorov, see Kendall, "Andrei Nikoaevich Komogorov"; on Kolmogorov and Soviet statistics, see: Mahalanobis, "*Some Impressions,*" 5; Zarkovic, "Note on the History of Sampling Methods in Russia," 338; and on the freeing up of certain technical sciences, see Ostrovitianov, "The Discussion on Statistics Summed Up," 321. Other Soviet mathematicians who endured similar fates include Romanovsky, Smirnov, and Slutsky, the last of whom abandoned statistics in favor of astronomy (Zarkovic, "Note on the History of Sampling Methods in Russia," 338). For similar phenomena in the Western world, see Porter, *Trust in Numbers*, 222–225, on high energy physics; Erickson et al., *How Reason Almost Lost its Mind*, 17–20, on game theory, whose applications were proscribed, and on cybernetics, which flourished.

40 Zarkovic, "Note on the History of Sampling Methods in Russia," 338.

numbers for the five-year plans, but they also had to suppress real economic and demographic data that contradicted the official picture."[41] In Blum and Mespoulet's formulation, "in the hybrid that was Soviet Statistics, the Soviet triumphed over the Statistics."[42] This dichotomy extended into the 1950s and was clearly visible to outside observers. During his 1957 visit to the PRC, the Indian statistician Prasanta Chandra Mahalanobis informed his Chinese hosts that since 1947 the Soviet Union had not attended any sample survey activities organized by the International Statistical Institute, adding that the Soviets just did not attach importance to mathematical statistics.[43] The separation of mathematical statistics from statistical work, of theory from practical application, was complete.

Confusion over the true definition and status of statistics, however, lingered until it was officially resolved in 1954. It was through deliberations at a national conference, held in Moscow from 16 to 26 March 1954, that what had for the most part already occurred in practice was given a final and official imprimatur. K. V. Ostrovitianov, vice president of the Soviet Academy of Sciences, subsequently published a summary of the conference.[44] Organized by the Soviet Academy of Sciences, the Central Statistical Board, and the Ministry of Higher Education, the purpose of the conference was "to discuss matters in dispute and to examine the subject and method of statistical science."[45] In all, 760

41 Gordin, "Statistique et revolution en Russie," 809.

42 Ibid., 808.

43 WJB 105-00530-06: 15.

44 Ostrovitianov's report was published in *Vestnik Akademii Nauk SSSR*, no.8 (1954): 3–12. All citations here are from a translation of the Russian original (Ostrovitianov, "The Discussion on Statistics Summed Up"). In his commentary to the translation, J. M. [*sic*] notes that "Ostrovitianov's main point, broadly stated, is that, at the present time at any rate, the name *statistics* should be limited to the discipline and practice which handle the quantitative aspects of socioeconomic phenomena; that this is the only statistical discipline deserving the status of science in its own right; and that statistical work on natural and technological phenomena is most conveniently regarded as a branch of mathematics, namely *mathematical statistics*" (Ostrovitianov, "The Discussion on Statistics Summed Up," 321). *Vestnik Statistiki* also published a detailed report on the conference: "Review of the Scientific Meeting on the Question of Statistics," *Vestnik Statistiki*, no. 5 (1954): 39–95. Ostrovitianov would visit China as part of a Soviet cultural delegation in October of that year; see "Пребывание советско й делегации деятелей культуры в Китае" (The Stay in China of the Soviet Delegation of Cultural Figures), *Izvestiia*, 9 October 1954, 3; "Банкет в честь советских специалистов в Китае" (Banquet in Honor of the Soviet Specialists in China) and "Банкет у Го Мо-жо" (Banquet with Guo Moruo), *Pravda*, 6 October 1954, 4.

45 Ostrovitianov, "The Discussion on Statistics Summed Up," 323.

invitations were issued throughout the various Soviet republics.[46] The majority of those invited were academic and practicing statisticians, and they were joined by smaller numbers of economists, engineers, mathematicians, philosophers, and medical men. In all, sixty people spoke at the conference and twenty submitted written contributions. Three principal points of view emerged during the conference. The first held that statistics is a universal science that studies phenomena of society and nature. The second posited that statistics is a social science concerned with method, a system of principles for collecting quantitative information in order to characterize social phenomena. The final view was that statistics is a social science that only studies social productive relations, that is, the economy.[47]

During the ten days of the conference, the first two views were discussed and summarily rejected. Proponents of the first claim, which was no different from the views in the "bourgeois liberal" West, were accused of a "gross error in attempting, with the help of a single science of statistics, to study by the same methods the phenomena of both nature and social science."[48] This error was not only a scientific error but it also had moral dimensions. As Ostrovitianov declared in his report on the conference:

> The comrades who regard statistics as a universal science which studies nature and society make of it some kind of science over and above the [socio-economic] classes, coldly indifferent to good and to evil, without any preference at all as between [socio-economic] classes and between social structures.[49]

In support of this claim, Ostrovitianov offered a criticism of the comparison made by some participants at the conference between the luminosity of stars and the categorization of the peasantry.

> What can there be in common between the grouping of stars according to their luminosity and, let us say, Lenin's grouping of the peasantry according to their [socio-economic] class characteristics? *Nothing, other than certain*

46 Invitations went out to (name of contemporary republic in parentheses): Leningrad (Russia), Kiev (Ukraine), Minsk (Belarus), Baku (Azerbaijian), Tashkent (Uzbekistan), Tbilisi (Georgia), Riga (Latvia), Erivan (Armenia), Alma-Ata (Kazakhistan), Kharkov (Ukraine), Sverdlovsk (Ukraine), and Irkutsk (Russia).

47 Ostrovitianov, "The Discussion on Statistics Summed Up," 323–324.

48 Ibid., 324.

49 Ibid., 326.

technical statistical devices used in the study of these totally different kinds of phenomena. The grouping of stars according to luminosity requires no class analysis at all, whereas in investigating the differentiation of the peasantry we must make a [socio-economic] class analysis of this phenomenon, proceeding from the postulates of historical materialism and political economy. Statistical devices and methods of investigation, the choice of characteristics for grouping, the combining of groups, were entirely without exception subordinated by Lenin to the task of [socio-economic] class analysis of the rural population (emphasis added).[50]

Ostrovitianov conceded that quantitative relations in their pure form—equally applicable to "inorganic nature, to organic nature and to the domain of the social sciences"—could be studied by mathematics, more specifically by the subdomain of mathematical statistics. But apart from this branch of mathematical statistics and the socio-economic science of statistics, there did not exist a third "non-mathematical but nevertheless universal 'general' statistics."[51] This distinction formed the basis for the rebuttal of the second claim, that statistics was a science of method. "Every social science," Ostrovitianov explained, "has not only its own subject, but also its own method of investigation." Furthermore, he noted that in political economy, the method employed was abstraction (historical materialism in the sphere of economic relations), and not, quoting Marx, a microscope or chemical reagents.[52] Thus, it was decided that only in the study of social phenomena did statistics reach the level of an independent science, with its own subject and methods of investigation.[53]

And so, following almost two weeks of deliberation and debate, the conference was in a position to offer a definition of statistics.

Statistics is an independent social science. It studies the quantitative aspects of mass social phenomena in full awareness of and therefore within

---

50 Ibid.

51 Ibid. "Thus, there is no scientific grounds for a universal statistics as an independent discipline which studies both nature and society. It would be an extremely meagre science made up of some statistical devices which are used in all branches of knowledge, and would hang somewhere between socio-economic and mathematical statistics."

52 It is indeed not clear how the act of abstraction can be compared to the use of a physical instrument, such as a microscope, or a substance, such as a chemical reagent, under the single category of method.

53 Ostrovitianov, "The Discussion on Statistics Summed Up," 327.

their qualitative aspect; it investigates the quantitative expression of the laws of social development in the specific conditions of place and time.[54]

Statistics thus was to serve, in the words of Lenin, as the "rock of exact and indisputable facts," the very foundation on which rested not only political economy but all other social sciences as well. In early June *Pravda* reported on the conference, explaining that "it [had] clarified major problems in the development of statistics as a social science."[55] The effect, in the end, was to ratify and reify a distinction that was already fairly well entrenched, that is between statistics and mathematical statistics.

The 1954 resolution had a signal influence on the wider PRC statistical community. News of the resolution and what was now the Soviet state-sanctioned definition of statistics, however, took time to spread; a complete Chinese translation of the deliberations was not published until early 1955.[56] But its impact was evident in the more confident tone and tenor of the critiques that began to appear in 1955.[57] Most prominent was an essay by Xu Qian and Liu Xin. In 1953, the two had published an essay critical of bourgeois statistics in the journal *Xin jianshe* (新建设; New construction).[58] In 1955 they published an updated version in the SSB's journal *TJGZTX*, explaining that their basic critique had been justified in light of the 1954 Moscow conference.[59] In the ensuing debate, which we shall look at more closely in the following chapter, statistics was circumscribed largely within the realm of application and divorced from mathematics.[60]

54 Ibid., 328.

55 "Problems in the Development of Statistical Science." *Pravda*, 2 June 1954, 3. Quote from a translated version published in *The Current Digest of the Russian Press* 6, no. 22 (14 July 1954). In 1954, the seventh issue of the *Bulletin of the Academy of Sciences of the Armenian SSR: Social Science* also published a report on the Moscow conference, including a discussion of follow-up activities in Armenia.

56 Editorial Note, "Sulian tongji kexue huiyi jueyi," 1–4, 45.

57 This is an impression based on interviews I conducted in Beijing with, among others, Renmin University Professor of Statistics Yuan Wei on 25 March 2011.

58 Xu Qian and Liu Xin, "Guanyu zichan jieji tongji lilun de pipan."

59 Ibid.

60 In her insightful discussion of science in China during the Cold War, Schmalzer, "Self-Reliant Science," 86, observes that the "overdetermined" dichotomy between applied and basic science has to be understood through the emphasis on self-reliance, which itself had at least two implications: the use of local methods and local independence from the center.

TABLE 3.1. Two Approaches to Statistics

Statistics	Socialist statistics
Statistics is a universal science	Statistics is a social science
Chance exists; randomness is part of both the natural and social order of things	There is no chance in the social world (there are only laws that can be ascertained)
All counting generates errors; methods can be devised to ascertain and control errors in any count	Everything can be counted accurately; the best count is a complete count
Key methods	
Law of large numbers	Exhaustive enumeration
Probability theory	Periodic reports
Variance, correlation, regression	Typical sampling
Random sampling	

Taming and Rejecting Chance

Trygve Lie's desire for "clear and systematically organized facts" thus echoed Lenin's evocation of statistics as the "rock of exact and indisputable facts." But although such a desire united the Universalist claim and its Marxist (Soviet-Chinese) critique, the contrasting understandings of statistics within each placed them at loggerheads when it came to questions of theory and method. The contradictory claims regarding the universality of statistics as outlined above were undergirded by two fundamentally different ontological positions as to how social reality could be ascertained via quantification (see Table 3.1). The first, as we have seen, by embracing uncertainty, accepted its own limits. Chance and error were seen as part of the natural order of things. Not everything could be counted, at least not accurately, and even more to the point, all attempts at counting generated errors. This did not mean that exhaustive enumeration or other nonprobabilistic methods were entirely abandoned.[61] Just that the goalposts had shifted: Intellectual energy was no longer channeled toward devising methods to count exhaustively, but instead to devising methods for studying variation, and eventually accurately determining, controlling,

61 Up to the end of the nineteenth century and beyond, many German and Italian statisticians continued to see statistics largely as a social science, and data collection in U.S. agriculture continued to rely on exhaustive enumeration well into the 1930s; my thanks to Sarah Milov and an anonymous reviewer for my *Osiris* essay (Ghosh, "Lies, Damned Lies, and [Bourgeois] Statistics") for pointing this out.

and minimizing the errors that resulted during any count. Probability, the law of large numbers, and general mathematical reasoning thus became crucial tools in devising new methods of counting and assessment.

This was in stark contrast to the second approach, which, fueled by the teleological inevitability of Marxist historical progression and by the confidence offered by dialectical materialism, was predicated on the absence of chance and uncertainty in the social world. Within such an ontological worldview there was no place for probability, the law of large numbers, or any kind of abstract mathematical thought. The social world had no chaos, only laws of production, population, and relations—all of which could be discovered via measurement. The result was an emphasis on complete enumeration—the counting of everything. In other words, just as the liberal tradition pivoted on an appreciation and eventual acceptance of uncertainty, chance, probability, and error in the world and in counting, so too was the Soviet and Chinese Socialist-Marxist critique based on the comfort of certainty and the desire to enumerate and account for everything.

In taming and rejecting chance, both traditions were interested in increasing overall control. But they had very different expectations about what they could control, and each approach taken to its extreme could also lead to perverse social and economic outcomes. For instance, the political economists Peter Katzenstein and Stephen Nelson have argued that the emphases on error, accuracy, and calculability in the West has led to a fixation on risk, frequently at the cost of acknowledging the importance of uncertainty. Unlike uncertainty, assessment of risk was based on past experience and was calculable, and therefore could be used as the basis for decision making. For this reason, uncertainty came to be discounted in the design of economic models, which over time contributed to an increasing disengagement from real-world actions.[62] To substantiate their claims about the importance of uncertainty, Katzenstein and Nelson analyzed the 2008 financial crisis, with its basis in

---

62 The distinction between uncertainty and risk in the sphere of economic activity was first independently articulated in the early 1920s by the economists Frank H. Knight (*Risk, Uncertainty, and Profit*) and John Maynard Keynes (*Treatise on Probability*). For claims that among American economists uncertainty has largely been written out in favor of risk, which is calculable, see Katzenstein and Nelson, "Worlds in Collision"; and Nelson and Katzenstein, "Uncertainty, Risk, and the Financial Crisis." A more "orthodox" depiction of risk conquering uncertainty can be found in Bernstein, *Against the Gods.* For a discussion of how risk came to be incorporated into economic thinking and capitalist practice in the United States during the

increasingly sophisticated mathematical models that were supremely confident in their ability to ascertain and account for risk.[63]

Socialist statistics' fetishization of complete enumeration would also lead to certain systemic cul-de-sacs. But the illusion of control that this method offered was extremely attractive. As Yugow noted in his review of Soviet statistics in 1947:

> Accounting of all economic phenomena on a national scale and the possibility of statistically following all economic phenomena in their complete economic cycle give Soviet statistics such great advantages in comparison with the statistics of other countries, that the forms, methods, and the results of the work of the statistical organs of the U.S.S.R. are highly instructive to all who work in the field of the theory and practice of economic statistics, even in cases when these data are not entirely exact and commensurable in particulars. The gathering of primary materials, the formulation of problems, and the effect of the plan upon economic factors in the U.S.S.R. provide an almost "laboratory" picture; in any case, they reflect the existing facts far more clearly and fully than in other countries, where the statistical study of national problems and those of individual branches is not only greatly complicated, in view of the technical incompleteness of the data, but is also deliberately distorted because of private interests, trade secrets, competition, etc.[64]

From a planning perspective, the promise of achieving data collection on a national scale, and thereby (in theory at least) ensuring that all manner of economic activity is accounted for, was hard to resist. It was also the kind of comprehensiveness that nonsocialist countries with their decentralized statistical apparatus found impossible to achieve.

In many parts of the world, and especially in much of what would become the Global South, these strengths of the Soviet system were seen as the basis for its economic performance. For instance, the Indian industrialists who drafted the Bombay Plan in 1944 wrote with admiration and aspiration:

---

nineteenth and twentieth centuries, see Mohun, *Risk: Negotiating Safety*; Levy, *Freaks of Fortune*; and Bouk, *How Our Days Became Numbered*.

63 On the role of risk and uncertainty management in precipitating two earlier financial crashes (the stock market crash of 1987 and the market turmoil that engulfed the hedge fund Long-Term Capital Management in 1998), see MacKenzie, *An Engine, Not a Camera*.

64 Yugow, "Economic Statistics in the U.S.S.R.," 244.

This [growth target for India] might appear to be too modest a goal for a planned economy to achieve, especially in view of the fact that in the U.S.S.R., within a period of 12 years since the beginning of the First Five-Year Plan, the national income is reported to have increased from 25 billion rubles to 125 billion rubles, i.e., five-fold.[65]

Even so, in India such high regard for planned economic growth did not extend to the adoption of Soviet socialist statistics. In China, however, adoption of Soviet socialist statistics in the years after 1949 was enthusiastic and comprehensive. Even as late as 1957, by which time tensions between China and the Soviet Union were already increasingly evident, Xue Muqiao, director of the SSB, noted in an editorial:

> [We must] seriously study the Soviet Union's advanced experience in constructing and managing a socialist economy, especially its experience in the planned management of the people's economy and in the establishment of planning and statistical work.[66]

In order to achieve this, Xue and the SSB relied heavily on Soviet statistical experts and the expertise and experience to which they laid claim. It was based on their advice that the PRC envisioned, and attempted to construct, a comprehensive statistical system.

Soviet Experts in China—The General Story

In his 1987 study of PRC economic development, Carl Riskin observed that, in a sense, Mao's slogan of "leaning to one side" anticipated the long-term unsustainability of Sino-Soviet cooperation. This did not mean, Riskin clarified, that Soviet contributions to Chinese development were insignificant.[67] Rather, it serves to caution against assuming that the PRC's enthusiasm for emulating the Soviet model was uncritical or unchanging over the course of

---

65 Thakurdas et al., *Memorandum Outlining a Plan of Economic Development for India*, 29. This was given as the reason for hoping for a 200 percent jump in national income within fifteen years (also quoted in Harris, "Appraisals of Russian Economic Statistics," 213). This is not to say that Indian statisticians and economists accepted Soviet statistical theory, but there is no doubt that they were impressed with what the Soviet Union had achieved in the three decades since its inception.

66 Xue Muqiao, "Woguo de jihua gongzuo," 7.

67 Riskin, *China's Political Economy*, 60.

the 1950s.[68] As much recent scholarship (discussed below) has shown, the real story is far more complex.

The general facts about Sino-Soviet cooperation during the 1950s are now fairly well known. In February 1950, the two nations signed the thirty-year Sino-Soviet Treaty of Friendship, Alliance, and Mutual Aid.[69] In the event, Soviet financial aid was rather limited. For instance, of the total investment in the First Five-year Plan (1953–1957), the Soviet share was only 3 percent.[70] But the aid took nonfinancial, nonmateriel forms as well.[71] From August 1952 to May 1953, a Chinese delegation visited the Soviet Union with the express purpose of drawing up China's First Five-year Plan with the help of Soviet experts.[72] Indeed, equally significant, if not more so, was aid that took the form of imparting technical expertise and experience.[73]

Until recently, this aid remained a relatively understudied subject. Only in the last twenty years or so has it begun to receive significant scholarly attention. The opening of the archives, in both the former Soviet republics and China, has allowed scholars to explore this relationship in much greater detail,

68 As Kirby, "China's Internationalization," has shown, the PRC's allegiance to the Soviet Union during the 1950s was entirely voluntary (unlike the case of many Eastern European states) and was fueled not only by a desire to emulate but also to contribute to the creation of a socialist world economy.

69 For an interesting analysis of the negotiations involved in the signing of the treaty during what was Mao's first foreign tour, see Yang Kuisong, "The Sino-Soviet Alliance and Nationalism."

70 Meisner, *Mao's China and After*, 112–113; see also Riskin, *China's Political Economy*, 74–77.

71 Naughton, in "The Pattern and Legacy of Economic Growth," was among the earliest scholars to point to the importance of technical knowledge-sharing within the broader aid provided to the early PRC by the Soviet Union.

72 The delegation was initially led by Zhou Enlai. Bernstein and Li, *China Learns from the Soviet Union*, 12.

73 Meisner, *Mao's China and After*, 113. Meisner notes that as many as 12,000 Soviet and East European experts were sent to China in the 1950s (a more recent estimate is discussed below). In return, some 6,000 students were trained in modern science and technology in Soviet universities, and 7,000 workers gained experience in Soviet factories. Another estimate is provided by Goikhman, "Soviet-Chinese Academic Interactions," who suggests that as many as 20,000 Chinese students, technicians, and scholars studied in or visited the Soviet Union in the 1950s. Shen Zhihua, *Sulian zhuanjia zai Zhongguo*, 405; and Naughton, "The Pattern and Legacy of Economic Growth," 232, estimate an even higher number, i.e., 38,000. The highest estimate of over 50,000 Chinese studying abroad in the socialist world is in Kirby, "China's Internationalization," 884.

focusing on the men, materiel, and technology that formed its basis.[74] These studies generally conclude that from 1949 to 1956 China emulated, if not always uncritically, the Soviet experience. The Chinese commitment to the Soviet Union's "advanced experience" began to weaken in 1956 and by the Sino-Soviet split in 1960, had been entirely repudiated.[75]

In recent years Soviet advisers and specialists in China have also attracted focused attention. Deborah Kaple was among the first scholars to address this subject by synthesizing scattered and frequently fragmentary Soviet archival materials and combining them with oral interviews with several of the men and women who had served as specialists in China.[76] Kaple discovered that the specialists were predominantly men, who remained in China from as little as a few months to a year, and who were not well supervised once they arrived in China. Most had a hard time establishing personal friendships with the locals. Conversely, the Chinese frequently complained about the quality of the specialists who were sent to China. Building and substantially expanding on Kaple's work, in 2003 Shen Zhihua published what remains the most detailed study of Soviet specialists in China.[77]

According to Shen, the twelve years from 1948 to 1960 can be divided into four distinct periods.[78] The first, from 1948–1949, was marked by a Soviet policy of wait and watch. The years from 1949 to 1953 witnessed the beginning of exchanges, but the experts were predominantly involved in either military affairs or in specific industrial sectors. Agreeing with this general assessment of limited engagement during this early phase, Izabella Goikhman has observed that until 1953 interactions were largely characterized by the exchange of publications. Even so, according to Shen's findings, about 5,000 Soviet experts, including those working with the armed forces, were in China during this period. It was only after 1953 that academic exchanges were institutional-

74 Among the first in this new wave of exploration was the edited volume by Westad, *Brothers in Arms*. The latest and most comprehensive work that covers not only the 1950s but also brings the story up to the present is Bernstein and Li, *China Learns from the Soviet Union*. The Sino-Soviet split has also received fresh treatment in the excellent multiarchival work by Lüthi, *The Sino-Soviet Split*.

75 Bernstein and Li, *China Learns from the Soviet Union*, 3.

76 Kaple, "Soviet Advisors in China."

77 Shen Zhihua, *Sulian zhuanjia zai Zhongguo*. Shen made extensive use of archives in both the Soviet Union and the People's Republic and was able to augment his findings with oral interviews.

78 For a general summary of this periodization, see Shen Zhihua, *Sulian zhuanjia zai Zhongguo*, 405–410.

ized, and the number of visiting personnel increased.[79] For Shen, the late 1952 slogan "completely study the Soviet Union" (全面学习苏联; *quanmian xuexi Sulian*) marked the beginning of this period. The four years starting from 1954 represented the high tide of Soviet experts in China. As many as 11,000 visited China to share their technical expertise and experience. For the visiting experts, this was also a period when the work environment was the best, and when Chinese eagerness to learn was at its most earnest. This was also, however, the period when finding China's own path began to be openly discussed in certain areas. By 1956, the first doubts about the Soviet model began to be articulated. Even so, in the field of science and technology, studying Soviet methods was still the pre-eminent strategy.[80] During the final phase, 1958–1960, Shen finds a gradual decline and then a sudden withdrawal of all experts in 1960. He calculates 2,000 Soviet experts were in China during these final years, producing a total of 18,000 Soviet experts over the course of the twelve-year period.

Soviet Statistical Experts

In spite of the varied and contingent nature of Chinese engagement with Soviet aid and expertise, statistics was one area in which the Soviet impact was seminal and long-lasting. Although it is difficult to assess how many of Shen Zhihua's estimated 18,000 Soviet experts were statisticians, we can identify several key figures who were assigned to important positions as consultants at the SSB or at major educational and cadre-training institutions, such as Renmin (People's) University.[81] The first mention of Soviet statistical experts is found in Wang Sihua's speech of 1950. In it he observed that the Soviet Central Statistical Administration was personally overseen by Molotov (莫洛托夫), which was suggestive of the importance the Soviets attached to statistics.[82]

79 Goikhman, "Soviet-Chinese Academic Interactions," 282.

80 Shen Zhihua, *Sulian zhuanjia zai Zhongguo*, 18.

81 Such information can likely be found among materials in the Archives of the SSB, but I was unable to gain access. Shen Zhihua, who maintains lists of Soviet experts in various sectors, does not possess a list of Soviet statisticians. Author's personal communications with Shen Zhihua, November 2012.

82 Wang Sihua, "Tongji gongzuo de renwu," 8. This is very likely a reference to Vyacheslav Mikhailovich Molotov (Вячеслáв Михáйлович Мóлотов, 1890–1986) of the 1939 Molotov-Ribbentrop pact on nonaggression between the USSR and the Nazi Third Reich. An important politician, Politburo member, and diplomat, until 1949 Molotov served as Soviet minister of foreign affairs. Oddly enough, the paragraph that mentions Molotov was excised from the

Wang was therefore appreciative of the help offered by Soviet statistical advisers as the NSB organized and carried out statistical work.[83] Some years later, Yang Jianbai, Wang's deputy at the NSB, reiterated the importance of Soviet expertise for the NSB's statistical work. In an article in the *People's Daily* he reaffirmed the need to take the Soviet Union as their teacher, adding: "The Soviet Union's advanced statistical experience is abundant and completely correct, and we must study her experience."[84]

The Soviet statistical advisers who assisted the NSB (and later the SSB) operated solidly within a tradition of socialist statistics. Their own experience in the Soviet Union had primarily focused on statistical work and not on research in any other discipline, and they brought their particular understanding of statistics to the PRC. They belonged to a group that would have wholeheartedly agreed with the conclusions drawn at the 1954 meeting in Moscow. As Xue Muqiao, the first director of the SSB, would recall in his memoirs, the Soviet experts were decidedly unenthusiastic about the use of sampling of any kind. In his words, "Soviet statistical experts did not promote the use of surveys, [but rather] advocated making comprehensive arrangements for statistical reports."[85] On one occasion, Xue and Li Fuchun attempted to have V. N. Starovskiy's book *Chouyang diaocha* (抽样调查; Sample surveys) issued for study by the State Planning Commission (SPC) and the SSB, but they were blocked by the Soviet statistical experts who were present in Beijing.[86] Thus, from the very beginning, the advice the Chinese received, and which they valued highly, was about the primacy of socialist statistics and the concomitant irrelevance of abstract mathematics.

---

version of the speech that was reproduced in the 1986 compilation of Wang's writings on statistics. Cf. Wang Sihua, *Wang Sihua tongji lunwenji*, 3.

83 Wang Sihua, "Sannianlai dongbei tongjiju," 8; and Wang Sihua, *Wang Sihua tongji lunwenji*, 45. For examples of how Soviet advice and experience were put into practice in the Northeast, see Wang Sihua, "Sannianlai dongbei tongjiju," 8; and Wang Sihua, *Wang Sihua tongji lunwenji*, 45–46.

84 Yang Jianbai, "Sulian zhuanjia bangzhu women," 3. Yang was secretary-general and deputy director of the NSB from 1951 to 1954, when he was transferred to the main offices of the SSB in Beijing. At the SSB, Yang specialized in national income accounting. In 1956 he was transferred to the State Planning Commission (SPC).

85 Xue Muqiao, *Xue Muqiao: Huiyilu*, 174.

86 Vladimir Nikanovich Starovskiy (1905–1975; 斯塔洛夫斯基) was the head of the Soviet Central Statistical Administration from 1940 to 1975. Xue Muqiao claimed in his memoirs (*Xue Muqiao: Huiyilu*, 174), that Starovskiy had personally presented him with a copy of this book during his 1954 visit to Moscow.

The first among a series of high-level Soviet statistical experts deputized by the Soviet Central Statistical Administration to serve in the PRC was A. I. Ezhov (Ru.: А.Е. Ежов; Ch.: 叶若夫/叶诺夫).[87] Ezhov arrived in the PRC in early 1950 and returned to the Soviet Union in February 1952. A deputy director of the Soviet Central Statistical Administration, Ezhov brought with him the prestige of his high office. He also played a key role in disseminating information about Soviet statistics to a wider audience, in both Russian and other languages.[88] As the first Soviet statistical expert in China, Ezhov quite possibly had the greatest impact in shaping PRC statistical work.[89] During his two years in the PRC, Ezhov consulted with the Statistical Office of the National Finance Committee, played a critical role in helping the NSB design statistical work, and was instrumental in offering advice and general guidelines for the establishment of a national statistical system.

Less is known about most of the men who followed Ezhov to the SSB, five of whom are listed in Table 3.2. Sergei Krotevich (Ru.: С. К. Кротевич; Ch.: 克拉特维奇), who spent a little over three years in the PRC, was the first among them. In addition to acting as a consultant on financial economics, he also played a key role in designing and carrying out the 1953 national population census. An expert by the name of N. P. Semidevkin (Ru.: Н.П. Семидевкин; Ch: 谢木杰夫金) arrived on 22 July 1955.[90] A specialist in infrastructure statistics, he stayed in the PRC for about three years.

87 There are several different versions of his name: in *DSJ* he is identified as Yenuofu (叶诺夫); in a Beijing Statistical Bureau document located in the Beijing Municipal Archives he is referred to as Yeruofu (叶若夫); and in an NSB publication he is listed as Yeraofu (叶饶夫). The last transliteration is the closest approximation of his actual name. Cf. *DSJ*, 2; BMA 002-020-00969: 6; and NSB, *Tongji gongzuo: Di san xuanji*, table of contents. On the general problems with transliterations of Russian names during the 1950s, see footnote 90 below.

88 Among his works translated into English are: (1) *Soviet Statistics*; (2) *Industrial Statistics*; and (3) *Organisation of Statistics in the U.S.S.R.*

89 In addition to his name appearing with regularity in published and archival materials, Ezhov was also the Soviet expert whose name was recalled most frequently by several of my interviewees. The possibility remains that this is because his name was relatively easy to pronounce, and therefore easy to remember.

90 The transliteration of Russian names during this period appears not to have followed a consistent method of concordance between the Cyrillic alphabet and Chinese characters. As a result, ascertaining the names has not been a straightforward task. In some instances, all I had to work with was the transliterated name in Chinese. In such cases, handbooks of Russian-Chinese names and terms, which I consulted to arrive at guesses about the names in Cyrillic, were less instructive than hoped. Input from Russian and Chinese speakers offered limited solutions. Cf. (1) Xin Hua, *Eyu xingming yiming shouce* (see, in particular, Appendix 5, which is a

TABLE 3.2. Soviet Statistical Experts at the SSB

Dates in the PRC	Name in Chinese	Name in Pinyin	Name in Cyrillic	Name in English
Early 1950–February 1952	叶诺夫/叶若夫	Ye nuo fu/Ye ruo fu	А. Е. Ежов	A. I. Ezhov
February 1952–July 1955	克拉特维奇	Ke la te wei qi	С. К. Кротевич	Sergei Krotevich
22 July 1954– (for about 3 years)	谢木杰夫金	Xie mu jie fu jin	Н. П. Семидевкин	N. P. Semidevkin
22 August 1955– (for about 2 years)	马克西莫夫	Ma ke xi mo fu	Г. М. Максимов	G. M. Maximov
1954–1956 (?)	拉格乔夫	La ge qiao fu	И. И. Логачев	I. I. Logachev
11 December 1957–11 June 1958	索包里	Su bao li	В. А. Соболь	V. A. Sobol'

Note: Table compiled from: *DSJ*, 2, 6, 30, 39, 55–56; and SSB, *Sulian tongji zhuanjia tanhua jilu xuanbian.*

There is also evidence of an I. I. Logachev (Ru.: И.И. Логачев; Ch.: 拉格乔夫) who was active in China from 1954 to 1956. The next expert to arrive was G. M. Maximov (Ru.: Г.М. Максимов; Ch: 马克西莫夫), who remained in China until 1957.[91] The last expert, V. A. Sobol' (Ru.: В.А. Соболь; Ch: 索包里) arrived on 11 December 1957 but spent less than one year in the PRC, returning to the Soviet Union on 11 June of the following year.

It is somewhat telling that information about Ezhov and the five statistical advisers who followed him is difficult to locate in both English- and Russian-language materials.[92] For instance, Robert Campbell's biographical dictionary of Russian and Soviet economists carries an entry for Sobol' (1896–1968) but not for any of the others. It states that Sobol' was educated at the Ural-Siberian Communist University and the Institute of Red Professors. After a brief time at GOSPLAN, the Soviet State Planning Commission, he was transferred to the Central Statistical Administration, where he ended his career. In addition

Russian-Chinese transliteration table); (2) Zhongyang Makesi Engesi Liening Sidalin zhuzuo bianyiju, *Zhengzhi jingjixue jiaokeshu*; and (3) Onoprienko, *Bibliograficheskii.*

91 Onoprienko, *Bibliograficheskii*, offers two possible candidates for who this might be: (1) G. Maximov, author of (with A. Isupov), "Natsional'nosti SSSR po dannym Vsesoiuznoi perepisi naseleniia"; and "Naselenie SSSR"; or (2) I. Maximov, author of (with D. Zhak and N. Koshliak) "Opyt makhanizatsii ucheta v kolkhozakh."

92 The bibliographic index of fifty years of articles in *Vestnik Statistiki* (Onoprienko, *Bibliograficheskii*), for instance, lists only a handful of articles authored by these figures.

to working on the balance of the national economy, Sobol' was perhaps most influential as long-term editor of *Vestnik Statistiki*. Campbell, however, is rather dismissive of his contributions, noting that "he was basically a bureaucrat who really had nothing much to contribute to economics."[93] Such fragmentary information accords with the suspicion that the experts deputized to the PRC were not always drawn from the elite strata of Soviet scientists. Even so, Soviet statistical experts appear to have been either directly or indirectly involved in three types of activities: consulting on the design and execution of statistical work, aiding in the publication of translations from Russian to Chinese, and directly teaching statistical cadres.

Consultation

The most basic form of consultation involved introducing elements of the Soviet statistical system and offering advice to Chinese statisticians on how to proceed with setting up their own statistical system. Thus, for example, in late 1950 Ezhov introduced to statistical cadres in Beijing the structure and key tasks of the Soviet system.[94] Such discussions usually involved a wide range of topics, including how to distinguish between agriculture and industry, whether handicrafts industries should be regarded as private industries, how statistical survey work should be organized, how data analysis should be conducted, the importance of original records, and why statistics was a science with a party spirit (党性的科学; *dangxing de kexue*).[95]

Soviet advisers were also direct participants in specific survey projects that were initiated by the SSB, and they served as advisers at particular enterprises. For instance, Sergei Krotevich offered key advice in 1953 as the SSB carried out the first complete modern census in China's history.[96] Data were collected

93 Campbell, *A Bibliographical Dictionary of Russian and Soviet Economics*, 387–388.

94 See Ezhov's December 1950 speech at a meeting on statistical work in Beijing. BMA 002-020-00969: 5–6.

95 BMA 004-011-00019: 2–12, 13–18. These materials are based on notes from a lecture delivered by Krotevich; the accompanying handwritten cover letter praised his exegesis and requested that the materials be widely disseminated.

96 The Qing government had attempted a complete enumeration in 1908, but the project floundered in the face of political exigencies and the eventual demise of the dynasty. The four decades leading up to the 1950s witnessed some surveys conducted by the Guomindang regime in regions under its control, but there was nothing on a national scale. During the 1940s, scholars in Republican China advocated the administration of a census and the establishment of a census bureau. Foreign scholars too noted the need for a standardized census. But the civil war between the Guomindang and the Communists blocked implementation of any of these recommenda-

across only four parameters: name, age, sex, and nationality.[97] Two and a half million central and local government workers took part in the recording, aided by volunteer census workers in the localities.[98] Final enumerations and tabulations were compiled by June 1954, and on 1 November 1954, the SSB reported that at midnight of 30 June 1953, the population on the Chinese mainland was 583.6 million.[99] This number surprised many observers who were familiar with the 1933 estimate by the League of Nations for Republican China of about 450 million.[100] Krotevich himself wrote a report on the census in Russian that was published in *Vestnik Statistiki* in 1955. A Chinese-language translation was published the following year by the Statistical Press.[101]

In similar fashion, Soviet consultants also spent time at designated factories and enterprises, helping set up or rationalize statistical work. For instance, in an article in the journal *TJGZ*, the Statistics Section of the Planning Office of Angang Steel thanked the Soviet statistical expert Cheburnikh (?) (Ru: Чебурних; Ch: 切布尔尼) for his wide-ranging assistance. The article explained that with his help the department was able to appreciate what was distinctive about socialist statistics; improve original records; simplify

---

tions. After the establishment of the PRC, the Communists made several attempts to derive a new population total for the country by collecting reports on population through administrative channels. These efforts resulted in a population total of about 566 million. On the 1908 census, see Lam, *A Passion for Facts*; on foreign views of pre-1949 demographic activities, see Jaffe, "A Review of the Census and Demographic Statistics of China," which notes that "at no time did the Government appear to realize the value of accurate population data as such, and the data actually collected have been very largely the by-products of inquires with more immediate administrative aims." On early PRC administrative attempts to determine the population, see Chung and Hsiao, "How We Made the Census."

97 Chung and Hsiao, "How We Made the Census," 9; on the question of nationality, see Mullaney, *Coming to Terms with the Nation*.

98 Chung Lin and Hsiao Lu, "How We Made the Census," 10.

99 SSB, "Communique of the Major Statistics from the First National Population Census," 62–65.

100 For a sense of the surprise and skepticism these results generated, see Aird, "China's Population, Census and Vital Statistics," 443; Aird, "Estimating China's Population," 67–72; Krader and Aird, "Sources of Demographic Data on Mainland China," 623–630; Coale, *Rapid Population Change in China*; and Cressey, "The 1953 Census of China," 387–388. It is possible that none were aware of the early 1950s' estimate of 566 million noted above. Many may also have assumed that the disruptions caused by the civil war served as a check on population growth.

101 See Krotevich, "Vsekitayskaya perepis' naseleniya 1953 g."; and Krotevich, *Zhongguo 1953 nian quanguo renkou diaocha*.

forms; establish monthly, ten-day, and daily reports; and initiate research and analysis.[102]

Another important consulting task was inspection of local statistical work. From November to December 1955 the Soviet specialists Maximov and Semidevkin were accompanied by SSB deputy director Sun Yefang (1908–1993) on a tour of Hubei, Hunan, Guangdong, Zhejiang, Jiangsu, and Shanghai, with the goal of understanding city- and province-level statistical work. Principal questions included how to strengthen local statistical work, especially under the conditions brought about by the high tide of cooperativization, and how to strengthen county-, district-, and village/township-level statistical work. On 18 January 1956, the team produced a report that included three key elements: First, it was severely critical of the work done by the various provincial bureaus, noting that regardless of how well the local or national bureaus worked, the provincial bureaus did not exhibit sufficient leadership. Their organization was incomplete, and their materials were either lacking in quality or late. Second, the team offered suggestions on how to strengthen province-, district-, and village-level statistical work. Third, the team offered advice on how to set up a united and scientific agricultural survey scheme.[103]

Publications

A second way in which Soviet experts contributed to statistical work in the PRC was by aiding with the translation of Soviet textbooks, organizing curricula at universities and training centers, and compiling collections of special articles.[104] The earliest compilations of the NSB's *DBTJGZ* frequently contained a section entitled "Introducing Soviet Experience," which included several translated articles. For most workers in the Northeast (and in the rest of the country as well), this was the first opportunity to read about Soviet methods through writings by Soviet statisticians. For instance, a relatively early critique of bourgeois statistics can be found in an article by the author N. 廖佐夫 (Ru: Н. Леозов(?); Ch: *N. Liao zuo fu*).[105] Each of the NSB's

102 Angang Steel Planning Office Statistics Section, "Ganxie Sulian zhuanjia dui women de bangzhu," 31–32.

103 *DSJ*, 43.

104 For more on Soviet advisers at major universities, such as Renmin, Tsinghua, and Peking, see Yu Shengqi, "Zai Sulian zhuanjia bangzhu xia de Zhongguo renmin daxue"; Chen Jun, "Sulian zhuanjia zai Beijing daxue"; Bai Sheng, "Sulian zhuanjia zai Qinghua daxue"; and Ge Jiali, "Zai Sulian zhuanjia de jiaodao xia."

105 N. 廖佐夫, "Xin Zhongguo tongji de gaikuang yu tongji gongzuozhe de renwu."

compilations from 1950 to 1952 carried articles by Soviet statisticians, including, in some cases, by experts who were present in the PRC as consultants.[106] Subjects ranged widely, including the meaning, role, and organization of economic statistics in the Soviet Union; survey methods in the Soviet Union; the Soviet experience in unifying and simplifying tables and reports; and discussions of statistics in people's democracies. At the same time, advice on how to deal with local statistical problems was also offered, focusing on the role of city and village statistics, the methods of statistical and economic analysis, and so on. By 1953, after several compilations had been produced, the Soviet authors were largely replaced by Chinese authors, and the main focus was primarily Chinese, not Soviet, experience in statistical work.

With the establishment of the national SSB in Beijing in August 1952, much of the translation and editing work was devoted to textbooks.[107] By September 1957, over 100 items had been translated. Their scope was extremely wide: from general outlines on the principles of statistics to every kind of specialized statistics, such as industry, agriculture, capital construction, trade cooperation, supplies, transportation, finance and economics, health and education, and so on. In addition to advanced texts for specialists, several books targeted basic-level statistical workers and were therefore at the middle-school or high-school level. Examples of such books, whose publication runs were more than 20,000 or 30,000 volumes, included *Tongji lilun* (统计理论; Statistical theory) produced by the Soviet Central Statistical Administration; *Jingji tongjixue jiaocheng* (经济统计学教程; Guide to economic statistics) by Petrov (Ru: Петров; Ch: 彼得洛夫); and *Gongye tongjixue jiaocheng* (工业统计学教程; Guide to industrial statistics) by Shaensky (Ru: Щаынски; Ch: 沙文斯基). Beginning in 1956, collaboration between the Soviet and Chinese Departments of Culture resulted in China's receipt of additional Soviet books. The Chinese were also able to directly approach many Russian authors and request that they write forewords for the Chinese editions of their books.[108]

Finally, starting in 1955, *Tongji chubanshe* (统计出版社; Statistical press) began producing compendia of translated essays and articles. The first of these was a collection of reports that Sergei Krotevich had authored through early 1954. In the first quarter of 1957, the press published a more ambitious volume containing records of selected lectures delivered by the Soviet experts Maxi-

---

106 See, for instance, Northeast Statistics Bureau, *Tongji gongzuo: Di yi xuanji*; *Tongji gongzuo: Di san xuanji*; and *Tongji gongzuo: Di si xuanji*.

107 The establishment of the SSB is discussed in chapter 5.

108 See Anon., "Banianlai woguo fanyi chuban le daliang Sulian tongjixue zhuzuo."

TABLE 3.3. Collected Translations of Statistical Materials

Series No.	Theme/Topic	Publication dates	Print run	Number of Essays
1	General	1956 May	7,100	13
2	General	1956 June	7,150	9
3	General	1956 September	7,140	14
4	General	1956 December	5,060	8
5	?	?	?	?
6	Agricultural labor productivity	1957 May	2,070	15
7	Bourgeois statistics	1957 July	1,570	8
8	Accumulation and consumption	1957 July	1,570	7
9	Sample surveying	1957 August	1,460	10
10	The history of the development and use of sampling theory	1958 February	1,260	4
11	The use of sample surveying in population research	1958 April	1,470	6
12	The theory of index numbers	1957 December	1,460	6
13	?	?	?	?
14	Bulletin of the Meeting of All-Soviet Statistical Workers[1]	1958 March	1,480	13

1. Referring to a meeting that took place in Moscow, 4–8 June 1957.
Source: Beijing: Tongji chubanshe.

mov, Semidevkin, and Logachev. Drawn primarily from conversations between 1954 and 1956, the book was divided into seven sections, each devoted to a particular kind of statistical work: comprehensive, balancing the economy, industry, agriculture, infrastructure, trade, and labor wages. Within each section, short essays explained Soviet practices in specific areas. For instance, an essay in the section on industry discussed the definitions of modern industry and handicrafts. Similarly, an essay in the section on agriculture explained the calculating and reporting practices on collective farms in the USSR. Most sections concluded with "Questions and Answers," which were typically an attempt to clarify the differences between Soviet and Chinese practices.[109]

Also beginning in 1956, a series titled *Tongji yiwen zhuanji* (统计译文专辑; Collected translations of statistical materials) was published. It consisted of at least fourteen issues (see Table 3.3). In its first year, each issue covered a wide variety of topics and was intended for wide circulation. By 1957, each issue focused on a specialized theme, with reduced circulation numbers

109 SSB, *Sulian tongji zhuanjia tanhua jilu xuanbian*; Zhu Hong'en, "Du 'Sulian tongji zhuanjia tanhua jilu xuanbian.'"

suggesting a more targeted audience. In 1958, its final year, two issues (nos. 10 and 11) were devoted to introducing materials from "bourgeois" countries, most notably the United States and the United Kingdom.[110]

Teaching

A final method of Soviet involvement was that of direct teaching via lectures and discussions. During interviews, the statisticians Ni Jiaxun (1931–) and Lin Fude (1925–2017) recalled attending lectures delivered by Soviet experts at Renmin University, in particular lectures by Ezhov.[111] Both Ni and Lin remarked that problems of translation and access made these experiences rather episodic and of limited value: The Soviet expert would lecture via an interpreter. Attendees were not permitted to ask any questions during the session. If the participants had questions, they were required to write them down and hand them to the expert's local liaison. The liaison would then select and translate several of the questions and deliver them to the expert, who would prepare responses. These responses would then be read at the beginning of the following lecture.[112] Such a discontinuous learning process, with no classroom interaction, arguably had limited pedagogical value for the participants. But the lectures may have had a wider impact, since they were typed and then circulated as reference materials for study sessions. Krotevich's lecture from August 1952 appears to be one such example. The cover letter to his circulated lecture stated that "his report is very good, and can serve as reference material for cadres in agricultural and industrial statistics during their vocational learning."[113] Finally, from time to time, larger seminars were organized by the

110 Essays in no. 10 were drawn from *The Journal of the Royal Statistical Society* and essays in no. 11 were drawn from *The Journal of the American Statistical Association*.

111 Based on interviews with statisticians Lin Fude and Ni Jiaxun, who were both active in statistical work and education in the 1950s. I met Lin Fude on 22 April 2011 in Beijing, and I had two meetings with Ni Jiaxun, both in Beijing, on 29 March 2011 and 7 September 2011. A statistical worker during the early 1950s, Ni attended Renmin University, where he obtained a bachelor's degree in 1960. He subsequently took up a faculty position at the university. It was through him that I learned of the identity of the two statisticians who visited India later in the year.

112 This description is based, in particular, on Lin Fude's recollections.

113 See BMA 004-011-00019: 1, 2–12. In another instance, the lecture by Semidevkin, "He qingnian tongzhi tantan tongji xuexi," at Xi'an Statistics School was published in *TJGZTX*. Reproductions of lectures could also include works by more hallowed figures: one of Lenin's letters to the Soviet Central Statistical Administration was translated and circulated in June 1953 for use as reference material in study sessions. See BMA 123-001-00416: 50–56.

SSB or by one of the universities. For instance, on 12 May 1955, the SSB organized a Beijing seminar at which Soviet experts and professors from Renmin University delivered lectures to statistical cadres in all the departments within the State Council.[114] Such exercises, however, appear not to have been a frequent feature of the work performed by the statistical experts.

A fourth, and final, way of drawing upon Soviet expertise also existed, though it took place in the Soviet Union rather than in the PRC. Although it is not clear how many Chinese students studied statistics in the Soviet Union, at least two senior Chinese statisticians visited there. From December 1954 to January 1955, Xue Muqiao visited as part of a Sino-Soviet Science and Technology Cooperation delegation. During his time in Moscow, Xue met with the director and deputy director (Ezhov, who had recently returned) of the Soviet Central Statistical Administration. He also met with members of the Planning Commission.[115] A little over one year later, in July 1956, Sun Yefang, one of the SSB's deputy directors, led a seven-person team of statisticians to the USSR. They met a number of officials, visited various bureaus, and studied statistical and planning activities. In total, the team spent over thirty days in the USSR and discussed a variety of subjects, including the calculation of gross output value, constant prices, and the reporting structure in the Soviet Union.[116]

Conclusion

Li Fuchun's 1951 dismissal of statistics as it had been practiced heretofore indicated a crisis on two levels, one ontological and the other epistemological. Into what categories should the world be divided? And what forms of knowledge can be deployed to ascertain these categories? In line with the discourse within the Soviet Union, as summarized by Ostrovitianov in 1954, the point of departure was the premise that there was no single universal reality for the known world; rather, different theories and methods were applicable to the different divisions—the social and the natural—of the world. These different theories and methods, in turn, promoted very different interests and led to very different social and economic outcomes in society. For Chinese thinkers, drawing upon a specific set of Soviet and Marxist views, the natural and social worlds were not the same and could not be understood using the same tools.

---

114 *DSJ*, 37.

115 Ibid., 34; Xue Muqiao, *Xue Muqiao: Huiyilu*, 174.

116 *DSJ*, 47.

The Chinese were exposed to these views, not only from reading Marx and Lenin, but also and even more importantly from Soviet statistical experts who were present in the PRC and from a vast array of translated material. The overall effect was to engender an overarching theoretical approach to all knowledge that prefigured what statistics could be. The practical impact of this theoretical distinction between the natural and social worlds was to idealize and attempt to construct a vast national network of statistical offices, from Beijing down to the county level, with the aim of capturing all necessary data. In Part II we will explore the implications of this ideal. For now, let us turn to the content of socialist statistics and how it gained primacy.

4

No "Mean" Solution

REFORMULATING STATISTICS,
DISCIPLINING SCIENTISTS

IN FEBRUARY 1957, Professor of Statistics Jin Guobao published a self-criticism in the journal *TJGZ*.[1] The focus of the self-criticism was his successful textbook, *Tongjixue dagang* (An outline of statistics), the most recent revised edition of which had been published in 1950.[2] An abridged version published in 1935 under the title *Statistics* had also become popular as a textbook in vocational schools.[3] In his opening paragraph to his self-criticism Jin explained that the ancients' (古人; *guren*) possessed an important principle: getting rid of hackneyed phrases (陈言务去; *chenyan wuqu*). By that principle alone, Jin confessed, *Tongjixue dagang* was beyond redemption. Copied entirely from Anglo-American bourgeois statistics, the book was full of stale words (陈言; *chenyan*); not one sentence was the product of new thinking or composed of original ideas. His self-criticism was thus an attempt to atone for the failures of the book and for his failure to correct his thinking in a timely fashion.

Such strong self-criticism is not entirely surprising, especially in 1957 when intellectuals became targets of thought reform.[4] Nor was the self-criticism by

---

1 Jin Guobao, "Dui 'Tongjixue dagang' de ziwo pipan."

2 Jin Guobao, *Tongjixue dagang*.

3 Jin Guobao, *Tongjixue*.

4 On the thought reform of intellectuals during the 1950s more generally, see Chen, "The Thought Reform of Intellectuals." For a recent work that focuses on the thought reform campaigns conducted against vagrants and the lumpen proletariat, see Smith, *Thought Reform and China's Dangerous Classes*.

Jin the only such self-criticism by a statistician in the mid-1950s.[5] These self-criticisms were themselves part of an extended series of debates within the academy, which from 1950 onward had sought, on the one hand, to establish a new definition of statistics, and on the other, to single out those statistical tools that were appropriate for a socialist society.[6] Jin's self-criticism, in particular, offers a useful route through which to access two related strands of inquiry. The first of these has to do with understanding consensus formation within the scientific community of early PRC statisticians. As shown in chapter 2, the pre-1949 world of Chinese statistics (and social science) was methodologically and epistemologically diverse. In that chapter I make the case for the centrality of the NSB in reshaping the nature and content of statistical work, indicating that changes in the practice of statistical work preceded their articulation within the theory and pedagogy of statistical science. Crucial to the latter transition was disciplining a vibrant and varied community of statisticians so that they would come to accept and espouse statistics as a Marxist-Leninist social science in the years after 1949.

The sense that practice preceded changes in pedagogy raises questions about how we conceptualize scientific change during and after political revolution. How did socialist statistics become the dominant paradigm during the 1950s? Recent work on the sociology and history of science has devoted much attention to studying how practitioners arrive at scientific consensus and closure. On the one hand, these studies stress "extra-scientific" factors, such as culture, power and funding, politics, and personal credibility.[7] Other works explore the cultures of knowledge societies, in particular the disunity of science made visible by a focus not on "the construction of knowledge but in the construction of the machineries of knowledge construction."[8] This machinery

5 At least two other self-criticisms were published: by Zou Yiren, "Guanyu zichan jieji tongji xueshu sixiang"; and by Wang Sili, "Pipan wo zai tongjixue zhuzuo zhong."

6 Articles critical of pre-1949 statistics and its practitioners were published in journals such as *DBTJGZ, TJGZTX, TJGZ, Xin jianshe* (New Construction), and *Zhongguo gongye* (China Industry), as well as in scientific theses written at Renmin (People's) University in Beijing. See also Zou Yiren, "Guanyu zichan jieji tongji xueshu sixiang," 28.

7 For a brief overview of this literature, see Shwed and Bearman, "The Temporal Structure of Scientific Consensus Formation"; see also Sismondo, *An Introduction to Science and Technology Studies*, 130–135.

8 Knorr-Certina, *Epistemic Cultures*, 3. Knorr-Certina's work builds on earlier approaches that focus on the work of scientists as actors, most notably works by Bruno Latour, Peter Galison, and Lorraine Daston.

is further broken down into empirical, technological, and social machinery, the argument being that it is their combination that makes possible the rise of "truth effects."[9] More recently, Sheila Jasanoff has made the case for the coproduction of social and natural orders, demanding that we no longer separate the "domains of nature, facts, objectivity, reason and policy from those of culture, values, subjectivity, emotion and politics."[10]

Yet others have focused primarily on the "social processes within core sets of practicing scientists who negotiate results," demarcate knowledge claims, and construct boundaries.[11] In a 1983 article Thomas Gieryn noted that such "demarcation is routinely accomplished in practical, everyday settings."[12] Gieryn developed his arguments into a book-length study of "boundary-work," which he also labeled a form of "cultural cartography," and identified three genres through which it usually operates: expulsion, expansion, and protection of autonomy.[13] Scholars of science and technology studies have offered related schemas for understanding how controversies are resolved, focusing in particular on the use of detailed critiques, the creation of new tests and procedures, the isolation of more-scientific or utilitarian positions, and the discounting of deviant data.[14]

At the heart of scientific closure therefore is negotiation. This negotiation, as Ted Porter has argued, is frequently conducted by a select group of experts to present a way to understand how science takes place.[15] Such negotiation can apply not only to questions of demarcation within science but also across the boundary between science and pseudo-science. Michael Gordin has explained that "individual scientists (as distinct from the monolithic 'scientific community') designate a doctrine a 'pseudoscience' only when they perceive themselves to be threatened—not necessarily by the ideas themselves, but by what those ideas represent about the authority of science, science's access to resources, or some other broader social trend."[16] Studying demarcation and boundary making thus allows us to "understand *science* and what scientists

9   Knorr-Certina, *Epistemic Cultures*, 11–13.
10 Jasanoff, *States of Knowledge*, 3.
11 Shwed and Bearman, "The Temporal Structure of Scientific Consensus Formation," 819.
12 Gieryn, "Boundary Work and the Demarcation of Science," 781.
13 Gieryn, *Cultural Boundaries of Science*, 15–17.
14 Sismondo, *An Introduction to Science and Technology Studies*, 130–132.
15 Porter, *Trust in Numbers*, 219.
16 Gordin, *The Pseudoscience Wars*, 2.

thought about their standards, their position in society, and their future" and to situate that understanding in a particular time and place.[17]

Gordin's advice to be attentive to time and place is especially salient in the case of China in the 1950s. Indeed, much of the scholarship cited above is predominantly about the practice of science in liberal societies. Jasanoff notes that certain scholars (such as Yaron Erzahi) have taken the argument further, claiming that "modern science provides the template for a particular form of politics: liberal democracy."[18] What then of other politico-ideological settings and contexts? In her work on the agricultural sciences in Mao-era China, Sigrid Schmalzer provides a useful counterpoint by expanding our lens beyond the liberal-democratic frame. "Whether state-socialist or liberal-capitalist," she notes, "every historical context produces ideologies that reflect and reinforce the perspectives and interests of the social groups that generate them."[19]

For China in the 1950s, the drastic political changes following the establishment of the PRC in 1949 brought with them an overarching new theoretical paradigm, which then had to be negotiated, both by the people promoting the paradigm and by those whose original positions were opposed to that paradigm. In a setting where the state and its ideology were not one among a set of players but rather reigned supreme, the consequent negotiation over statistics had a degree of path dependence. But we "must resist becoming so jaded that we fail to take seriously the political ideology that in fact meant a great deal to, and had very real consequences for, people on the ground."[20] The concerns over demarcating a clear boundary between the social science of statistics and the natural science of mathematical statistics that animated Jin Guobao and his critics had human as well as material implications. They led to a restructuring of intellectual communities and to a re-evaluation of key statistical concepts, the effects of which reverberated throughout Chinese statistical work in the 1950s.

This process of demarcation involved at least three principal elements: disciplining a generation of statisticians to accept the new definition, creating textbooks to train younger generations, and expunging all residue of

17 Ibid., 3, 14.

18 Jasanoff, *States of Knowledge*, 32.

19 Schmalzer, *Red Revolution*, 15.

20 Ibid., 14. Schmalzer goes on to suggest that "instead of mounting the critique against the fact that ideology is present, we may mount it against the role it plays in social oppression, cultural imperialism, environmental destruction, or other negative forces" (ibid., 15).

"bourgeois" influence. The first half of this chapter unpacks these attempts at demarcation by focusing on the nearly decade-long sequence of publications that sought to criticize Anglo-American "bourgeois" statistics and its practitioners. In the second half of the chapter, the focus shifts to the changes in statistical education and to the debates over the content of statistics. In broad terms, the consequences of this transition are well known: As a social science, statistics dealt with the social world rather than the natural world. As a result, it was bifurcated from what was seen as the abstract and formal theorizing of mathematical statistics, which was, in turn, banished to departments of mathematics.[21] Comparing new textbooks with Jin Guobao's *Tongjixue dagang* (An outline of statistics) permits us to see what this meant from a curricular perspective: what was retained, what was jettisoned. In many instances, the distinctions ran deeper. To merely reject the law of large numbers and general probability theory was far from enough. Even fundamental concepts such as measures of central tendency (especially the mean), indicators, and time series would come under the microscope, as earnest attempts were made to excise the last vestiges of any polluting bourgeois influence.

Jin Guobao and the Road to (Self-) Criticism

In the years immediately after 1949, most Chinese statisticians were under increasing pressure to adjust to a new overarching theoretical paradigm, which stood in fundamental contradistinction to the one with which they were familiar. In order to attempt to answer the question of how this occurred, it is important first to disaggregate the community of statisticians.

In oral history interviews with the statistician Lin Fude and the demographer Wu Cangping (1922–), they both provided contrasting views on how they adapted to the new statistics of the 1950s. Both had come of age and been trained in the 1940s, and as such prior to 1949 they had a relatively catholic approach to statistics. In the years that followed—Lin informed me—he and other statisticians had to recalibrate their ideas of statistics and bring them into closer alignment with the new definitions. Wu, in contrast, while accepting that such recalibration did take place, observed that the change was often only rhetorical or symbolic; the actual statistical ideas remained the same. Both pointed out that the tangible result was that methodology and statistical work

---

21 Zou Yiren et al., "Guanyu shuli tongji zai shehui jingji tongji kexue zhong."

were privileged over research, and theory and application stood estranged from each other.[22]

While Lin Fude and Wu Cangping may have successfully negotiated the transition, as the youngest generation of statisticians trained before 1949, they were hardly in positions of influence; that is, positions from which they could have an immediate and widespread impact on the state of the discipline and on the education and training of future statisticians and statistical cadres. That space was occupied by an older generation of statisticians who had been active during the preceding decades, especially in the 1930s when the social survey movement was at its peak. This generation had published textbooks that were still widely used. Its members had served in government, in education, in banking, and as researchers in a private capacity. In particular, because they were authors of important and popular textbooks, curtailing their activities and reformulating their views became crucial to the disciplinary reformulation of statistics. Prominent among members of that generation was Jin Guobao.[23]

Today, Jin Guobao is widely regarded as one of the founders (奠基人; *dianjiren*) of modern statistics in China.[24] To a large extent, such recognition is based on his activities prior to 1949. A professor of statistics and a one-time banker and senior government official, in the decades before the CCP victory he was arguably China's most famous statistician as well as its most prominent teacher of statistics. There were multiple editions of his textbook, *Tongjixue dagang* (An outline of statistics), which was widely regarded as "the most influential statistics textbook in the country during that time."[25] It was precisely this perceived influence that made Jin one of the targets of concerted criticism in the early PRC.

Jin Guobao was born on 12 October 1893 in Tongli town of Wujiang county, Jiangsu province. His early education was at Shanghai's Fudan Public School. After graduation, he served as an English teacher at Wujiang Middle School.

---

22 I interviewed Lin Fude and Wu Cangping in Beijing on 22 April 2011 and 11 March 2011, respectively.

23 Biographical details for Jin Guobao are drawn primarily from Shi Tai, "Tongji mingren lu: Jin Guobao"; Xu Youchun, *Minguo renwu dacidian*; and Gong Jianyao, *Shijie tongji mingren zhuanji*, 402–407.

24 See, for instance, the entry for Jin Guobao in Xu Youchun, *Minguo renwu dacidian*.

25 Shi Tai, "Tongji mingren lu: Jin Guobao," 61. By 1948, the book was in its ninth edition. I have not been able to locate actual publication details for the 1950s, but a thirteenth "revised" edition was published in 1961, which suggests three possible editions during the 1950s. It appears to have been republished as recently as 2014, this time by Shanghai sanlian shudian.

During this period, he frequently contributed essays to the journals *Shishi xinbao* (时事新报; New bulletin of current affairs) and *Dongfang zazhi* (东方杂志; Eastern miscellany). Using the nom de plume Jin Lüqin, he was the first person to translate Lenin into Chinese, publishing Lenin's "Bolshevik's Statement of Demands" in the first issue (September 1919) of the journal *Jiefang yu gaizao* (解放与改造; Liberation and reform).[26]

In 1921 Jin traveled to America to study statistics at Columbia University on a scholarship from the South Seas Tobacco Company. He received a master's degree in 1923, after which he returned to China and taught statistics successively at Shanghai Chinese Public School, Fudan University, and the University of Commerce. During these years, he was also a section chief in the Ministry of Finance. In 1928, on the orders of the president of the newly founded Academia Sinica, Cai Yuanpei (1868–1940), he spent over a year traveling in Europe and the Americas, researching statistical activities and gathering a variety of statistical materials.

Upon his return to China, Jin became directly involved in government, banking, and statistical education. In May 1929, he was appointed director of the Nanjing Finance Bureau and served as a special member of the Capital Construction Committee. In March of the following year, he was among those who joined D. K. Lieu to establish the Statistical Society, which was to be a forum for the promotion of private academic discussions. He was elected the society's deputy chairman and was selected, along with Franklin L. Ho (He Lian, 1895–1975), Chen Zhongsheng, and others, to represent the society at the International Statistical Institute's nineteenth biennial conference in Tokyo, held from 15 to 25 September 1929.[27] Prior to that trip, Jin had also visited Japan from April to July to study local finance. Upon his return to

26 This journal apparently only existed until 1922; Liang Qichao was its one-time editor.

27 It is not clear if a delegation from the society attended the conference. The official bulletin of the conference only lists three invited Chinese attendees: Chang-Heng Chen (member of the Legislative Yuan), Warren H. Chen (chief of the Bureau of Statistics under the Ministry of the Interior), and D. K. Lieu (director of the Bureau of Statistics in the Legislative Yuan). In another issue of the bulletin, seven papers (*mémoires*) on China are listed as being presented at the conference. Three of these were authored by the attendees listed above. Four other papers—it is unclear whether they were presented in person or in absentia—were authored by Chungshen S. Chen (chief of the International Trade Division under the Ministry of Industry, Commerce, and Labor), Peter Wei Lin (division chief of the Bureau of Statistics in the Legislative Yuan), C.C. Chang (of the Bureau of Statistics in the Legislative Yuan), and John Lossing Buck (of the University of Nanking). Topics covered in these papers included population, the composition of the Chinese National Government, statistical work in China, agriculture, and rural

China, he moved to Shanghai and reentered the world of finance. During the next several years he held various managerial, auditing, accounting, and research positions at the Communications Bank and at the Central Bank.[28] He would eventually rise to senior positions, such as director of accounting and chairman of the board, at several of these institutions. After 1937, Jin also held professorial appointments at Shanghai University of Commerce and at Jinan University.

Among Jin's various professional identities, the one that would persist into the 1950s following the Communist takeover was that of a respected professor and prolific author of monographs and textbooks on statistics and economics. In addition to *Tongjixue dagang*, he authored tracts on subjects such as England's tax structure, currency reform, index numbers, industrial and economic problems, and Keynesian economics. Early in the decade he was appointed professor at Fudan University, where he also served as the director of special training courses on statistics and trade. During these years he also concurrently held professorships at Shanghai Institute of Finance and Economics (SIFE) and at the Institute of Economics of the Shanghai Academy of Social Sciences. He continued writing in the 1950s, and authored at least four new books on statistics, including an abridged version of *Tongjixue dagang*, entitled *Tongjixue* (Statistics).[29] He was sixty-nine years old when he passed away in Shanghai on 15 February 1963.

It was Jin's *Tongjixue dagang*, which because of its content and widespread and continued use after 1949, became one of the early targets of criticism during the 1950s. *Tongjixue dagang* was published in 1935 by two different presses: first, as part of a series by the *Guoli shangxueyuan chubanshe* (国立商学院出版社; National School of Commerce Press), and second, by the Commercial Press. By 1950, it had gone through multiple editions.[30] Jin's views on statistics and its links to mathematics, however, had been in evidence three decades earlier. In the introduction to *Xinlun tongji* (New theory of statistics) (1925), his first textbook on statistics and published shortly after his return from Columbia University, he wrote:

---

variability. For more, see Institut International de Statistique (1932), 17, 23–24, 34, 44; for the papers, see Institut International de Statistique (1930), 18–220.

28 See the entry in Xu Youchun, *Minguo renwu dacidian*.

29 The other three books were: *Gaoji tongjixue*; *Gongye tongjixue yuanli*; and *Shehui zhuyi zhidu xia de guomin shouru yanjiu*.

30 See Song Tao, *20 shiji Zhongguo xueshu dadian: Jingjixue, xiace*, 785.

The study of statistics is subtle and broad. If one is without mathematical awareness and a scientific mind, he cannot speak about it. And yet, it is used frequently and very widely; there is no one who does not use it. Indeed, scholars who care not to learn it are immediately struck by dizziness when they come across its charts and tables; and the less said about [their reaction to] advanced mathematical principles and refined and meticulous analysis, the better.[31]

Ten years later, the first sentence of *Tongjixue dagang* was more direct:

Statistics is the study of using counting or estimation to represent the dynamic-ness and static-ness of *social or natural* phenomena and to carry out an analysis of the relationship between numbers (emphasis added).[32]

In thinking about statistics as fundamentally mathematical in nature and universally applicable to the social and natural worlds, Jin's views were very much a reflection of the prevailing wisdom of the time. Like Jin, a significant number of Chinese statisticians (and economists and social scientists) active in the 1930s had been trained in the United States or Western Europe, or were students of those who had been trained abroad.[33] They had served in government, and in many instances they were active participants in what Tong Lam has called the Social Survey Movement of the 1930s.[34] At this time, a well-defined socialist critique of statistics had yet to be articulated in China or elsewhere. The espousal of statistics as a universal science by Jin and his contemporaries, many of whom were also authors of prominent textbooks, had no coherent counter-discourse. This view was reflected in the several textbooks they authored shortly after 1949, all of which enjoyed widespread popularity:

---

31 Quoted in Shi Tai, "Tongji mingren lu: Jin Guobao," 60.

32 Jin Guobao, *Tongjixue dagang*, 1.

33 For the role of Anglo-American training and the larger influence of the Anglo-American academy in shaping disciplines such as demography, economics, and sociology in Republican China, see Lam, *A Passion for Facts*; Trescott, *Jingji Xue*; and Chiang, *Social Engineering and the Social Sciences in China*. According to Trescott, *Jingji Xue*, 59, the Economics Department of Columbia University produced the most Chinese graduates in the social sciences during the first half of the twentieth century and was particularly distinguished by its heavy emphasis on statistics. For a discussion of Marxist social sciences during this period, see Chiang, *Social Engineering and the Social Sciences in China*, 136–183. For an alternate view that considers developments in economic thought during the Republican years based on the rubric of the totalizing governmentalizing logic of capitalism, see Thompson, "The Birth of the Chinese Population."

34 Lam, *A Passion for Facts*.

Jilin-based professor of statistics Wang Sili wrote *Tongjixue tonglun* (统计学通论; General theory of statistics) (1951); Zhang Zhihong wrote *Nongye tongjixue* (农业统计学; Agricultural statistics) (1953); and Zou Yiren authored *Gongye tongji* (工业统计; Industrial statistics) (1952).[35] By publishing books that promoted what was to quickly become a theoretically unsound understanding of statistics, these authors attracted the immediate attention of New China's socialist statisticians. As one of these socialist statisticians went on to explain, the hazards such books posed were indeed quite grave:

> [They] will not only impede the progress of statistical science but will also cause ideological confusion among statistical workers, which will lead statistical work to suffer losses; they therefore [will eventually] hinder development of the enterprise of socialist construction.[36]

Thus, in very short order after 1949, Jin Guobao, many of his fellow statisticians, and the textbooks they had composed, all came to represent the most obdurate of foes to those seeking to enforce a new understanding of statistics.

Patterns of Critique and Self-Critique

Obdurate foes Jin and his fellow statisticians may well have been, but they were also the type of higher intellectual whom the CCP prized. For the CCP, their value lay in possessing skills and knowledge that were sorely needed. But such groups were also the least susceptible to dogma. As Harriet C. Mills noted in a 1959 article,

> Communist policy toward this strategic group [of higher intellectuals] in the past decade has consistently aimed at securing most effective utilization of its knowledge. Zigzagging steadily toward this goal, the Communists have now attacked, now united, now criticized. Meanwhile, they are recruiting their own Red intellectuals, but have not yet had time to train a new group both Red and expert.[37]

This then was the challenge: to not wholly discredit the statisticians but to reform them so that they could continue to contribute from within the system

---

35 Song Xuguang, "Xin Zhongguo jingji tongji fazhan."
36 Gui Shizuo, "Ping Zou Yiren, Jin Guobao liangwei xiansheng," 24.
37 Mills, *Thought Reform*, 496.

and to be held up as exemplars of the potential of remolding.[38] A lack of adequate numbers of Red experts in statistics remained a challenge throughout the decade, meaning that this older generation of statisticians never lost their value.

The criticisms of the 1950s can also be understood as a younger generation's attempts to supplant an older generation. The people responsible for "zigzagging" toward the goal of effectively utilizing statisticians belonged, for the most part, to a middle generation, between the Jin Guobaos, on the one side, and the Lin Fudes and Wu Canpings, on the other. While it is unclear if either Lin or Wu directly participated in the criticisms (Lin co-authored a textbook in the mid-1950s along socialist lines), many of their colleagues did, and they all shared certain general characteristics. Most were born in the 1910s or 1920s, had been trained in economics or statistics (sometimes abroad), and by the late 1930s had thrown in their lot with the CCP. Many had received further training in Yan'an and at party schools elsewhere and, as such, were particularly receptive toward socialist statistics. As they took on administrative and pedagogical positions at key institutions following 1949—in some cases they had already been involved in administering CCP-controlled areas prior to that—they turned their enthusiasm towards propagation. Notable among these institutions were the Northeast Statistics Bureau (personnel included Yang Jianbai, Wang Sihua, and Wang Jianzhen), the State Statistics Bureau (Xue Muqiao, Wang Sihua, and Wang Jianzhen), People's University (Lin Fude), and Beijing University (Hu Daiguang).[39]

Newly established journals of the PRC provided the principal channels through which this younger generation attacked Jin Guobao and his colleagues. Of particular prominence were journals that focused specifically on statistics, most notably the NSB's *DBTJGZ* (1949–1954) and the SSB's *TJGZTX* (1953–1956), *TJGZ* (1957–1958), and *JHTJ* (1959–1960).[40] Other venues where statistics was debated included *Xin jianshe* (New construction)

---

38 On thinking about the Chinese revolution as a redistribution of skills and knowledge, see Eyferth, *Eating Rice from Bamboo Roots.*

39 The senior statistical leadership, people like Xue Muqiao, Li Fuchun, Sun Yefang, Wang Sihua, and so forth, did not directly participate in the criticism, though they must have sanctioned it.

40 *TJGZ*, known as *TJGZTX* from 1953–1956, was the SSB's principal journal. A monthly from 1953 to 1955, it became a fortnightly beginning in 1956. In 1959 it was published under a new name, *Jihua yu tongji* (计划与统计; Planning and statistics), but ceased publication in 1960. It was revived in the late 1970s, under the name *Zhongguo tongji* (中国统计; China statistics).

and *Caijing yanjiu* (Journal of finance and economics) (1956–1960, 1980–). By the middle of the decade, student theses written at Renmin (People's) University also began to tackle the subject of socialist and bourgeois statistics.[41]

The pattern of these criticisms and self-criticisms did not follow the larger patterns of national thought reform campaigns, in which newspapers were a key element. Harriet Mills described such national campaigns by noting that they generally began "with a series of articles and editorials in newspapers."

> Since newspaper reading is a political obligation in China and items of the day are often taken up in study groups, a subject which has received more than usual attention will begin to be discussed. Thus, a demand is created for further study, for which materials and instructions are soon forthcoming. The campaign, which may last several months, is launched.[42]

As noted, most criticisms of bourgeois statistics, however, were published not in newspapers but in specialized journals, which were unavailable to the general public. Therefore, the readership and target audience for study sessions were more narrowly conceived: statistical cadres, students, and teachers. In addition, unlike nationwide campaigns, such as the three-antis (1951), the five-antis (1952), or the anti-Rightist movement (1957–1959) that usually had a clear start and end date, the criticism of statistics persisted throughout the decade, gathering steam along the way.[43] The interlocutors in these debates no doubt took some of their cues from national developments, but they were also subject to the internal and international dynamics of the world of statistics.

Based on these general observations, we can divide the years from 1949 to 1958 into four periods (Table 4.1).[44] The relative paucity of critical articles during the first phase, lasting from 1949 to 1951, suggests that attention was primarily devoted to establishing a basis for PRC statistical work. Once this

41 Zou Yiren, "Guanyu zichan jieji tongji xueshu sixiang" (1956), 28.

42 Mills, *Thought Reform*, 492.

43 On the slippage between the normative start and end dates of campaigns and policies and their actual implementation in practice, see Diamant, "Policy Blending, Fuzzy Chronology and Local Understandings."

44 The periodization here is reconstructed from the following materials: (1) Liu Chang and Zhang Yun, *Ershi shiji Zhongguo*, 119–127; (2) *TJGZTX* and *TJGZ*; (3) The serial anthologies produced by the NSB that consisted of materials culled from the NSB's monthly journal *DBTJGZ* (1950–1954; the journal was discontinued in mid-1954); and (4) other journals, such as *Xin jianshe*, *Jingji zhoubao*, and *Caijing yanjiu*. It should be pointed out that this is not a comprehensive list of all publications attacking "bourgeois statistics"; it does, however, reflect some of the general trends during the decade.

TABLE 4.1. Frequency of Articles Criticizing "Bourgeois" Statistics, 1949–1958 (including self-criticisms)

	1949–1951	1952–1954	1955–1957	1958
Number of Articles	2	6	10	1

foundation was in place in the Northeast, a sustained analysis of existing statistical theory and pedagogy could be undertaken. Indeed, through the end of 1951, criticisms of "Anglo-American" or "bourgeois" statistics were rather limited. At the national level, the first book to be criticized was Gou Shisheng's *Tongjixue yuanli* (Principles of statistics). Critical essays were published in 1950 by He Sheng and in 1951 by Tie Hua and Liu Xin.[45] These initial criticisms appear not to have been very detailed. At the more regional level, some criticisms were published by the NSB beginning in 1951, especially following the start of the three-antis campaigns later in that year.[46]

Developments in late 1951 and 1952 provided a clearer focus for the critiques, which expanded both their targets and their frequency. The first of these developments was the publication of new editions of textbooks by Jin Guobao, Zou Yiren, and other statisticians. If these new editions offered a significant challenge to the rise of socialist statistics, then the preparations for, and establishment of, the SSB in 1952 and the selection of Renmin University as the main training site for the (social science) of statistics that same year provided the twin loci through which this challenge was met. Both these developments, as well as the nationwide reorganization of educational institutions (院系调整; *yuanxi tiaozheng*), encouraged a more sustained analysis of the state of textbooks and academic discourse. Not surprisingly, therefore, 1952 witnessed a large number of articles published on the subject, including articles by Yang Jianbai and Wang Jianzhen in the NSB's *DBTJGZ*.[47] In that same year, Guo Gengji published an essay criticizing Jin Guobao's *Gongye tongjixue yuanli* (Principles of industrial statistics) and Zou Yiren's *Gongye tongji* (Industrial statistics).[48] In his essay, Guo stated that in spite of claiming to have

45 He Sheng, "Ping 'Tongjixue yuanli: Di yi juan'"; Tie Hua and Liu Xin, "Lun tongjixue de duixiang."

46 In particular, a series of articles by Wang Jianzhen, reprinted in the compilations of *DBTJGZ*, in January 1952, October 1952, and January 1953.

47 For Yang Jianbai's essays, see, for instance, NSB, *Tongji gongzuo: Di si xuanji.*

48 Guo Gengji, "Fandui sanbu zichan jieji tongji sixiang dusu."

changed their texts to reflect Soviet statistical theory, these authors were guilty of still following Anglo-American statistical theory. Additional essays on the hypocrisy of capitalist index numbers and the dangers of bourgeois statistical thought were also published during 1952.[49]

In early 1953, Xu Qian and Liu Xin published what was up to that point the most sustained and detailed analysis and criticism of the state of statistical theory in the PRC. Their essay, entitled "Criticism of Bourgeois Statistical Theory," appearing in the first and second issues of that year's *Xin jianshe*, expanded on the earlier criticisms of several statisticians, including Gou Shisheng, Jin Guobao, and Zou Yiren. In their essay, Liu and Xu offered specific examples to demonstrate the problems with bourgeois statistics (to be discussed in the following section).[50] While one would have expected the essay to have served as a clarion call for additional critiques, 1954 was an oddly quiet year. One possible reason was the focus on the First Five-year Plan (1953–1957).

The third period, from 1955 to 1957, was marked not only by a quantitative jump in the number of articles but also by a qualitative shift in their content. Three features, in particular, characterize this period: the impact of the 1954 Soviet resolution on statistics, the emergence of self-criticisms, and the role of the SIFE. Let us first consider the 1954 resolution. A meeting in Moscow in the spring of 1954, which had categorically adjudicated in favor of defining statistics as a social science, had a signal influence on the wider PRC statistical community. News of the resolution and the now Soviet state-sanctioned definition of statistics, however, took some time to filter through to China; a complete Chinese translation of the deliberations was not published until early 1955.[51] But its impact was evident in the more confident tone and tenor of the critiques that began to emerge in 1955.[52] Close on the heels of publication of the Soviet resolution, Xu Qian and Liu Xin published an updated version of their 1953 article.[53] It retained its earlier title, but this time it was published in the SSB's journal *TJGZTX*. A brief editorial comment noted that it had been

49 Fang Bingzhu, "Cong zhishu shang kan zichan jieji tongji sixiang de xuwei benzhi"; and Feng Jiqing, "Xijing tongji sixiang zhong de zichan jieji yidu."

50 Xu Qian and Liu Xin, "Guanyu zichan jieji tongji lilun de pipan" (1953). The authors noted that the essay was drafted in December 1952.

51 See Editorial Note, "Sulian tongji kexue huiyi jueyi."

52 This is an impression based on the interviews I conducted with, among others, Renmin University professor of statistics, Yuan Wei.

53 Xu Qian and Liu Xin, "Guanyu zichan jieji tongji lilun de pipan" (1955).

updated based on the conclusions arrived at during the 1954 Moscow confer-ence. In the ensuing months, additional essays were published by Hu Dai-guang and Wang Jianzhen; the former demanding the eradication of the perni-cious influence of bourgeois statistics and the latter offering an analysis of the failings of bourgeois theories of index numbers.[54]

The year 1956 was particularly fruitful in terms of criticisms. Early in the year, Li Zhenzhou published an essay criticizing agricultural statistics because of their rightist conservative thought.[55] Li was followed by Wang Zhengxian, who critiqued the capitalist thought present in educational materials pub-lished at the start of the 1950s, and by Shi Qing, who focused on the statistics of infrastructure.[56] Zhang Zhihong and his *Nongye tongjixue* (Agricultural statistics) were critiqued by Zhang Minru.[57]

The year was also significant for it witnessed the first self-criticism by a statistician. Early in the year, Zou Yiren, who had been one of the subjects of targeted criticism at the beginning of the decade, published a self-criticism.[58] Gui Shizuo responded at the end of the year with an essay in which he ap-proved of and welcomed Zou's contrition, but he felt that in many instances the self-criticism did not go far enough. For good measure, in the same piece he returned to Jin Guobao's *Gongye tongjixue yuanli* (Principles of industrial statistics) and offered additional criticism.[59] Perhaps in response to the in-creased stridency during the previous year, 1957 witnessed additional self-criticisms. Jin Guobao published a self-criticism early in the year. As noted, he did not discuss his book on industrial statistics or any of his other works, but instead focused on his most successful textbook, *Tongjixue dagang*.[60] A third self-criticism was published several months later by Wang Sili, the

54 Hu Daiguang, "Ganqing zichan jieji tongji sixiang de liudu"; Wang Jianzhen, "Dui zichan jieji zhishu 'lilun' de pipan."

55 Li Zhenzhou, "Chedi pipan yu kefu women nongye tongji gongzuo."

56 Wang Zhengxian, "Guanyu jiben tongjixue yuanli jiaocai de yixie yijian"; Shi Qing, "Pipan jiben jianshe tongji gongzuo."

57 Zhang Minru, "Dui 'nongye tongjixue' yishu de jidian yijian."

58 Zou Yiren, "Guanyu zichan jieji tongji xueshu sixiang de ziwo pipan" is the first self-criticism that I have been able to locate. Zou (1908–1994) was a Paris University–trained math-ematician and statistician who had taught at Fudan University, Chongqing University, and Shanghai Business University. In 1952, he was transferred to the Shanghai Institute of Finance and Economics (SIFE). In addition, he also held appointments at the Shanghai Academy of Social Sciences and Jiaotong University. For more on Zou, see Shi Tai, "Zou Yiren."

59 Gui Shizuo, "Ping Zou Yiren, Jin Guobao liangwei xiansheng."

60 Jin Guobao, "Dui 'Tongjixue dagang' de ziwo pipan."

author of books such as *Tongjixue tonglun*.[61] Late in the year, Zou Yiren published a short history of the evolution of statistics, and, in line with his self-criticism during the previous year, he described statistics as, after all, a social science.[62]

These self-criticisms followed a common pattern. Each author began by observing that he had been trained within the Anglo-American system, in some cases had also spent time abroad, and was completely in thrall to bourgeois vices, such as formalism and idealism (唯心主义; *weixin zhuyi*). After 1949, even though he may have paid lip service to the slogan of learning from the Soviet Union, he still regarded Marxist-Leninist statistical theory as too simplistic. Only after extended study of Marxist-Leninist thought during the following five to six years had he come to realize the true object of statistics (the social world) and its theoretical basis of dialectical materialism (辩证唯物主义; *bianzheng weiwu zhuyi*).[63] The process can be summed up by the following three-point schema:[64]

1. He now understood that statistics was a social science that researched large-scale social phenomena. It was different from all the natural sciences.

2. Statistics studied the quantitative aspects of such large-scale social phenomena. Therefore, it was also different from all other social sciences.

3. Finally, statistics took into account the qualitative aspects of numerous social phenomena in order to study their quantitative aspects; that is, it took into account the interconnectedness of society. In so doing, it was again different from bourgeois statistics.

A final feature that distinguished the 1955–1957 years was the role that SIFE came to play, both as a site to which several recalcitrant "bourgeois" statisticians were relocated and as the venue where those same statisticians could perform their contrition and restitution via structured discussion and criticism sessions. The SIFE's origins can be traced to Nanjing Higher Normal School, which was established in 1917. Departments dealing with commerce and business had migrated to Shanghai in 1921, and by 1932 they had become

61 Wang Sili, "Pipan wo zai tongji xue zhuzuo zhong de zichan jieji sixiang."

62 Zou Yiren, "Lun tongji shi yimen shehui kexue."

63 Jin Guobao, "Dui 'Tongjixue dagang' de ziwo pipan," 28.

64 Wang Sili, "Pipan wo zai tongjixue zhuzuo zhong de zichan jieji sixiang," 27.

the National Shanghai School of Business. In August 1950, the school's name was changed to SIFE. During the following three years, the business and finance departments of twenty schools in eastern China were merged with it.[65] As a result, by the mid-1950s it was home to many of the major pre-1949 figures in the world of statistics, some of whom we have already encountered: Zhu Junyi (the last director of statistics under the Guomindang), Jin Guobao, and Zou Yiren, among others.

By the mid-1950s, many of the statisticians at SIFE began organizing informal discussions and meetings focusing on the various foibles of bourgeois statistics. One such meeting in 1956 engaged in criticism of bourgeois index numbers. Organized by the Statistics Department at SIFE, the meeting and its outcome were originally described in a report published in the first issue of *Caijing yanjiu* during the second half of the year.[66] It included the views of statisticians such as Zou Yiren (the journal's first editor and whose self-criticism had just been published), Zhu Junyi, and Jin Guobao. The report was reprinted in abridged form in *TJGZ* later in the year.[67] The group at SIFE remained active into 1958, when, motivated by the tumult of the Anti-Rightist movement, they published another lengthy report on an informal meeting that discussed the status and use of mathematical statistics within a socialist society. Among the many involved in this discussion were the usual suspects: Wang Sili, Jin Guobao, and Zou Yiren.[68]

The different periods through which we have viewed the growing intensity and changing nature of criticisms permit some generalization. What is evident is that even with a top-down dominant paradigm, consensus formation was a process that had to be negotiated. For its target audience—statistical cadres and students—this negotiation took place in the pages of various statistics and economics journals, thereby producing the veneer of a community of scientists in pursuit of "true knowledge" and arriving at a consensus through debate and discussion, that is, approximating the idealized workings of a general scientific community. The testing of boundaries was in clearer evidence during the early 1950s, when none of those being criticized felt it necessary to respond in print. As the space for alternate views shrank to near nonexistence after 1955, staying above the fracas was no longer an option. As more textbooks

65   http://www.shufe.edu.cn/41/list.htm, accessed December 26, 2018.
66   Tang Qinghong, "Zichan jieji tongji zhishu lilun pipan."
67   SIFE, "Pipan zichan jieji tongji zhishu de fan kexuexing he fandong benzhi."
68   Zou Yiren et al., "Guanyu shuli tongji zai shehui jingji tongji kexue zhong."

based on the new paradigm were published, especially by the SSB and by Renmin University Press, and as more people were trained in socialist statistics, self-criticisms offered the only route to salvaging one's reputation and position. Although it is impossible to parse an individual's stated acceptance of the new paradigm from his true views, very quickly the entire exercise became as much about performance as it was about belief. In this way, by 1956 the burden of performing criticism shifted from New China's socialist statisticians to those whom they sought to reform—the likes of Jin Guobao, Wang Sili, and Zou Yiren.

Before we proceed to a discussion of how these processes affected the actual content of statistics, it is worth asking if any of the obvious candidates were successful in evading criticism. Perhaps the most famous case is that of Chen Da, the eminent demographer and sociologist. Chen's career can be neatly divided into two parts. He received a doctorate in sociology from Columbia University in 1923 and from 1923 to 1951 he taught at Tsinghua University in Beijing and at National Southwestern Associated University in Kunming, Yunnan province. During that time, he led several sociological studies, authored numerous books and articles on labor, migration, and population, and held a number of important administrative positions. He was also a vocal advocate of improved demographic data, in particular the establishment of a national census.[69] From 1938 to 1948 he directed the Institute of Census Research, where he developed a census program and carried out surveys in and around Kunming. After 1949, Chen's activities become untraceable, in stark contrast to his previous activities.[70] Yet, as a practitioner who used statistical methods but did not write textbooks or seek to teach statistics he remained uninteresting to the exponents of socialist statistics.[71] Thus, even though Chen was targeted during the Anti-Rightist movement of 1958, he escaped the attention of socialist statisticians.

69 See, for example, Chen, *Population in Modern China.*

70 In 1951, Chen lost his position at Tsinghua University because sociology was scrapped as a discipline.

71 Gong Jianyao's biographical essay, *Shijie tongji mingren zhuanji,* 377–381, devotes only one paragraph to Chen's post-1949 activities, noting that he delivered one paper in 1957 using data from the 1953 national census. The same skewed pattern can be observed in Boorman, *Biographical Dictionary of Republican China,* Vol. 1, 235–239. Boorman notes that Chen participated in a conference on population research in Beijing. His paper, "New China's Population Census of 1953 and its Relation to National Reconstruction and Demographic Research," was the main paper presented at the conference. In many ways, Chen's case is symptomatic of the side-lining of several major pre-1949 figures.

The Content of Critique

Unlike the nature and timing of the self-criticisms, which leave open the interpretation that they may have been primarily about performance and survival, the impact of the imposition of socialist statistics on the content of statistical science is more readily discernible. At a general level, criticisms of bourgeois statistics were repetitions of arguments already encountered in the previous chapter. Bourgeois statistics conflated the natural and social worlds as a common unit of analysis.[72] It lacked comprehensiveness, and as a consequence it lacked objectivity.[73] It existed to exploit workers, and it helped capitalists and imperialists increase their profits.[74] Among such broader grievances, perhaps the most trenchant and specific charge was that it placed formal mathematical doctrine at the very center of statistics. By doing so, it failed to distinguish between the causal mechanisms behind functions (common to the natural world) and the multifaceted relations captured by correlation (common to the social world).[75] As a result, it did not accurately take into account the economic content and reality of what was being studied, nor did it adequately account for the relatedness of events.[76] Or, put another way, it failed to adequately take into consideration the inter-relatedness of the quantitative (量; *liang*) and qualitative (质; *zhi*) aspects of social phenomena.[77] As Jin Guobao explained in his self-criticism, probability theory was based on the independence of events, such as the toss of a coin or the throw of dice, but "events in society are interrelated and mutually constrained and definitely not independent [of each other]."[78] And so, many critiques concluded dismissively that by detaching itself from the real world, bourgeois statistics was nothing but a game of numbers (数字游戏; *shuzi youxi*).[79]

72 Xu Qian and Liu Xin, "Guanyu zichan jieji tongji lilun de pipan" (1955), 30.

73 Ibid., 32; Zou Yiren, "Guanyu zichan jieji tongji xueshu sixiang de ziwo pipan," 26; Tang Qinghong, "Zichan jieji tongji zhishu lilun pipan," 16; Hu Daiguang, "Ganqing zichan jieji tongji sixiang de liudu," 23.

74 Zou Yiren, "Guanyu zichan jieji tongji xuexhu sixiang de ziwo pipan," 25.

75 Wang Sili, "Pipan wo zai tongjixue zhuzuo zhong de zichan jieji sixiang."

76 Wang Jianzhen, "Dui zichan jieji zhishu 'lilun' de pipan," 40.

77 Zhang Zhiji, ed., *Dazhong tongji*, 3–4.

78 Jin Guobao, "Dui 'Tongjixue dagang' de ziwo pipan," 29. Also see Wang Jianzhen, "Dui zichan jieji zhishu 'lilun' de pipan," 40.

79 Xu Qian and Liu Xin, "Guanyu zichan jieji tongji lilun de pipan" (1955), 21, 30; Hu Daiguang, "Ganqing zichan jieji tongji sixiang de liudu," 25; Jin Guobao, "Dui 'Tongjixue dagang' de ziwo pipan," 31.

These broader critiques filtered into two specific areas where their impact is clearly discernible: textbook content and the evaluation of specific statistical concepts. In the early years from 1949 to 1953, new educational materials including textbooks were produced either by the NSB or by the Statistics Department of Renmin University. These materials usually consisted of compilations of reports, speeches, and lectures by leaders such as Wang Sihua and Di Chaobai (1910–1977) as well as by Soviet experts. By 1951, the compilations had taken the shape of sector- or activity-specific handbooks. For instance, in 1951 the first in a series of reference materials was published for trade and industrial statistics.[80] Over the following two years, reference materials were published on statistical theory, agriculture, statistical work in the Northeast, national income, and much else.[81]

By 1953, Tongji chubanshe, the publishing division of the SSB, was already active and would rapidly become a major producer of didactic material. In addition to general textbooks, the press published books that discussed specific methods in various fields, shared experiences, reprints of lectures, and much else. Provincial- and city-level statistics bureaus established their own presses and followed suit. By the mid-1950s, general introductory textbooks were being produced for students at all levels by several different organizations. Representative examples of such books targeting beginners and nonspecialist cadres are listed in Table 4.2.[82]

These new volumes differed from the pre-1949 textbooks with regard to both their content and their focus. In his 1954 report on Soviet statistics, Ostrovitianov had noted that the controversy over the true nature of statistics had prompted Soviet "economists and statisticians to be afraid of making use of mathematical methods in investigating social phenomena. This, evidently, is why the methods of mathematical statistics are almost entirely neglected in the drafts of the textbook prepared by the Moscow Institute of Economics and Statistics."[83] A similar pattern is discernible in the production of textbooks and

80 Renmin University of China, Department of Statistics, *Maoyi tongjixue cankao ziliao—yi*; Renmin University of China, Department of Statistics, *Gongye tongjixue cankao ziliao*.

81 Renmin University, Department of Statistics, *Tongji lilun zhong de jige wenti*; Renmin University, Department of Statistics, *Nongye tongjixue cankao ziliao*; Renmin University, *Tongjixue cankao ziliao—yi*; and Renmin University, Department of Trade Statistics, *Jingji tongjixue cankao ziliao—yi*.

82 Specialized textbooks and translations of Soviet textbooks became increasingly available during the course of the decade (Zu Ting'an, "Siben gongye tongji shuji").

83 Ostrovitianov, "The Discussion on Statistics Summed Up," 329.

TABLE 4.2. Introductory Statistics Textbooks

Title	Authors/Editors/Compilers	Publisher	Date	Comments
Dazhong tongjixue (People's statistics)	Zhang Zhiji and Zou Yiren, eds.	Lixin kuaiji tushu yongpinshe	January 1953	Revised editions in May 1954 and 1956. Self-study materials for common cadres; Middle-school level
Tongjixue yuanli jiangyi: Shangxia ce (Lectures on statistical principles—Vols. 1 and 2)	Renmin University Department of Statistics	Zhongguo renmin chubanshe	August 1956	This was the 4th edition (the first edition appeared in 1952). The 9th printing; including this print run, produced a total of 60,038 copies
Tongji lilun yiban wenti jianghua (Common problems in statistical theory)	Xu Qian, Jiang Zhao, Lin Fude, Zheng Yao, Liu Xin, Zhou Fugong	Tongji chubanshe	March 1956	First edition; by January 1957 there had been six print runs, with a total 59,910 copies

related educational material in China. Table 4.3 offers a comparison of the tables of contents of the above three books against that of Jin's *Tongjixue dagang*.

The first striking difference is the overwhelming emphasis on the collection, grouping, and aggregation of data in PRC-era textbooks, where chapters devoted to these activities usually follow an introductory chapter (ch. 2, 3, 4 in two cases; ch. 3, 4, 5 in one case). In contrast, early chapters in *Tongjixue dagang* discuss statistical tables and charts, but a discussion of actual collection and collation is relegated to a single chapter (ch. 20) at the very end. This inversion and change of emphasis are indicative of the much more practical orientation of PRC-era textbooks. The purpose of statistics was to produce data that could inform national construction. Statistical work, therefore, was a principal task, enjoying prominence over everything else. In general, the chapters in the three PRC-era textbooks can be divided into four categories: (1) Introduction; (2) Methods of collecting, collating (grouping), sorting, and summarizing data; (3) Exposition of key analytical tools (average numbers, indices, etc.);

TABLE 4.3. Comparison of Statistics Textbooks

	Tongjixue dagang (An Outline of Statistics)	*Dazhong tongji* (People's Statistics)	*Tongjixue yuanli jiangyi: Shangxia ce* (Lectures on Statistical Principles—Vol. 1 and Vol. 2)	*Tongji lilun yiban wenti jianghua* (Common Problems in Statistical Theory)
Title				
Year	September 1934	1956	First ed. in 1952, 4th edition in 1956	1956
Author	Jin Guobao	Zhang Zhiji, ed.	Renmin University Statistics Department	Xu Qian, Jiang Zhao, Lin Fude, Zheng Yao, Liu Xin, Zhou Fugong
Publisher	Shangwu yinshuguan	Xin zhishi chubanshe	Zhongguo renmin chubanshe	Tongji chubanshe
Print run	?	?	60,000 (up to the 4th ed.)	59,910 (up to the 6th ed.)
Chapter				
1	Introduction	The Idea of Statistics (Introduction)	The Object and Method of Statistics	The Object and Method of Statistics
2	Statistical Tables	The Principle Method of Collecting Data: Statistical Reports	Statistical Surveys	The Organization of Statistics in the PRC
3	Statistical Charts	A Subsidiary Method of Collecting Data: Specialized Surveys	Statistical Grouping	Statistical Observations
4	Average Numbers	Birds of a Feather: Grouping and Sorting	The Organization and Technology of Statistical Summaries and Statistical Tables	The Aggregation and Grouping of Statistical Data
5	Variance	Making Comparisons: Relative Numbers	Comprehensive Indices	The Organization and Technology of Summary Statistics: The Statistical Table
6	Probability and Error Normal Curve	What is the Average? —Average Numbers	Dynamic Series	Absolute and Relative Numbers

and (4) Graphical representation of statistical results. Even the chapters on analytical tools focused only on various forms of average and index numbers (ch. 5–9, 5–8, and 6–9 respectively), which were the principal tools used to process rates of change, ascertain the level of plan implementation, and carry out related analyses.

The most salient change, however, is the absence in the PRC volumes of any chapters or content focusing on "mathematical" statistics. In contrast, Jin Guobao devoted individual chapters in *Tongjixue dagang* to topics such as Variance (ch. 5), Probability and Errors (ch. 6), Correlation (ch. 10), and Time Series (ch. 11–14).[84] Most of these topics were predicated on the acceptance and employment of probability theory and the law of large numbers. Their omission in the PRC textbooks raises the question of the exact status of mathematics within the new socialist statistics. Chinese statisticians were quick to point out that they did not completely reject mathematics. Rather, they insisted that what was wrong was to take mathematical statistics as the basis of all statistics, but the use of mathematics to pursue statistical calculations was perfectly reasonable.[85] Thus, having set probability aside, what was left to decide was what particular methods were best suited for use within a socialist society and economy. These debates usually centered on four concepts: Measures of Central Tendency, that is, Average Numbers; Indices and Index Numbers; Time Series; and Sample Surveys. We will return to Sample Surveys in Part III of this book. For now, let us take a closer look at the first two concepts, both of which elicited fractious but also incisive criticism.[86]

Measures of Central Tendency

Among the most important concepts that came up for debate and discussion were measures of central tendency. Of the five principal measures of central

84 By the 1950s random sampling had become a major statistical method globally and was being widely taught; more on this in Part III.

85 Zou Yiren et al., "Guanyu shuli tongji zai shehui jingji tongji kexue zhong," 57. Ostrovitianov, "The Discussion on Statistics Summed Up," 329, included a similar exhortation in his 1954 report: "However, the use of the best mathematical methods, where possible and appropriate, in studying social phenomena is neither shameful nor un-Marxist. What should not be done is simply to substitute abstract mathematical formulae for politico-economic analysis."

86 One additional topic that came up for sporadic discussion was the study of business cycles (商情预测; *shangqing yuce*), a prominent topic but one that was easy to dismiss in a socialist setting because it studied market forces, was predicated on speculation, and used the tools of probability analysis (Jin Guobao, "Dui 'Tongjixue dagang' de ziwo pipan," 31–32).

tendency—the mode, the median, the arithmetic mean, the geometric mean, and the harmonic mean— the latter three generated the greatest amount of debate.[87] Among the key points of interest were the importance of grouping and the problems of bias and weights.[88]

The problem of grouping brought together the socialist concern for melding theory with economic realities on the ground. In his self-criticism, Jin Guobao cited an example from Marx's own writings to illustrate how average numbers could easily mislead:

> Imagine a situation in which among ten workers each worker earns 2 shillings every week, five workers each earn 5 shillings every week, and another five workers earn 11 shillings every week. Thus, the total weekly income of these 20 workers is 100 shillings, or 5 pounds. Now, imagine if total wages rose by 20 percent, so that now the total weekly wage stood at 6 pounds. If we were to calculate the new average wage, we would expect to see everyone's wages rising by 20 percent. But, in reality, the wage hike only affected the two five-person teams, one of which saw weekly wages go up to 6 shillings, and the other went up to 14 shillings. Thus, half of the total workforce saw no improvement in their income, even though the average income for all had indeed increased.[89]

As this example indicated, calculating the average wage hike actually served to distort class distinctions. A similar analysis was made of American per capita GDP calculations, which did not take into account the differences between the bourgeoisie and the proletariat, thereby covering up the class contradictions of capitalism.[90] In textbooks, the importance of grouping was framed in a more positive light: Attention to group averages permitted the identification of advanced workers as well as those lagging behind.[91] In this way, Chinese statisticians sought to stress that it was absolutely critical to calculate not only total averages but also group averages.[92]

---

87 Wang Sili, "Pipan wo zai tongjixue zhuzuo zhong de zichan jieji sixiang," 28–29.

88 Bias could be caused by weights (权偏误; quan pianwu) or by the model itself (型偏误; xing pianwu). Wang Jianzhen, "Dui zichan jieji zhishu 'lilun' de pipan," 41.

89 Jin Guobao, "Dui 'Tongjixue dagang' de ziwo pipan," 29.

90 Xu Qian and Liu Xin, "Guanyu zichan jieji tongji lilun de pipan" (1955), 33.

91 Xu Qian et al., Tongji lilun yiban wenti jianghua, 97–98.

92 Jin Guobao, "Dui 'Tongjixue dagang' de ziwo pipan," 30; Xu Qian and Liu Xin, "Guanyu zichan jieji tongji lilun de pipan (1955)," 30, 32–33; and Wang Sili, "Pipan wo zai tongjixue zhuzuo zhong de zichan jieji sixiang," 28, also emphasized the importance of grouping.

The problem of bias was used to determine which type of average was best suited for making economic calculations. Most discussions on this subject initially offered a simple example to illustrate that for any set of numbers the arithmetic mean (AM) was larger than the geometric mean (GM), which was in turn larger than the harmonic mean (HM). For example, given the three numbers—400, 300, and 200—the corresponding mean values would be:[93]

$$AM = (400 + 300 + 200)/3 = 300$$

$$GM = \sqrt[3]{(400 \times 300 \times 200)} = 286$$

$$HM = 3/(1/400 + 1/300 + 1/200) = 277$$

But, as Wang Sili explained, these numbers did not represent any real-life situation; the example merely served to highlight that the three methods yielded different results. Given such an example, therefore, each type of mean could theoretically be correct. If, however, examples with actual economic content were considered, then it would become apparent that in one instance it was the AM that was the only correct choice, and in another instance, only the HM was the correct choice. Consider the following two cases that were discussed by Wang Jianzhen:[94]

Case I: Three types of food products 甲, 乙, 丙 have production and pricing characteristics listed in columns B and C (Table 4.4).

If all three products were to be considered together, what would be the combined average price of each ton? According to Wang Jianzhen, the only suitable method to arrive at such a price was to use the arithmetic mean:

$$\text{Average Price} = \Sigma BC / \Sigma B$$

$$\text{Average Price} = (500 \times 400 + 1000 \times 300 + 2500 \times 200) \\ /(500 + 1000 + 2500) = 250 \, 元$$

93 Example drawn from Wang Jianzhen, "Dui zichan jieji zhishu 'lilun' de pipan," 41; other instances of the relationship between AM, GM, and HM are listed in Tang Qinghong, "Zichan jieji tongji zhishu lilun pipan," 13; and Jin Guobao, "Dui 'Tongjixue dagang' de ziwo pipan," 29–30.

94 Wang Jianzhen, "Dui zichan jieji zhishu 'lilun' de pipan," 41.

TABLE 4.4. Production and Pricing of Three Food Products

A Product	B Quantity (tons 吨)	C Price/ton (元)	D Share (%)
甲	500	400	20
乙	1,000	300	30
丙	2,500	200	50

Case II: If the only information available was the price of each product and its relative share in monetary terms, that is, columns C and D, then to answer the same question, only the HM would suffice:

$$\text{Average Price} = \sum D / \sum \tfrac{D}{C}$$

$$\text{Average Price} = (20 + 30 + 50) / \left(\tfrac{20}{400} + \tfrac{30}{300} + \tfrac{50}{200}\right) = 250 \, 元$$

In neither case was the GM the appropriate choice. And yet, that is precisely what bourgeois statistics would have preferred and what had been preferred by the various bureaus of the Nationalist government prior to 1949.[95] As Wang Jianzhen and other Chinese statisticians explained, bourgeois statistics considered the GM to be superior to the AM and the HM because it lay between the other two and therefore was putatively less biased.[96] Indeed, in a paper read at the eighty-second annual meeting of the American Statistical Association in 1920, the noted mathematical economist Irving Fisher (1867–1947) had observed: "As to bias, it can be shown that the arithmetic average has an upward bias, i.e., has in the very nature of the arithmetic process a natural tendency to give results too large, that the harmonic has a downward bias, that the geometric, median, and mode have no inherent bias in either direction."[97]

The tendency in the PRC, therefore, was to ignore the GM. For instance, the chapter on average numbers in *Dazhong tongjixue* (People's statistics) did not even discuss the GM, listing only the four other measures of central

95 See Li Deyin's statement in Tang Qinghong, "Zichan jieji tongji zhishu lilun pipan," 17. Also see the discussion of index numbers below.

96 Wang Jianzhen, "Dui zichan jieji zhishu 'lilun' de pipan," 41; Jin Guobao, "Dui 'Tongjixue dagang' de ziwo pipan," 30; Tang Qinghong, "Zichan jieji tongji zhishu lilun pipan," 13.

97 Fisher, "The Best Form of Index Number," 536.

tendency and devoting the greatest space to the AM.[98] But the GM was not entirely dismissed. The SSB's *Tongji lilun yiban wenti jianghua* (Common problems in statistical theory) discussed it, but explained that its use was extremely limited.[99] As Hu Daiguang explained elsewhere, the GM could be used to calculate rates of growth only in those instances where the individual rates represented change in the same direction (i.e., consistently up or consistently down) over the entire time period under study.[100] In all other instances, the GM was nothing but an example of mere mathematical formalism. Put differently, the GM for annual growth rates A_1, A_2, A_3, A_4, A_5, was determined only by the first and last rates, A_1 and A_5. Therefore, the GM could not tell us anything about A_2, A_3, or A_4, or the fluctuations they may represent.[101] Thus, it was in this form and this form only—unidirectional change—that we see the GM being used to calculate the combined annual growth rate in industry and agriculture.[102] One implication of the rejection of the GM in favor of the AM is that it leaves open the possibility that our sense of the changes in the 1950s is based on numbers that may possess an upward bias.[103]

Indicators and Index Numbers

The mean is a specific type of indicator, and the distinction drawn between its three types was part of a wider discussion on indicators and index numbers. Broadly speaking, an index number is "a quantity which shows by its variations the changes over time or space of a magnitude which is not susceptible to direct measurement in itself or to direct observation in practice."[104] Examples of index numbers include those for wholesale prices, labor productivity, business activ-

---

98 Zhang Zhiji and Zou Yiren, eds., *Dazhong tongjixue*, 41–51 (chap. 5).

99 Xu Qian et al., *Tongji lilun yiban wenti jianghua*, 113.

100 Hu Daiguang, "Ganqing zichan jieji tongji sixiang de liudu," 26–27.

101 $\sqrt[4]{(A_2/A_1) \times (A_3/A_2) \times (A_4/A_3) \times (A_5/A_4)} = \sqrt[4]{(A_5/A_1)}$. For a more detailed explanation of this line of reasoning, see Jin Guobao, "Dui 'Tongjixue dagang' de ziwo pipan," 30.

102 SSB, *Weida de shinian*, 90. The table in question lists the average annual rate of increase in output of major products (steel, pig iron, coal, electric power, etc.) for three time periods: 1950–1952, 1953–1957, and 1950–1958. For each time period, the method used is the GM. The data used to calculate these rates are in the tables on pp. 84–89.

103 For a related discussion on the upward bias in reconstructed retail price indices for this period, see Wang, *Economic Policies and Price Stability*, 4–8.

104 "Important features in the construction of an index number are its coverage, base period, weighting system, and method of averaging observations. The term can also be applied to a series of values which are standardized by being referred to a basic period or area." Dodge, ed.,

ity, physical volume of production, cost of living, and many more. The calculation of index numbers was considered absolutely essential to socialist statistics,
and it is here that the distrust of formal mathematical statistics became trans-
parent.[105] In a book on industrial statistics, A. I. Ezhov, deputy director of the
Soviet Central Statistical Administration and the first Soviet statistical expert
in China (1950–1952), had stressed the superiority of socialist statistics by observing that "Industrial statistics of all major capitalist countries does not,
owing to the absence of a centralized governmental statistics, possess a number
of important indicators and many indicators are estimated indirectly."[106]

The central challenge in arriving at a good index number is to be able to
account for changes in all variables; for instance, in the case of price indices,
this would mean accounting for changes in price and in quantity. Two of the
most basic price indices are known after their respective inventors, Hermann
Paasche (1851–1925) and Étienne Laspeyres (1834–1913). In the Paasche Index,
prices at time $= n$ (p_n) and time $= 0$ (p_0) are weighted by quantity produced
over time n (q_n).

$$P_P = \frac{\Sigma\, p_n q_n}{\Sigma\, p_0 q_n}$$

In contrast, in the Laspeyres Index, prices at time $= n$ (p_n) and time $= 0$
(p_0) are weighted by quantity at time $= 0$ (q_0).

$$P_L = \frac{\Sigma\, p_n q_0}{\Sigma\, p_0 q_0}$$

The only difference in the two formulas is the period from which the quantity
values are drawn: The former uses period "n" quantities; the latter uses base
period quantities. While numerous indices, such as these and many others,
have been proposed and tests have been devised to check their quality, the
problem of over- or under-estimation remains unresolved.[107] In an attempt to
resolve this problem, in 1921 Irving Fisher proposed what he called the "two

The Oxford Dictionary of Statistical Terms, 197. A contemporary Chinese definition can be found
in Xu Qian and Liu Xin, "Guanyu zichan jieji tongji lilun de pipan" (1955), 34.

105 Ibid.

106 Ezhov, *Industrial Statistics*, 2.

107 For more, see Dodge, ed., *The Oxford Dictionary of Statistical Terms*, 227–228, 299; and
the entries in *Wolfram MathWorld*, http://mathworld.wolfram.com/topics/StatisticalIndices.
html, accessed 14 December 2018.

supreme tests" that could help determine the formula for an ideal index: first, that it "should *work both ways* as to the two factors, prices and quantities"; and second, that "the formula should *work both ways* as to time."[108] In mathematical terms, Fisher's ideal index formula took the square root of the product of the Paasche and Laspeyres indices:

$$P_B = \sqrt{P_L P_P}.$$

Irving Fisher was regarded as the foremost American economist by several of his most accomplished peers, including Joseph Schumpeter, James Tobin, and Milton Friedman. He was also a brilliant statistician and among the pioneers in the subfield of econometrics.[109] Among the many areas in which he made seminal contributions was the study of index numbers.[110] Fisher also had a direct link to Chinese statistical practice during the 1920s and 1930s. During his time as a PhD student at Yale, economist Franklin L. Ho worked with Fisher in developing index numbers. Upon his graduation in 1926, Ho returned to China and took up a position as professor of public finance and statistics at Nankai University in Tianjin. At Nankai, Ho put to good use the $500 gift he had received from Fisher to organize the Committee on Social and Economic Research.[111] In the early 1930s, this committee would become the Department of Economics. Among Ho's first research projects was the compilation of historical price indices, which he calculated by applying Fisher's ideal index to over sixty years of price and quantity data from the Imperial Maritime Customs Service.[112] These indices, as well as others that Ho and his

108 Fisher, "The Best Form of Index Number," 534. The equivalent Chinese terms are 时间互换测验 and 因子互换测验. See Renmin University, Department of Statistics, *Tongjixue yuanli jiangyi: Xia ce*, 89.

109 See, for instance, Schumpeter, *Ten Great Economists*, 223; Tobin, *Essays in Economics*; and Friedman, *Money Mischief*, 37.

110 Fisher, "The Best Form of Index Number"; and Fisher, *The Making of Index Numbers*.

111 As Ho recalled: "I prepared a memorandum for Chang Po-ling in which I proposed the establishment at Nankai of a University Committee on Social and Economic Research, independent of any of its colleges and having as its principal field of research the finding and measuring of facts bearing upon social, economic, and industrial problems in China." The $500 gift from Fisher was combined with CN$5,000 from the university budget and another CN$4,000 from the China Foundation. See Reminiscences of Ho Lien (Franklin L. Ho), Oral History, 1966, in Chinese Oral History Project, Rare Book and Manuscript Library, Columbia University, New York, 58–59.

112 Ho, *Index Numbers of the Quantities and Prices of Imports and Exports*, 4.

colleagues calculated, were published between 1934 and 1936 in *Nankai Index Numbers*.[113]

For China's socialist statisticians after 1949, however, in promoting the concept of an ideal index formula (理想公式; *lixiang gongshi*) Fisher was conflating two distinct problems: what type of mean to use and what type of weights to use. According to the Chinese socialist statisticians, Fisher began by assuming that the problem was a mathematical one. He completely ignored the economic content of the various indices.[114] Fisher thus became the embodiment of the failings of bourgeois statistics as a whole.[115] The failings of the two tests as well as his "eight big formulas" were dissected in detail by several interlocutors.[116] The differences they sought to articulate can be summarized in the following two sets of equations:[117]

$$\frac{\Sigma\, p_1 q_1}{\Sigma\, p_0 q_1} \times \frac{\Sigma\, q_1 p_0}{\Sigma\, q_0 p_0} = \frac{\Sigma\, p_1 q_1}{\Sigma\, p_0 q_0}$$

$$\sqrt{\frac{\Sigma\, p_1 q_0}{\Sigma\, P_0 q_0} \times \frac{\Sigma\, p_1 q_1}{\Sigma\, p_0 q_1}} \times \sqrt{\frac{\Sigma\, q_1 p_0}{\Sigma\, q_0 p_0} \times \frac{\Sigma\, q_1 p_1}{\Sigma\, q_0 p_1}} = \frac{\Sigma\, p_1 q_1}{\Sigma\, P_0 q_0}$$

p_0= Price at time 0 p_1= Price at time 1

q_0= Quantity at time 0 q_1= Quantity at time 1

Both sets of equations offer a way to disaggregate overall change over time, which is represented by the formula $\Sigma\,(p_1 q_1)/\Sigma\,(p_0 q_0)$ on the right side of each

113 Fong, "Recollections of Early Research on Chinese Economy," 70, 72. See also Trescott, *Jingji Xue*, 263–264 (on the proliferation of index number calculations across China); Chiang, *Social Engineering and the Social Sciences in China*, 84 (on Franklin Ho's obsession with calculating indices).

114 Tang Qinghong, "Zichan jieji tongji zhishu lilun pipan," 19; Jin Guobao, "Dui 'Tongjixue dagang' de ziwo pipan," 31; Xu Qian and Liu Xin, "Guanyu zichan jieji tongji lilun de pipan (1955)," 34.

115 Wang Jianzhen, "Dui zichan jieji zhishu 'lilun' de pipan," 41.

116 See, for instance, Tang Qinghong, "Zichan jieji tongji zhishu lilun pipan,"17–18; Jin Guobao, "Dui 'Tongjixue dagang' de ziwo pipan," 28, 30–31; and Wang Jianzhen, "Dui zichan jieji zhishu 'lilun' de pipan," 41. Also see Renmin University, Department of Statistics, *Tongjixue yuanli jiangyi: Xia ce*, 89–90.

117 For the first equation, see Renmin University, Department of Statistics, *Tongjixue yuanli jiangyi: Xia ce*, 71; for the second equation, see Ho, *Index Numbers of the Quantities and Prices of Imports and Exports*, 4.

equation. The first instance uses the product of Laspeyres' and Paasche's price and quantity indices to arrive at overall change. In the second instance, the same result is generated by multiplying Fisher's ideal price and quantity indices. The former uses the AM method, and the latter use the GM method.[118] The Chinese strongly rejected the second method, dismissing it as a game of mathematical formalism (一套形式主义的数学游戏; *yitao xingshizhuyi de shuxue youxi*).[119]

The Chinese critique of Fisher's ideal index was certainly a reasonable and powerful one. And while Fisher may have served as a productive and justifiable focal point for the wider critique of Anglo-American bourgeois index number theory, despite all their deliberations, China's socialist statisticians did not propose any suitable or sophisticated alternatives.[120] Constant prices of industrial products were produced with 1952 and 1957 as base years, but "proper price indices for different expenditure components" were not created.[121] Given such an absence, the most frequently used types of index numbers were much simpler, focusing on year-on-year change or taking one particular year as the base year and tracking nominal changes in future years based on that year. The following two tables from *Weida de shininan* (Ten great years) provide an example of each such case.[122]

Table 4.5 provides index numbers for annual industrial and agricultural production from 1950 to 1959. For each year (row), the previous year (row immediately above) is taken to be the base year (i.e., 100). Thus, we can see that total industrial and agricultural output in 1957 was 107.8, or 7.8 percent greater than that in 1956. The primary driver of this increase was industrial growth (111.4, or 11.4 percent) and not agricultural growth (103.5, or 3.5 percent). In Table 4.6, in contrast, all comparisons were made taking 1949 as the base year (=100). Thus, the growth in modern industry by 1957 was just over 7 times (703.1) that in 1949.

118 For an example using the HM method, again acceptable to the Chinese, see Renmin University, Department of Statistics, *Tongjixue yuanli jiangyi: Xia ce*, 75–78.

119 See Renmin University, Department of Statistics, *Tongjixue yuanli jiangyi: Xia ce*, 89.

120 Discussions incorporated critiques and assessments of a wide variety of index numbers, such as composite index, aggregate index, relative index, average index, structural index, comparative index, strength index, dynamic index, chain indices, and so on.

121 As Xu, "The Establishment, Reform, and Development of China's System of National Accounts," 446–447, notes, "there was no consumer price index and fixed assets investment price index."

122 SSB, *Weida de shinian*, 16, 80.

TABLE 4.5. Gross Value of Industrial and Agricultural Output

	工农业总产值 三, 指数 (以上年为100)			%
	工农业 总产值	工业 总产值	其中: 手工业产值	农业 总产值
1950 年	123.3	136.4	156.4	117.7
1951 年	118.9	137.9	121.3	9.4
1952 年	121.1	130.3	119.1	115.3
1953 年	114.4	130.2	124.7	103.1
1954 年	109.4	116.3	114.7	103.3
1955 年	106.6	105.6	96.8	107.7
1956 年	116.5	128.2	115.6	104.9
1957 年	107.8	111.4	114.2	103.5
1958 年	148.0	166.2	—	125.0
1959 年计划	120.0	125.6	—	110.0

Source: Reproduced from State Statistics Bureau, *Weida de shinian*, 16.

TABLE 4.6. Continuous Growth of Modern Industry

	现代工业的比重不断提高		
	现代工业的 总产值 (亿元)	指数 (%) (以1949年为100)	现代工业总产值 占全部工业总产 值的比重 (%)
按1952年 不变价格计算			
1949 年	79.1	100	56.4
1950 年	103.9	137.5	56.9
1951 年	159.1	201.0	60.4
1952 年	220.5	278.6	64.2
1953 年	288.1	364.1	64.5
1954 年	339.8	429.5	65.4
1955 年	370.8	468.6	67.6
1956 年	503.4	636.2	71.6
1957 年	556.3	703.1	70.9
按1957年 不变价格计算			
1957 年	496.7	—	—
1958 年	872.7	12.4 倍	74.6

Source: Reproduced from State Statistics Bureau, *Weida de shinian*, 80.

Conclusion

Neither of these transitions—the disciplining of a generation of statisticians and the changes in content—occurred overnight, even though that was very much the idea behind the re-organization of schools and disciplines in 1952.[123] Genuine concerns about capitalism, such as the existence of incentives to dress up data to fool people and to favor private interests, were conflated with the perceived weaknesses of specific statistical methods. To return to Schmalzer's point about paying attention to the context within which ideologies operate, ideology performed a dual role for statisticians in the 1950s. On the one hand, it gave them uncluttered insights into the failings of Western statistical claims. Indeed, they were not wrong to point out that bourgeois statistics often failed to meet its own stated goals. Yet ideology also constrained them, forcing them to shift the emphasis away from what kinds of data were needed—which of course remained important—to what kinds of methods were appropriate. As the struggle over reformulating statistics unfolded, a space was created to mount a robust critique of Western methods, but this occurred without a corresponding search for creative solutions.

What the reversion to simpler methods of indexing did achieve, however, was a more accessible statistics. It made the statistics that did exist much easier to comprehend, producing a sort of mass-line effect. Nonetheless, the narrower range of statistical theory that was taught beginning in the 1950s significantly limited the abilities of several generations of practicing statisticians. The strong boundaries established between the social science of statistics and mathematical statistics left both communities of scientists (and practitioners) mutually ignorant.

Particularly revealing in this regard is the case of the statistician Xu Baolu (1910–70).[124] Born in 1910[125] in Beijing, Xu did his undergraduate work at Tsinghua University, and graduated with a degree in mathematics. He then

123 The readjustment of universities and faculties was carried out in an attempt to Sovietize the existing educational structure. In the humanities and social sciences, the starkest change was in the delegitimization of such disciplines as sociology and anthropology, both of which ceased to exist.

124 For more on Xu Baolu, see Anderson, Chung, and Lehmann, "Pao-Lu Hsu 1909–1970"; Jiang Xuepei, "Beijing daxue jinian zhuming shuxuejia, tongjixuejia, Xu Baolu jiaoshou"; and Chen and Zheng, "Academic Achievements of Professor P. L. Hsu."

125 There appears to be some disagreement about the date of his birth. One biography cites 1 September 1910, and another cites 1909.

spent four years at University College, London, earning a PhD in 1938 and a ScD in 1940, both in Mathematical Statistics. Upon his return to Beijing, he was appointed professor in the Department of Mathematics at Peking University. Barring a two-year stint in the United States from 1945 to 1947, he spent the remainder of his career at Peking University. Xu's major area of research was probability theory and mathematical statistics; today he is identified as the first person in China to engage in research in these areas.[126] He counted among his friends E.A. Pearson and Jerzy Neyman and "he was the first Chinese who was internationally recognized in the area of probability and statistics."[127] These are precisely the areas that socialist statistics refused to include within its purview. And yet, none of the criticisms targeting bourgeois statistics ever touched Xu for the simple reason that he was based in a Department of Mathematics. This dichotomy between statistics and mathematical statistics cut both ways—it constrained what one knew of the other—and remained influential into the 1980s.

126 See Jiang Xuepei, "Beijing daxue jinian zhuming shuxuejia, tongjixuejia, Xu Baolu jiao-shou," 39.

127 Chen and Zheng, "Academic Achievements of Professor P.L. Hsu," 1; Anderson, Chung, and Lehmann, "Pao-Lu Hsu 1909–1970," 468. A survey of Chinese contributions to statistics notes: "It was Professor Pao-Lu Hsu (许宝騄) who first introduced modern statistics to China in the 1940s. He was also the first Chinese statistician to win a renowned international reputation in his field." See Chen, Fang, and Yang, eds., The Development of Statistics, 1.

PART II

Seeing like a Socialist State

5

The Nature of Statistical Work

"[We are] able to reflect production, but [can only offer] a poor reflection of livelihood."[1]

FROM A VERY EARLY STAGE, the idealized vision of statistical work that had been espoused in the writings of Wang Sihua and claimed to have largely been achieved by the NSB had to face up to the reality of expansion across a vast, complex, and largely agrarian landscape. Through the work of the NSB and, in particular, the writings of Wang Sihua, a model of statistical work had been articulated that valorized extensiveness (across economy and society), completeness (of the statistical system), and objectivity (about social facts). This model was predicated on the primacy of exhaustive enumeration; other methods of data collection, notably surveys, were of secondary importance and only to be used in a supplementary or as-needed basis. The idea that such a valuation of methods was the only correct choice was based on a concerted and unambiguous theoretical understanding of the social world and how to account for it. By 1951, state and party leaders were in a position to start thinking about expanding the system in the Northeast throughout the country to establish a unified national statistical system.

As the emphasis moved from articulation of an ideal to national implementation, the ensuing six years witnessed the establishment of the SSB in Beijing and the rapid expansion of a nationwide network of subsidiary statistics offices at the provincial, city, and county levels. These bureaus and offices organized their work schematically around four key tasks: collection (收集; *shouji*),

---

1 BMA 063-001-00088: 5.

collation/ordering (整理; *zhengli*), research (研究; *yanjiu*), and supply (提供; *tigong*).[2] Among them, collection and supply were considered the more fundamental tasks.[3] The six-year period also included within it China's First Five-year Plan (1953–1957), thereby inextricably linking, at least in a putative sense, the twin regimes of statistical and planning work.

The nature of statistical activity during this period not only mirrored broader bureaucratic and governance-related trends frequently identified with the mid-1950s but by constraining them also gave them their shape.[4] The first of these key trends was the increasing degree of complexity across the economy as new factories and industries were set up, agriculture was re-organized and subsequently nationalized, and a new system of remuneration was put in place, all against the background of sustained economic growth. For statistics, this meant an expansion of the types of data to be collected as well as an expansion into the more virgin territory of performing statistical analyses and research based on the collected data. Complexity also brought into focus issues of organization and coordination across the various bureaus and organs. Another key trend during the mid-1950s was the push toward greater decentralization. Decision-making power, especially as it pertained to implementation of policies decided at the central level, was frequently devolved to the provinces or lower levels of administration.[5] In statistics, this broader push toward decentralization came up against the centralizing tendencies of the statistical apparatus. A resolution was attempted, via the principle of "unified leadership, dispersed management" (统一领导, 分级管理; *tongyi lingdao, fenji guanli*),

2 See, for instance, Xue Muqiao's closing remarks at the Sixth National Meeting on Statistical Work (BMA 133-001-00090: 83, 3 of 7 of document). Xue cited a similar list of tasks earlier during the same meeting, when delivering his report on experience gained in statistical work over the course of the First Five-year Plan. See Xue Muqiao, "Di yige wunian jihua qijian woguo tongji gongzuo de chubu jingyan he jinhou renwu." Numerous other instances of similar listings exist. For instance, in his self-criticism Zou Yiren listed the same four tasks. See Zou Yiren, "Guanyu zichan jieji tongji xueshu sixiang de ziwo pipan," 26. Such categorization is, of course, not unique to the PRC or to socialist statistics. Jin Guobao, in his *Tongjixue dagang*, 8, offered the same breakdown of the procedural methods associated with statistics.

3 Xue Muqiao, "Di yige wunian jihua qijian woguo tongji gongzuo de chubu jingyan he jin hou renwu," 5.

4 For a general discussion, Schurmann, *Ideology and Organization*, remains an invaluable resource.

5 Decentralization was a feature of planning work; determination of its extent has been a source of some controversy among economists (Perkins, "Research on the Economy of the People's Republic of China," 353).

which allowed local party and government administrations to execute their work as they saw fit and without too much oversight.[6] This tension between centralization and decentralization, in turn, brings into focus the two forms of political and executive authority—horizontal and vertical—that characterized both the organization and prosecution of statistical work.[7] Finally, a third key feature of the period was the realization that agriculture was absolutely critical to the economic fortunes and socialist goals of the PRC. Best articulated in statements that simultaneously reiterated the primacy of industry while also elevating agriculture to near equal status, such a rebalancing of emphasis forced the SSB to confront the unreliability of their agricultural data and the unsuitability of the methods used to collect those data.

Statistical work was thus carried out within the larger unfolding context of increasing complexity, changing (economic) focus, and struggles over administrative devolution. As work became more complicated and demanding, engagement with and discussions about statistical activities grew more sophisticated—methods were expanded, analyses were undertaken, and the results of such analyses were circulated among leaders and bureaucrats. However, as the following chapter will show, a key problem remained—the system that had been set up incentivized the overproduction of reports, and the state had little or no capacity to handle the resultant excess reports. This meant that in the interests of maintaining timeliness (及时性; *jishixing*), which the director of the SSB, Xue Muqiao, would repeatedly state was of higher importance than correctness (正确性; *zhengquexing*), there was an ever-increasing disjuncture between local data and regional and national data. Whereas local data were often of reasonably good quality, the need to report the data to higher levels in a timely fashion meant that estimation was permitted.[8] As a result, the higher up data were reported, the more estimation was likely used, rendering provincial and eventually national data victim to gross inaccuracies.

Furthermore, given their reliance on the systematic collection, collation, and upward reporting of data, statisticians had no independent or systematic means to check the accuracy of the numbers.[9] At the same time, such a system was particularly prone to delays. Archival reports of statistical work are

6 BMA 133-001-00090: 83.

7 For a general discussion of decentralization as well as vertical and horizontal forms of governance, see Schurmann, *Ideology and Organization*.

8 See, for instance, BMA 002-020-00746: 12.

9 The best methods were to do spot-checks or one-time surveys. See Zhang Zhiji, ed., *Dazhong tongji*, 9–13.

full of complaints about factories, work units, and agricultural cooperatives all falling behind reporting schedules. Delays combined with the excess issuance of reports thus created an absolutely chaotic situation by the mid-1950s, leading Wang Sihua to lament that statistical work was unable to address the needs of the State Planning Commission (SPC) and the demands of the nation's leaders.[10]

The Creation of a National Network

Prior to September 1952, when the SSB (the national statistical agency) was set up, the only bureau (局; *ju*) for statistical work in the PRC was the NSB headquartered in Shenyang. This does not mean that no statistical work was carried out during the first three years of the PRC. Much of this work was coordinated through finance and economics committees and statistical departments (处; *chu*) and branches (科; *ke*), which collected data or prepared estimates based on pre-existing data. In similar fashion, regional and provincial governments possessed statistics divisions, as did various ministries. Most statistical reports were compiled through their efforts.[11] The key national level coordinating institution was the Central Finance and Economics Committee, which was set up on 21 July 1949. Composed of six departments, it was led by Chen Yun (1905–1995), who was supported by such figures as Bo Yibo (1908–2007), Ma Yinchu (1882–1982), Li Fuchun, and Xue Muqiao.[12]

It was this Central Finance and Economics Committee that initiated discussions about a national statistical system. In July 1951, Li Fuchun, a deputy director of the Committee, chaired a national-level conference on financial and economic statistics to begin discussions on how to implement the Northeast model on a national scale. Held in Beijing from 16 to 20 July, this conference in later years would be retroactively dubbed the First National Meeting on Statistical Work.[13] Eighty-five delegates, joined by ninety-one observers,

10 See Wang Sihua, *Wang Sihua tongji lunwenji*, 100–101.

11 For instance, a Work Department of the CCP Central Finance and Economics Committee was set up in July 1948, headed by Dong Biwu (1886–1975) and with Xue Muqiao as secretary. A Department of Customs Statistics was set up in early 1950. See *DSJ*, 1–2.

12 See *DSJ*, 2.

13 In all, there would be five additional such annual national meetings on statistical work, all held in Beijing: December 1952, February 1954, February 1955, February 1956, and September 1957. For details, see *DSJ* and SSB, *Tongji gongzuo zhongyao wenjian huibian*, 3 vols.

gathered to address for the first time the statistical needs of the country.[14] Echoing Stalin decades earlier, in his speech Li Fuchun observed that without statistical data the tasks of national construction, planning, and economic work could not be undertaken. He went on to make clear that China would learn socialism from the Soviet Union and would take Marxism-Leninism as the guiding principle, criticizing Anglo-American capitalist statistics along the way. But the key task facing the conference was the issue of establishing a national statistical system, something that even two years after its establishment, the PRC still did not possess. It was Wang Sihua who proposed setting up a national statistical agency, which would centrally guide statistical work throughout the country.[15]

The details of such a national system were articulated by the main Soviet statistical expert and adviser, A. Ezhov.[16] In an address delivered on the penultimate day of the conference, Ezhov pointed out that although at the regional levels there had been much statistical work, China still lacked national statistics under a unified leadership. Therefore, the most urgent task was to link the various statistical activities under one unified structure. Ezhov identified nine key tasks. The most important was the absolute necessity of one unified seamless system that linked the local to the central levels and the ministry to the work unit, with clear lines of command and oversight. To man this system, a large body of trained cadres was needed. The NSB was to serve as the model for all other provincial- and city-level bureaus throughout the country as they went about setting up such a system.[17]

It took nearly another year for the recommendations of the national meeting to be put into effect. On 7 August 1952, Xue Muqiao was appointed head of the newly established SSB.[18] The bureau began operations the following month, on 8 September. In a report delivered to the Central Finance and

14 Discussions ranged from how to ascertain guiding principles and specific tasks, what methods to employ, organizational issues, and the training and cultivation of cadres.

15 See *DSJ*, 6; Wang Sihua, *Wang Sihua tongji lunwenji*, 225.

16 As discussed in Part I, Ezhov, a deputy director of the Soviet Central Statistical Administration, arrived in China sometime in early 1950 and presumably returned to the Soviet Union in February 1952.

17 *DSJ*, 5–6. Other tasks included producing *Tongji gongzuo shouce* (统计工作手册; statistical work handbooks), simplifying and cleaning up the chaotic issuing of forms, collating pre-existing data for useful comparisons, etc.

18 See *DSJ*, 7. Wang Sihua, *Wang Sihua tongji lunwenji*, 225, was appointed one of the bureau's deputy directors and put in charge of industrial statistics.

Economics Committee, Xue noted that the bureau had established a secretariat, and departments for summary/comprehensive, industrial, agricultural, trade, capital construction, and communications statistics.[19] In addition, it also set up sections devoted to labor wage statistics, distribution of goods, and the world economy. Departments devoted to culture and education and to social statistics had yet to be set up. The plan was to employ about 250 people across these offices (excluding those working part-time).[20]

Some months after the establishment of the SSB, the Second National Meeting on Statistical Work took place in Beijing from 15 to 22 December. It included an agenda-setting address by the famed military commander and recently appointed vice chairman of the central government, Zhu De (1886–1976). For the most part, Zhu's speech covered issues already introduced by Ezhov. One notable difference was in how Zhu framed the importance of statistical work. In this, his speech was in agreement with Wang Sihua when he was in the Northeast: Statistical work was necessary because it was the basis of a planned economy. Zhu thus reiterated Wang Sihua's call for a united statistical apparatus. His exhortation was subsequently repeated by Xue Muqiao in his own speech to the delegates.

One of the primary tasks identified at the Second National Meeting on Statistical Work was the need to set up regional-, provincial-, and city-level statistics offices as part of a unified national network. In a letter dated 14 April 1953 and delivered to all regions, provinces, and cities, Xue Muqiao noted that by the end of March five regions and eighteen provinces and cities had reported the successful establishment of statistics offices.[21] But the status of statistics offices in the South-Central Region as well as in twenty-three provinces and cities remained unknown. Xue proceeded to accept the SSB's culpability in causing some of these delays: Organizational guidelines about a national statistical network had, after all, only been disseminated in early

---

19 DSJ, 7–8. For a chart depicting the seven departments in 1952, see DSJ, 492.

20 In a survey conducted the following month, it was established that there were 354 statistical workers in the main offices of the six administrative regions and Inner Mongolia. Another 766 statistical workers were on the various provincial-level finance committees. Thus, the total was 1,120 workers, as follows: in the North 98, Northeast 388, East 240, South-Central 168, Southwest 111, Northwest 95, and Inner Mongolia 20. See DSJ, 9.

21 These were: Northeast, North, East, Southwest, Northwest, and Inner Mongolia, Beijing, Tianjin, Hebei, Shaanxi, Suiyuan (绥远) (a former province in China; today part of Inner Mongolia), Guangdong, Hubei, Hunan, Henan, Shanghai, Jiangsu, Anhui, Chongqing, Yunnan, Xi'an, Shanxi, Shenyang, and Luda. See BMA 002-020-00670: 1.

March. Even so, Xue urged the recalcitrant to rapidly establish statistics offices.[22]

As shown in Table 5.1, Xue's exhortation met with direct results. By the end of June 1953, all but Liaoning and Qinghai had set up regional-, provincial-, or city-level statistics offices.[23] According to the sanctioned size of the various provincial bureaus at the time of establishment, 1,275 statistical workers were distributed across twenty-four bureaus. The average bureau size was about fifty-three workers.[24] City-level bureaus, however, were often larger than provincial bureaus: Shanghai had ninety-six statistical workers and Beijing sixty-five. Most bureaus were organized into departments, offices (室; shi), branches, and groups (组; zu).[25] As a result of the restructuring of government organs that took place in 1954 (in the wake of the First National People's Congress), the SSB was identified as the first among twenty organizations directly subordinate to the newly formed State Council (which replaced the Government Administrative Council of the 1949–1954 Common Program).[26] This is the national structure that remained in place until the summer of 1958.

By 1956, the SSB main office in Beijing had expanded to thirteen key branches employing 645 personnel.[27] Two of these were offices dedicated to administration and research. Six divisions (司 ; si) handled statistics related to comprehensive/summary data, industry, agriculture, capital construction,

22 BMA 002-020-00670: 1.

23 The Liaoning Statistical Bureau was formed by the merging of the Liaodong and Liaoxi statistics bureaus in August 1954. The Liaodong and Liaoxi Statistics Bureaus had been set up separately in May and August 1950 respectively, under the aegis of the NSB. Evidence from the directive, the first document in BMA 002-020-00670, is not always consistent with the dates reported in *DSJ*. For instance, the BMA document notes that Anhui had already set up a statistical bureau by March 1953. *DSJ*, however, lists it as occurring in June 1953.

24 These numbers and estimates are based on information culled from *DSJ*, 7–35. It should be noted that this was the sanctioned size and not necessarily the actual number of personnel employed.

25 I have not found uniform guidelines delineating a department from an office and a branch; considerable variation likely existed. For instance, in 1957 the Henan Statistics Bureau made the following recommendation: Cities with populations over 0.5 million should set up a bureau, those with a population between 0.2 and 0.5 million should set up a department, and those with a population of less than 0.2 million should set up an office. See BMA 133-001-00091: 14.

26 Schurmann, *Ideology and Organization,* 183–185. For a copy of the actual government notice, see BMA 153-001-01221: 1–3. In Schurmann, the order on the list indicates the relative importance of the organ. However, I have not found anything to substantiate this.

27 Huang, "The Statistical Agency in China's Bureaucratic System," 61.

TABLE 5.1. Establishment of National, Regional, Provincial, and City-Level Statistics Bureaus

Date established	Province/city	Authorized staff	Notes
7 August 1952	National		SSB Established; Director: Xue Muqiao
8 September 1952	National		SSB Starts Operations
12 January 1952	National		Statistical Bureaus (局), Departments (处), and Branches (科) set up in various bureaus and organs of the central government
26 November 1954	National		SSB first bureau listed under reorganized State Council
4 April 1950	Northeast	?	
January 1953	Northwest	80	
1 February 1953	North China	60	Stopped functioning on 5 August 1954, when large administrative regions were canceled.
13 March 1953	Southwest	70	Located in Chongqing
March 1953	East	110	
1 May 1953	South-Central	80	Located in Shanghai. Stopped work by end of year.
December 1952	Inner Mongolia	45	The Suiyuan (绥远) Bureau was set up in March 1953, and later incorporated into the Inner Mongolia Bureau
January 1953	Hebei	70	
January 1953	Hebei	60	
1 February 1953	Guangdong	57	Built on work done by the province's Survey and Statistics Committee (调查统计委员会) and the South China Finance Committee Planning Bureau (华南财委计划局统计处)
9 February 1953	Yunnan	55	
4 March 1953	Zhejiang	65	
12 March 1953	Shanxi	55	
14 March 1953	Shandong	90	
15 March 1953	Henan	66	
1 April 1953	Shaanxi	50	
April 1953	Heilongjiang	80	
April 1953	Jiangxi		
24 April 1953	Guizhou	50	Prior organ: The Statistics Office of the Provincial Finance Committee (省财委统计室)

Date	Name	Number	Description
25 May 1953	Sichuan	80	
May 1953	Jilin	55	Established by abolishing the Statistics Department of the Jilin Provincial People's Economic Planning Commission (吉林省人民经济计划委员会统计处)
May 1953	Hunan	50	
May 1953	Xinjiang	15	The existing Statistics Department (统计处) was upgraded to a Bureau (统计局). Branches and Offices (科室机构) began to be established starting June 1957.
9 June 1953	Anhui	65	Built on foundation of the Statistics Department of the Anhui People's Government Finance and Economic Committee (皖省人民政府财经委员会统计处)
June 1953	Fujian	45	In December 1957 became part of the Fujian Planning Committee (福省计划委员会); in October 1959 it was again separated).
June 1953	Guangxi		In 1958 it was changed to the Guangxi Zhuang Autonomous Region Statistics Bureau (广西壮族自治区统计局).
June 1953	Ningxia	99	Ningxia merged with Gansu on 23 September 1954
July 1953	Gansu	50	
August 1954	Liaoning	124	Liaodong and Liaoxi Bureaus had been set up separately in May and August 1950. They were merged into the Liaodong Bureau in 1954.
January 1955	Qinghai	29	Previously the Statistics Division of the Provincial People's Government Planning Committee (省人民政府计划委员会统计处)
1950?	Shenyang		
1953?	Chongqing		
30 January 1953	Beijing	65	Set up on the foundation of the Statistics Section of the Beijing Municipal Finance Committee (北京市财委统计科)
5 March 1953	Tianjin	65	Formed by combining: (1) the Statistics Department of the Tianjin Municipal Finance and Economic Committee (天津市财政经济委员会统计处), (2) the Statistical Section of the Ministry of State-owned Industry under the Tianjin Municipal Committee (天津市委国营工业部统计组), and (3) the Department of Statistics within the Research Office of the Tianjin Municipal People's Government (天津市人民政府研究室统计科).
March 1953	Xi'an		
1 April 1953	Shanghai	96	

Source: DSJ, 1–35.

trade, and the distribution of supplies. In addition, there were five departments dedicated to transportation, labor wages, culture, education, and compilation (publication).[28] By the following year, the total number of statistical cadres was said to be between 190,000 and 200,000.[29] Outside of regional-, provincial-, and city-level offices, statistical activity was undertaken by a mix of full-time and part-time staff. This was especially the case at the local levels. For instance, statisticians at the factory level frequently doubled as accountants.

By 1957, the fourfold description of statistical work—collection, collation, research, and supply—reflected both the *de jure* and *de facto* situations.[30] Of these four, collection, collation, and supply were part of the original tasks listed by Wang Sihua in his description of the work of the NSB, where research at that time was largely aspirational. But by the end of 1957, it had come to occupy a normatively equal status. In reality, however, statistical workers perceived the importance of these activities in unequal ways. In 1957, Xue Muqiao noted that some workers preferred performing supervisory (监督检查; *jiandu jian-cha*) or research (研究分析; *yanjiu fenxi*) tasks over the more basic tasks of collection and collation, thereby failing to recognize that without these latter tasks it would be impossible to undertake the former.[31] As he explained:

> If, in order to supervise or [carry out] research, one were to become relaxed about the collection and collation of statistical work, this is akin to destroying the foundations of one's own work; exactly like the scholar who, having lost his pen, takes a sword to a [literary] competition at a colleague's home[32]

The overemphasis on control was at least partially contingent on the early experience gained in the Northeast. In 1951, Li Fuchun had observed that sta-

---

28 *DSJ*, 492.

29 SSB, *Re'ai women de tongji gongzuo*, 3, 10 (cited hereafter as *RWTG*); SSB, *Tantan tongji ganbu de zixue*.

30 See, for instance, Xue Muqiao's speech at the Sixth National Meeting on Statistical Work, in BMA 133-001-00090: 83.

31 Xue Muqiao, "Di yige wunian jihua qijian woguo tongji gongzuo de chubu jingyan he jinhou renwu," 6. A truncated version of this report is located in BMA 133-01-00090: 66–80 (第六届全國統計工作會議文件 [一]), (it is missing the final two sections in the published version (on 如何运用各种调查方法 [略] and 统计工作的今后任务 [略]). The report was also reproduced in SSB, *Tongji gongzuo zhongyao wenjian huibian*, Vol. 3. In similar fashion, collection (here identified as survey) was described as the basis of all other statistical work in the introductory textbook Zhang Zhiji, ed, *Dazhong tongji*, 9.

32 Xue Muqiao, "Di yige wunian jihua qijian woguo tongji gongzuo de chubu jingyan he jinhou renwu," 6.

tistical work should first focus on the industrial sector. Later, with additional experience and capabilities, the work could expand to other sectors. The industrial sector, given its localization in space and its discrete nature (as compared to villages or cooperatives), permitted a focus on control (监督作用; *jiandu zuoyong*) and vertical leadership (垂直领导; *chuizhi lingdao*) that proved elusive in other sectors.[33]

Collection

Collection (收集; *shouji*; the term 搜集 [*souji*] was also frequently used) was the most basic statistical task performed by the SSB and its subsidiary organs. For Chinese statisticians and their Soviet advisers, the best way to count was to count exhaustively. This understanding generated a hierarchy of methods used for data collection. The most important, because it was the most fundamental, was the Complete Enumeration Periodic Reporting System. The periodicity of these reports ranged from ten-day (旬; *xun*) cycles, to monthly, quarterly, and annual reports. Among these, the annual report was the most important because it was used most directly to arrive at planning targets and to check their success and to summarize the state of economic and social affairs.[34] Reports issued at shorter frequencies were more useful at the provincial levels and below. Of secondary importance, but in widespread use, were censuses. Finally, unlike the first two methods, which were based on a complete enumeration, surveys, which only counted a selected sample, were used sparingly.[35]

Over the course of the 1950s there was a general trend toward increasing complexity, both in terms of the nature of the data collected and in terms of the methods used to collect that data. At the start of the 1950s, the emphasis was on setting up the periodic reporting system and conducting basic assessments of industry. If one-time specific sets of data were needed, then a census was the chosen form to collect such data.[36] For example, even before the establishment of the SSB, a census of state-owned and public-private industries was carried out in 1950. In 1952, the year the SSB was established, censuses of industrial production and employment were completed. A population census

33 BMA 133-001-00090: 82.

34 Xue Muqiao, "Di yige wunian jihua qijian woguo tongji gongzuo de chubu jingyan he jinhou renwu," 6.

35 For a similar hierarchy of methods, see Zhang Zhiji, ed., *Dazhong tongji*, 9–13.

36 A listing of major censuses (普查; *pucha*) carried out in the 1950s is provided in Wu Hui, "Zhongguo de tongji diaocha fangshi," 34–35.

in 1953 was followed by censuses of private enterprises and the handicrafts industry in 1954. Private activities, in particular, were tracked using the census method. By the mid-1950s, several additional censuses were announced with this specific goal in mind. For instance, an urgent notice issued on 20 October 1955 ordered that four types of surveys were of critical importance and had to be completed by the fourth quarter of the year. These were surveys of merchants, private workers in key industries, national employment, and material inventories.[37]

The periodic reporting system, however, remained the key method for collecting data.[38] And by 1956, an integrated statistical system extending down from the SSB in Beijing allowed for unprecedented levels of penetration. The SSB, headquartered in Beijing, was at the apex of this system. Below the SSB, statistics offices at the provincial level coordinated work across roughly 2,200 counties. Under the counties, village administrations coordinated the statistical work of about 750,000 village cooperatives. Unlike in the Soviet Union, where a similar unified chain of command was free from oversight by other party and government organs, in the PRC a statistical unit at any given level was also simultaneously under the authority of the party committee at that level. In other words, a provincial bureau took its orders not only from the SSB in Beijing but also from the provincial people's committee, which also had a say in its organization. Thus, the provincial people's committee could make its own requests for data from the provincial statistics bureau. A statistical worker at each level usually had more than one set of leaders to answer to.[39] It was hoped that such an arrangement would make the statistical system flexible enough to meet both local and national needs.[40] But it also generated problems of coordination and duplication, as the following section will show.

Given the goals of extensiveness and completeness, the management of reports was a major challenge for the SSB. The emphasis of the Second National Meeting on Statistical Work in December 1952 was on united leadership (统一领导; tongyi lingdao), which gave all decision-making authority regarding forms and schedules to the SSB in Beijing. The result was a proliferation

---

37 BMA 002-020-00746: 22–23.

38 "Among all the various methods of investigation, the complete enumeration periodic reporting system is the basic and [most] important method" (Wu Hui, "Zhongguo de tongji diaocha fangshi," 32).

39 Xue Muqiao, "Di yige wunian jihua qijian woguo tongji gongzuo de chubu jingyan he jinhou renwu," 8.

40 A description of this system can be found in Xu Qian et al., Tongji lilun yiban wenti jianghua, 18–19.

中央人民政府國家統計局制定
1 9 5 3 年 11 月

一九五三年國營、地方國營
及公私合營大型工業企業基本年報
（甲類表式）

（下表由綜合機關填寫，供匯總時分類用）

1.企業編號	
2.地 區	
3.主管系統	
4.經濟類型	
5.工業部門	
6.現代工業或工場手工業	

企業負責人簽章＿＿＿＿＿＿ 計劃統計部門負責人簽章＿＿＿＿＿＿
寄出日期：1954 年＿＿＿月＿＿＿日

表號：工年 0 1

企 業 概 況

1.企業詳細名稱＿＿＿＿＿＿＿＿＿＿＿＿＿＿
2.企業詳細地址＿＿＿＿＿＿ 電報掛號＿＿＿＿ 電話號碼＿＿＿＿
3.企業主管機關名稱：①年底直接管轄本企業的機關＿＿＿＿＿＿
　　　　　　②年底主管本企業的最高機關＿＿＿＿＿＿
4.經濟類型＿＿＿＿＿＿ 公私合營企業中公股佔＿＿＿＿＿＿ %
5.本企業主要產品：①＿＿＿＿ ②＿＿＿＿ ③＿＿＿＿
6.本企業為現代工業或工場手工業＿＿＿＿。
7.本企業在 1 9 5 3 年內實際開工日數＿＿＿＿日
　各班工作小時數 1.＿＿＿小時； 2.＿＿＿小時； 3.＿＿＿小時。
8.企業固定資產：　（單位：百萬元）

甲	合 計		其中：工業生產固定資產	
	原 值	淨 值	原 值	淨 值
	1	2	3	4
1952年12月31日固定資產				
1953年12月31日固定資產				

9.本企業在 1 9 5 3 年變動情況簡歷（如改組、改變領導系統、生產方向轉變、遷移、合併、分立等）：

製表人簽章＿＿＿＿＿＿ 審核人簽章＿＿＿＿＿＿

FIGURE 5.1. Summary Sheet from the Basic Annual Report for State-owned, Local State-owned, and Private-Public Large-Scale Industrial Enterprises for the Year 1953

of forms such as annual reports, for which the SSB in Beijing was the sole recipient. From the center's perspective, this was a world of neatness, simplicity, and legibility. All data would be collected via exhaustive enumeration at the enterprise level and would be received via successive rounds of upward reporting and collation as they traveled from the unit (单位; *danwei*) and county levels up to the provincial bureau levels and eventually to the SSB.

Consider the example of the 1953 *nian guoying, difang guoying ji gongsi heying da xing gongye qiye jiben nianbao* (Basic annual report for state-owned, local state-owned, and private-public large-scale industrial enterprises for the year 1953), which encapsulates this idealized vision.[41] Figure 5.1 is the summary sheet from this report.

Several centralizing and comprehensive features, which were meant to make an industrial enterprise "legible" to the SSB, are readily apparent: its basic identifying attributes, such as ID number and geographic and administrative location, industry division, key products, total labor input, and gross

41 BMA 018-011-00054: 80–91.

value of industrial output for 1953 and 1952 at current and fixed prices. The summary page was followed by twelve additional pages that presented further details on various activities of the industrial enterprise: total output value and commodity output value, output (quantity) of principal products, labor and wages, training of new workers, allocation of resources, consumption of raw materials, transportation, worker welfare, and so on. By collecting such annual reports from all industrial units, the SSB hoped to capture the total reality of industrial production in the entire country.

To arrive at the relative simplicity of the annual reports, each industrial enterprise had to track its activities far more frequently (daily, ten-day, monthly, and quarterly) with a much wider array of reports. It soon became clear that this situation had led to numerous contradictions between lower and upper levels of the statistical structure as well as across the same administrative level. Furthermore, as the Henan Statistics Bureau reported, enthusiasm ran high during the first year of work: A general impatience (急躁情绪; *jizao qingxu*) resulted in demands that were too high, too many, and too urgent. Several years later, the national policy of "more, better, quicker, cheaper" (多, 好, 快, 省; *duo, hao, kuai, sheng*) of the Eighth Party Congress (1956) generated additional tensions, because statistical work tended to focus on "better" and "quicker" at the expense of "more" and "cheaper."[42] A similar flip-flopping between simplification and complication was also reported by the Jiangxi Statistics Bureau, which explained that provincial commercial statistics underwent four instances of simplification and reversal between 1953 and 1957.[43]

It was in the field of agricultural statistics, however, that the SSB met its most intractable foe. For much of the decade, agriculture occupied a secondary, subsidiary position to industry within the economy. As Carl Riskin has noted, at the "most general level, the principal themes of the [First Five-year] Plan did not include agricultural production."[44] And yet, it was the surplus generated in agriculture that was an essential source of the investment needed to drive the growth of heavy industry. The efficacy of the periodic reporting system as a means of assessing agricultural output, and thus the surplus, was, however, constantly undermined by the scale of the agricultural sector, including variability in terrain, produce, and seasons. Take, for instance, a March 1955

42 BMA 133-001-00091: 16.

43 BMA 133-001-00091: 21. Henan province observed that simplification was successful to varying degrees based on the level of administration. For instance, simplification at the county level and above was quite successful. Below the county level, especially at the village level, the reverse was true. See BMA 133-001-00091: 11.

44 Riskin, *China's Political Economy*, 56.

telegram delivered to planning and statistical bureaus across the nation in which Xue Muqiao bemoaned the quality of the 1954 annual reports on agriculture. Xue observed that when compared with figures from earlier years, both the yield and the area sown showed massive upward and downward fluctuations for agricultural staples, such as grains, cotton, oilseeds, peanuts, and sesame. He concluded that "such large volatility (波动幅度; *bodong fudu*) in area sown and in the production figures for various crops amply illustrated that statistical work in agriculture still lacked a solid foundation and was in critical need of strengthening."[45] Even the use of typical sampling, which was meant to supplement the periodic reporting system, generated problems, since most cadres were unfamiliar with which methods to use, how to select typical cases, and other basic principles.[46]

Part of the problem was due to a lack of personnel. Unlike in industrial units, many villages and cooperatives did not possess a dedicated statistician. And even if they did, this person, in many cases, could only work part-time. The village clerk, whose regular duties became increasingly onerous as life in the countryside went through successive waves of cooperativization and eventual collectivization, was often responsible for statistical work also. In this situation, some clerks refused to perform statistical duties, claiming that such duties were not formally listed under their responsibilities and that such work was an "additional burden" (额外负担; *e'wai fudan*), which they could choose not to undertake.[47] By the mid-1950s, agricultural statistics had become a particular source of embarrassment for the SSB.

Even as a constant battle was fought to standardize and simplify the overall numbers and types of forms, there was a growing realization that the scope of data to be collected had to be expanded. A 1957 report on industrial statistics poignantly noted: "[We are] able to reflect production, but [can only offer] a poor reflection of livelihood."[48] That same year, Xue echoed the report's sentiments, stating: "In the past we spent more time thinking about production and construction statistics and did not emphasize [enough] surveys of people's lives."[49] Indeed, already by the end of 1954, this need to better understand changes (and improvements) in people's livelihood had precipitated a turn toward systematically collecting data on rural and industrial family income

45 BMA 002-020-00746: 9–11.

46 BMA 133-001-00091: 14.

47 BMA 133-001-00091: 20.

48 BMA 063-001-00088: 5. The Chinese original is: 因而能反映生产，不好反映生活 (*yin'er neng fanying shengchan, buhao fanying shenghuo*).

49 BMA 133-001-00090: 85.

and expenditures, purchasing power, wages, population, prices, and much else previously outside the purview of the SSB.[50] Exhaustive enumeration, ill-suited to these tasks, had to be augmented by newer, more sophisticated representative methods. Coupled with the challenges due to the re-elevated status of agriculture, where exhaustive enumeration via the periodic reporting system had already proven ineffective, 1955 and 1956 witnessed a broader engagement with nonexhaustive forms of counting, that is, survey sampling.[51] Even so, per the dictum of centralized leadership and dispersed management, the SSB could only offer regulations and suggestions on how to carry out the work.[52] The actual work was to be done by local units at the provincial-, city-, county-, and village-levels and coordinated by the various bureaus at each level.[53]

The *Jumin jiaji diaocha* (居民家计调查; Family livelihood survey of residents) planned by the Beijing Statistics Bureau in 1955 reflects these themes of expanded scope, new methods, and coordination.[54] The purpose of the survey was fourfold: first, to understand the material and cultural living standards of the people; second, to organize and plan for the circulation of goods; third, to plan for the circulation of currency; and fourth, to prepare cost of living indices for workers. Accordingly, the survey would measure income and expenditures (cash and in-kind) at every level of society, purchasing power, rates of consumption of major commodities, and changes in disposable income and savings.[55] These new types of statistical data were to be collected from all

50 BMA 133-001-00091: 13–14, 17 (Henan province report). On the 1954 Agricultural Household Survey, see *DSJ*, 35.

51 Xue Muqiao, "Di yige wunian jihua qijian woguo tongji gongzuo de chubu jingyan he jinhou renwu," 1–2; BMA 133-001-00091: 13–14.

52 A 1953 directive put the SSB in charge of all statistical regulations (Xue Muqiao, "Di yige wunian jihua qijian woguo tongji gongzuo de chubu jingyan he jinhou renwu," 9).

53 BMA 133-001-00090: 84. For an example of the SSB sending out guidelines for conducting a particular set of surveys, see BMA 002-020-00746: 22–23.

54 BMA 004-010-00805: 1–3. Starting in 1955, there was a nationwide campaign to collect such data. This was the kind of more intrusive but also more complicated statistical activity in which the SSB (and the state) began to engage during the First Five-year Plan (*DSJ*, 35). In earlier years, smaller-scale surveys exploring some of these questions had been attempted by individual bureaus. For example, in 1951 the Labor Bureau had organized about fifty people to survey more than 200 workers families; and in 1952, the City Finance Committee organized over 120 people to survey 700 households.

55 BMA 004-010-00805: 4–9, "Preliminary Comments on the Establishment of Household Accounts." The rest of the discussion on the survey draws from this document.

members of society, who were classified into four groups: workers, independent workers, private industrial and commercial workers, and peasants and other grassroots people.

If the survey had already indicated an expansion of the types of data to be collected, then the method—typical sampling—was also a departure from the exhaustive enumeration of the periodic reports. To ensure that the selected samples were indeed representative, the guidelines suggested using proportional sampling. In other words, urban and rural residents were separated into various groups based on selected characteristics, such as income and occupation. Once this was done, the relative weight of each group in the total population was used to compose the sample. This meant 250 survey households were divided into four groups: a little over 170 were classified as workers (Group 1), a little over twenty as independent workers (Group 2), twenty-five as private businesses (Group 3), and twenty-five as agricultural (Group 4).

A general lack of experience with such survey work led to a cautious, staggered implementation scheme. All preparations to survey Group 1 were to be completed by 1 January 1955 ; the three remaining groups were to be completed by 1 April 1955.[56] Furthermore, different methods were recommended to survey urban and rural households: urban families would complete the form with the help of a surveyor, whereas rural families would be interviewed by a surveyor on site.[57] The survey was tasked with collecting data across four parameters: family population and property, sources and categories of monetary and in-kind income, expenditures (both commodity and noncommodity), and daily cash savings.

Given its wide ambit, the survey required additional coordination across various government bureaus, both vertically and horizontally. Ten bureaus were asked to provide personnel to form a central team and a survey team. The responsibility of the central team was to guide the survey team in preparing and executing the survey. The combined strength of both teams was 58 persons, as outlined in Table 5.2. Twenty million yuan were allocated for the survey and included a special outlay for rewarding those households being surveyed. The instructions noted that based on experience in other cities, such rewards were necessary for the success of the survey.[58]

56 If this was felt to be too onerous, then Group 4 could be delayed until 1 July.

57 The problems of survey design and the need for trained interviewers are fairly universal problems. Lieu ("Collecting Statistics in China") focused on these problems in a pre-1949 Chinese context and described the solutions attempted.

58 The 20 million yuan was further divided into the following categories: 9 million yuan

TABLE 5.2. Composition of the Central and Survey Teams

City Bureau/Unit	Central	Survey
Planning Committee	1	1
Statistics Bureau	1	1
Commerce Bureau	1	12
People's Bank (Beijing branch)	1	14
Cooperatives	1	12
Grain Bureau	1	3
Labor Bureau	1	2
Federation of Trade Unions	1	1
Women's Federations	1	1
Public Security Bureau	1	1
Total	10	48

Indications are that such surveys remained ad hoc in nature. No regular schedule dictated their execution, nor did they have a dedicated team of personnel charged to carry them out. A team was put together only when instructions were received that a certain kind of data were needed. The surveyors' lack of experience also meant that the surveys were poorly executed, and hence their data were deemed unreliable.

Reconciliation and Correction

By the mid-1950s, collection of statistical data had expanded to include heretofore neglected subjects and also begun to employ methods that required the development of capacity and expertise in new areas, such as systematized survey sampling. Ensuring the quality of this data proved to be just as daunting as the actual collection. At each annual statistical meeting, reports on measuring quality were shared and discussed, but the SSB also intermittently approached the provincial bureaus to obtain a more real-time sense of the problems of organization and administration.[59] For instance, a June 1953 telegram from the North Region Statistics Bureau to the statistics bureaus of Hebei, Shanxi, Suiyuan, Beijing, and Tianjin sought responses to the following three questions:

---

for 3,000 account books, 7.5 million yuan as payment to the 250 households, and another 3.5 million yuan for other office expenses (BMA 004-010-00805: 4–9).

59 This was in addition to the reports it received during the annual national meetings on statistical work.

1. What kind of data do the bureaus provide to local leaders? How is this data supplied? (Do not only report on successful experiences; also point out problems and their possible solutions.)
2. What other data do local leaders ask for? What data are [currently] possible to supply?
3. Among data currently supplied, what kinds are regular/periodic and what kinds are of a temporary nature?[60]

As a result of numerous such inquiries and the debates and discussions they generated, we can classify data quality problems into three interrelated categories: overproduction/duplication/chaotic reporting, reconciling statistical and accounting data, and delays.

Overproduction/Duplication/Chaotic Reporting

A 1957 survey of a glassworks and a machine factory by the Nanjing Statistics Bureau is illustrative of the nature and scope of the problem (see Table 5.3). At both factories, not only were too many reports generated but they were also generated for several different clients, the statistics bureau being only one. Roughly half the number of forms in each factory (192 out of 351, or 55 percent in the glassworks factory; 107 out of 242, or 44 percent, in the machine factory) were produced for internal use. Among these, a further distinction was drawn between those used at the shop level and those used at the industry or unit level. The remainder was produced for outside submission. Among these, those sent to statistics offices accounted for 41.6 percent and 46.6 percent of the total respectively. Thus, the corresponding shares of total forms claimed by the statistics bureau were 19 percent (66 out of 351) and 26 percent (63 out of 242), respectively. The share of forms actually issued by the SSB in Beijing was even smaller: 12 percent and 14 percent respectively.[61] As a result, the bulk of the requests constituted a duplication of data.[62] On the basis of such

60 BMA 002-020-00670: 32. Another telegram (BMA 002-020-00670: 30) asked for provincial and city bureaus to report problems and their experiences in cleaning up false (illegal) reports.

61 More numerical and textual examples can be found in Si Gengsheng, "Fanlan chengzai de diaocha tongji biao."

62 That the SSB was particularly mindful of this duplication is evident in an August 1961 interview with a statistical worker in a county-level industrial and mining bureau (Li, "Statistics and Planning at the Hsien Level in Communist China"). Sometimes duplicate data were an outcome of genuine need. At other times, Li was told "some agencies liked to obtain all types

TABLE 5.3. Excess Forms at Two Nanjing Factories

	Nanjing Glassworks			First Machine Factory		
	Forms	Proportion	% of total	Forms	Proportion	% of total
Used within the industry	192	100	55%	107	100	44%
Used at the factory level	78	40.6	22%	47	43.9	19%
For outside submission	159	100	45%	135	100	56%
Planning	34	21.4	10%	32	23.7	13%
Financial accounts	50	31.4	14%	24	17.8	10%
Statistics	66	41.6	19%	63	46.6	26%
Technical accounts	9	5.6	3%	16	11.9	7%
Total	351	—	100%	242	—	100%

Note: In arriving at the count, the same type of form was counted twice if it was used for monthly and quarterly reports. The table is a faithful reproduction, except for the two "% of total" columns, which are not part of the original table produced by the Nanjing Statistics Bureau (BMA 133-001-00091: 57–58).

surveys, provincial bureaus requested that the center clarify and standardize only those reports it needed and not try and standardize all reports. The other tables could then be based on local requirements without having to conform to a national standard.

Attempts at simplification were also often ineffective. One 1957 survey of three bureaus—education, transport, and public health—in Jiangxi observed that two out of the three had failed to simplify and instead had increased their output of forms. In another survey, it was found that 22 offices and organs spread across 8 industrial enterprises produced 477 types of forms with 25,771 indicators.[63] Made repeatedly clear through such surveys was the point that what was of relevance to local agencies was not necessarily of relevance to the center.[64] The picture on the ground thus was quite different from that envisioned by the SSB in Beijing. Every administrative level complained of a flood of papers and numbers, and little capacity to make any sense of them. Excess

of information regardless of whether the information would be of use to them or not." Li also noted that "without prior permission from the bureau, other government agencies and offices were not allowed to send investigators to the factories and mines under the bureau's control for the purpose of collecting statistical information" (119).

63 BMA 133-001-00091: 27.

64 See, for instance, BMA 133-001-00091: 9; see also Si Gengsheng, "Fanlan chengzai de diaocha tongji biao," 3, which provides an exposition of the kind of detailed data collected by local bureaus.

issuing (滥发; *lanfa*) and chaotic issuing (乱发; *luanfa*) were the terms frequently used to describe such a state of affairs.[65]

The SSB tried to tackle this problem in September 1953 by going public. Earlier discussions about excess issuing and duplication had only circulated internally via memos, telegrams, and reports. But on 7 September 1953 *RMRB* published two directives that addressed the issue head-on. The first, "Instructions Regarding the Cleaning-up of Existing Statistical Survey Forms and the Cessation of their Reckless Issuing," was issued by the State Administrative Council (SAC) and carried the authorial imprimatur of Premier Zhou Enlai. The second was an SSB document that had been vetted by the SAC, entitled "Temporary Methods to Formulate and Approve Statistical Forms."[66] *RMRB* also published an accompanying editorial entitled "Wipe out Chaotic Phenomena from Statistical Survey Work."[67] These three documents served as a rallying cry to reduce duplication and mitigate chaos.[68]

Solutions that sought a gradual decrease in excess and chaotic issuing continued to be sought during the following years.[69] In some instances, such as instructions for statistical work in 1954, documents even contained claims that the "serious problem of excess reports by and large has been conquered."[70] And yet, by 1956, no long-term improvements had occurred. On the contrary, as the SSB expanded data collection into new areas, the *RMRB* bemoaned that the still excessive numbers, complicated content, constant emendations, and urgency all meant that it was finally "time to reduce statistical reports."[71]

---

65 See, for instance, BMA 133-001-00091; BMA 008-011-00054; BMA 002-020-00670: 28, 30–32; BMA 002-005-00088; and BMA 133-001-00043. Also see *DSJ*, 23 (Item 59) and Xue Muqiao, "Di yige wunian jihua qijian woguo tongji gongzuo de chubu jingyan he jinhou renwu," 18.

66 *RMRB* 7 September 1953, 1, 2. Both documents were released on 5 September (BMA 123-001-00416: 47).

67 *RMRB*, 7 September 1953, 2. Another article, published five days later, provided numerous examples from various provinces and economic sectors to bolster the claim of "too many, too complicated, [and] too chaotic" (Si Gengsheng, "Fanlan chengzai de diaocha tongji biao," 3).

68 The Beijing Statistics Bureau recognized the importance of these documents, and "hoped that statistical bureaus and offices at all levels will organize study sessions for all statistical cadres [so that the instructions] may earnestly be put into practice" (BMA 123-001-00416: 47).

69 BMA 133-001-00091: 11.

70 BMA 004-010-00807: 29.

71 Jiang Ling, "Zhonglei duo, xiangmu fan, biangeng duo, yaoqiu ji shi jianshao tongji baobiao de shihou le."

Indeed, the situation was so serious that internal memoranda issued a year later deemed the reports to be useless as soon as they were produced (随发随废; *suifa suifei*).[72] Unless action was taken, it was felt that such an unchecked spread of reports would lead to unmitigated disaster (泛滥成灾; *fanlan chengzai*).[73]

The possibility of an unmitigated disaster was very much on Xue Muqiao's mind in September 1957 when he observed that the problem of excess issuing remained unresolved. Given the demanding nature of exhaustive counting (全面调查化; *quanmian diaochahua*), Xue insisted that only the SSB should issue demands for periodic reports. If other levels of the statistical system needed their own data, they were to resort to one-time surveys, typical surveys, or annual reports, and not monthly or quarterly reports.[74] Using some of the campaign language of the time—the Hundred Flowers movement had just ended and the Anti-Rightist movement was in full sway—Xue offered three principal reasons for the problem: micromanagement, the twin tendencies of subjectivism (主观主义; *zhuguan zhuyi*) and bureaucratism (官僚主义; *guanliao zhuyi*), and inadequate oversight by the statistical organs.[75]

Reconciling Data

A second major problem was the presence of wildly disparate numbers for putatively the same observation or activity. Wang Sihua had touted standardization as a key characteristic of socialist statistics, but that ideal had to be reconciled with the reality on the ground. One reason for different numbers can be traced to confusion over what exactly was being measured. In 1953, for instance, a Beijing Statistics Bureau bulletin (sent to all industrial units in Beijing and copied to various central government organs) required that all entries in the annual report as well as in the monthly periodic reports be made according to the Catalog of Industrial Products. This would help avoid the problem

72 BMA 133-001-00091: 10.

73 BMA 133-001-00091: 30.

74 Xue Muqiao, "Di yige wunian jihua qijian woguo tongji gongzuo de chubu jingyan he jinhou renwu," 9.

75 Ibid., 18–19. Resorting to such sloganized reasons was of course hardly new and can be traced to the early 1950s and the campaigns against bureaucratism. Mao's essay "Solve the Problem of the 'Five Excesses,'" in *Selected Works of Mao Tse-tung*, Vol. 5, 89–91, raised similar concerns.

of leaky reporting (漏报; *loubao*), which was defined as consisting of not reporting goods produced for internal consumption or goods that factories considered unimportant. Similarly, there was sometimes confusion over the units of measurement: should one report tons (吨; *dun*) or bags (袋; *dai*)?[76] Even four years later, in 1957, ensuring that production was reported according to the Catalog of Industrial Products remained a challenge.[77]

But the challenges of standardization also reflected a more deep-rooted and systemic problem. We have already noted that data were collected by numerous agencies. In order to ensure that the data were properly reconciled, two systems of reporting were used: single-track reporting (单轨制; *dan guizhi*) and double-track reporting (双轨制; *shuang guizhi*). The former was used for data in fields where a high level of professionalization already existed, such as rail and air transportation, postal services, education, and public health. The business offices of these services were expected to report statistical data directly to the statistical bureaus at the same level and also to submit reports to their own next-higher levels.[78] In contrast, the dual-track system was used for the more important and much larger types of data, such as those relating to industrial and agricultural production, fixed capital investment, retail sales, and labor and wages. According to this system, data were collected by the twin tracks of the statistical apparatus and the various levels of the relevant ministry. The two sets of data thus produced could then be compared, with the normative expectation that they could be reconciled.

In practice, however, because of varying work and administrative conditions, the procedures of data collection were rarely standardized. In fact, in most cases a third source of data also existed, that is, the finance bureau.[79] But instead of solving, this merely exacerbated the problem. In other words, inconsistencies across statistics, accounting, and business reporting constantly brought into question the quality and reliability of the data being reported. The archives abound with bulletins, regulations, suggestions, and instructions

76 BMA 008-011-00054: 41–42.

77 BMA 063-001-00088: 5. By this time, the catalog consisted of about 230 items: over 180 intermediate goods and over 40 consumer goods.

78 Wu Hui, "Zhongguo de tongji diaocha fangshi," 32.

79 Dual tracking to control for the quality of data was used in other settings as well. For instance, in 1956 a related dual-track system was used at the Beijing Pharmaceutical Company. In this case, the two tracks were the statistical and the fiscal (accounting) (BMA 35-2-203: 35). Another example (with a chart) for public-private commercial enterprises, catering services, and so forth, can be found in BMA 133-001-00008: 67.

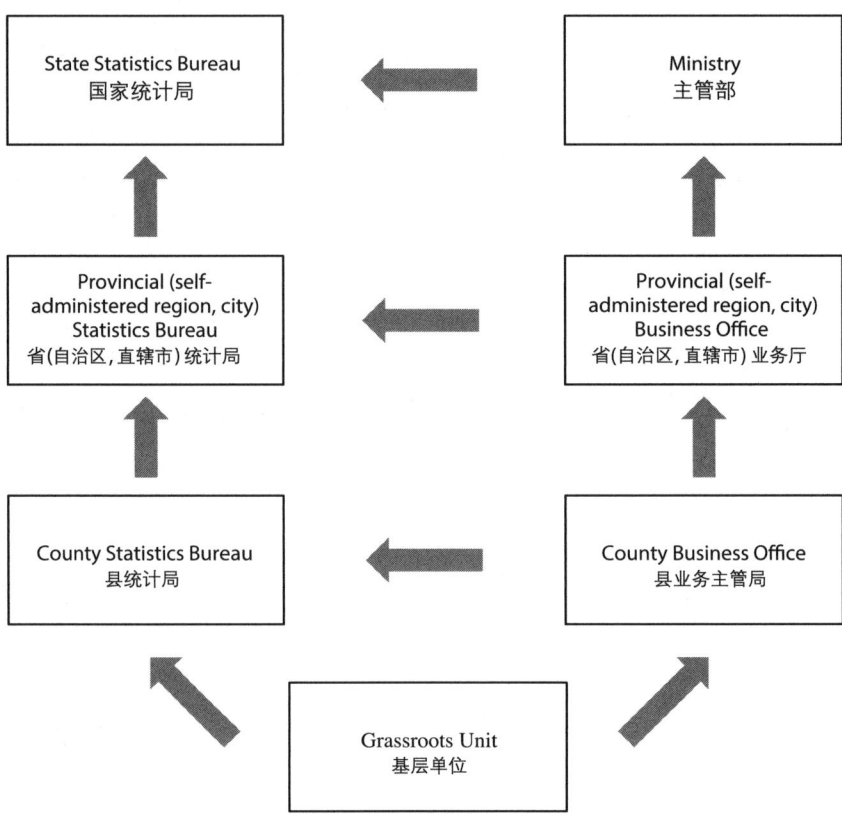

FIGURE 5.2. Systems of Data Collection. (Original schematic chart in Wu Hui, "Zhongguo de tongji diaocha fangshi," 33).

on how to grapple with this problem.[80] Many of these discussions stressed the generally superior reliability of statistical data, observing that accounting and business reporting was relatively vague because it was prone to variations in local conditions and local practices.[81] Yet others held the exact opposite view, asking not only for the abolition of the dual-track system but also for the removal of local-level statistical units and sole reliance on business/industry reports.[82] The problem was serious enough that on 20 June 1956 the SSB and

80 See, for instance, BMA 119-001-00118: 76–80; BMA 133-001-00065; BMA 004-010-00255; BMA 004-016-00268; BMA 133-001-0091: 32; and BMA 004-010-00618.

81 See, for instance, BMA 133-001-00065: 20.

82 For instance, BMA 0133-001-00091: 24.

the Ministry of Finance jointly released "Common Responses to Questions about Inconsistencies in Accounting and Statistics," which would, it was hoped, provide guidelines to begin to address the problem.[83]

Delays

Excess issuing, duplication, and the need to reconcile data contributed to yet another chronic problem: delays. A case in point is the 1954 Annual Report on Agricultural Production. In order for the national version of this report to be prepared in a timely manner, provincial reports, also compiled on the basis of periodic reports, were to be received by the end of January. But as late as 22 February, only sixteen out of twenty-nine provinces and cities had submitted their reports.[84] In a similar fashion, a 1955 notice from Xue Muqiao requested that all provincial- and city-level bureaus reconcile their reports in a timely manner, observing that only a handful of provinces had actually begun such work.[85]

Unclear procedures and multiple delivery deadlines contributed to such delays. For example, when reporting commercial statistics, the SSB regulations called for all forms to be filed within thirty days after the end of the quarter. But the Commerce Bureau's regulations called for the reporting to be done within twenty-six days, and the Finance Bureau, in turn, demanded the reports be delivered within forty-five days.[86] Such multiple deadlines did little to ensure timeliness on the part of any of the relevant bureaus.

As in the case of collection, the agricultural sector presented the greatest challenge for the reconciliation of data. We have already noted the secondary position of agriculture within the economy during much of the 1950s. In 1953 Mao had observed:

> In the present transition period we cannot introduce unified and planned production in agriculture, except on the state farms, and interfere too much with the peasants. We can only guide agricultural production and coordinate it with industrial production by our pricing policy and necessary

---

83 *DSJ*, 46–47.
84 BMA 002-020-00746: 9.
85 BMA 002-020-00746: 24.
86 BMA 133-001-00091: 18–19.

and practicable economic and political work so that it can be integrated into the national economic plan.[87]

It was as a result of such views, which acknowledged that the agricultural sector could not be administered with the same degree of control as other parts of the economy, that some proposed making county-level planning and statistics offices the main unit for reporting agricultural data. The rationale was that it was at the county level that local conditions required local control.[88] An even more drastic scheme suggested transferring responsibility for agricultural statistical work to the Agriculture Bureau.[89] But this was tempered by other less radical points of view, which recommended that the SSB and the Agriculture Bureau work together to reach a consensus on methods, requirements, and management.[90]

Solutions

By 1957 different solutions were suggested for each of the different levels of government: provincial, city, county, and village. At the provincial level it was felt that neither a system that granted too much centralized control to the province nor a system that let authority devolve down to the various departments was appropriate.[91] The problem at the provincial level included limited manpower and unfamiliarity with the work of the various departments, which resulted in much ado for very few results (忙乱尽缓; *mangluan jinhuan*) and did not mitigate the problem of the chaotic issuance of reports. If, however, authority to issue and manage reports were to be granted to individual departments, then, due to their numbers and variety, a chaotic situation (混乱现象; *hunluan xianxiang*) would also be unavoidable. The solution, as proposed by the Henan Statistics Bureau, was for the SSB and the various departments to issue a common set of periodic forms that were to be used by all.[92]

---

87 "Solve the Problem of the Five Excesses," 19 March 1953, in *Selected Works of Mao Tse-tung*, Vol. 5, 91.

88 BMA 133-001-00091: 2.

89 BMA 133-001-00091: 45–46.

90 BMA 133-001-00091: 70–71, 75.

91 This was, of course not the only opinion. In a federalist move, some called for greater power for the provinces, while others argued for increased centralization. See BMA 133-001-00091: 34, 41, and 26–27. For other solutions, see BMA 133-001-00091: 55–56 (for Nanjing); 58–61 (on simplification).

92 BMA 133-001-00091: 9.

At the city and special administrative levels (专; *zhuan*), it was felt that forms should only be issued after all city-level statistics departments had reached a consensus. This meant granting significant autonomy to city-level statistics bureaus. The logic was that city-level governments were close to and familiar with local conditions and hence had a better sense of on-the-ground realities. If their expertise and experience was not brought into play, the result would likely be a chaotic issuance of reports. At the same time, if certain departments had a strong foundation in statistical work, then they could be permitted to issue reports on their own, as long as there were still periodic checks.[93]

At the county levels, one-time reports or surveys usually outnumbered periodic reports. Since county offices were frequently situated far from the provincial offices, they exercised greater freedom and could respond more freely to short-term requirements for data, issuing new forms per need (随用随发; *suiyong suifa*). This reality, and the fact that time was always in short supply, meant that most forms reported from the county level up to the province levels were not approved. One proposed solution was to grant the county offices authority to approve one-time surveys but to require them to seek approval from the provincial offices for periodic reports. As for the village levels, most provinces felt that statistical work at this level was poorly administered. There were too many villages, and their tables and reports were generated in an erratic fashion. Most reports were frequently useless as soon as they were produced (随发随废; *suifa suifei*). The majority of the reports did not have a shelf life of more than one day and rarely made it to the county levels for approval. Accordingly, even as late as 1957 there was a pressing need to better oversee village-level statistical work by introducing a simplified periodic reporting system.[94]

Research

In his report to the September 1957 Sixth Annual Meeting on Statistical Work in Beijing, Xue Muqiao explained that the people's economy was extremely complicated because of the intertwining of a variety of intricate factors. It was the task of statistics and the statistician to explain and simplify this situation by providing sound research and analysis of statistical data. Therefore, research

93 BMA 133-001-00091: 10.
94 BMA 133-001-00091: 10.

was the third major service performed by statistics. The most important element of such research and analysis was the production of statistical indicators. As Xue noted:

> When conducting investigations, in order to make calculations easier we must simplify complicated phenomena by converting them into abstract [but easier to understand] statistical indicators.[95]

The calculation of indicators lay at the heart of most kinds of analyses, ranging from studies to comprehensively balance various sectors of the economy, to calculate relations among various factors, to increase savings, and to measure the successful implementation of plans.[96] In general, however, such work was neglected because of the primary emphasis on establishing a basic framework for collecting and collating data.[97] As the Anhui Statistics Bureau observed in 1957:

> Over the past several years, our work in this area [analysis] has been extremely inadequate. Our statistics bureau has done a much better job than other bureaus in mastering [the collection of] abundant data, but we have not put this data to good use.[98]

Given the overwhelmingly large number and types of forms, most bureaus had no time for analysis.[99]

Until about 1955, most research and analysis were left to the SSB in Beijing, to which provincial- and lower-level bureaus delivered annual reports. The SSB in Beijing, in turn, collated and analyzed the data to produce annual bulletins (公报; *gongbao*) that summarized economic growth and progress in plan implementation. These bulletins sought to present as comprehensive a picture of the economy as possible, including information, in absolute and relative terms, about changes in industry, agriculture, livestock, infrastructure, transportation, communications, investment, commerce and trade, prices, labor, education, publications, and public health.[100]

---

95 Xue Muqiao, "Di yige wunian jihua qijian woguo tongji gongzuo de chubu jingyan he jinhou renwu," 7.

96 See, for instance, BMA 133-001-00091: 12, 35–36.

97 BMA 133-001-00091: 52.

98 BMA 133-001-00091: 35.

99 As observed by the Henan Statistics Bureau, BMA 133-001-00091: 12.

100 See, for instance, SSB, *Guanyu yijiu wusan niandu guomin jingji fazhan he guojia jihua zhixing jieguo de gongbao*; SSB, *Guanyu yijiu wusi niandu guomin jingji fazhan*; SSB, *Guanyu 1956*

By the mid-1950s two developments had taken place that expanded the capacity of the SSB to carry out research and analysis. The first was the establishment of a small-scale computing facility at SSB headquarters. This foray into mechanized counting permitted the SSB to collectively summarize the data from the annual reports and censuses.[101] By 1957, Xue Muqiao was able to encourage those provinces where statistical work was particularly onerous to consider the gradual mechanization of their calculation work.[102] The mid-1950s also witnessed the creation of dedicated research sections within provincial- and city-level bureaus that were responsible, for example, for executing the household livelihood survey referred to above. Now let us look at an example of the type of research that was carried out by more local-level units.

In 1956, the Beijing Statistics Bureau began producing and circulating *Tongji fenxi jianbao* (北京市统计局统计分析简报; statistical analysis briefings).[103] Each issue was several pages long and offered a basic analysis of recent trends in a particular sector of the economy. Each briefing was organized under one of eight categories: summary, capital construction, commerce, industry, agriculture, communications (including transportation), [consumer] goods, and labor. Topics ranged across tracking agricultural production and urban incomes, forecasting plan completion in various bureaus during the first quarter of 1956, construction work completed during the first half of 1956, the arrival of capital construction equipment for local industries during the first half of 1956, the result of attempts to reduce costs, the ahead-of-time attainment of targets for the sale of retail goods, and so on. On the whole, the analyses drew upon actual figures and generated simple indicators of growth.

Each briefing was sent to two sets of recipients. The primary recipients included various city-level leaders and related offices, such as the city committee and its general office, and the city planning commission. The reports were

---

niandu guomin jingji jihua; SSB, *Guanyu fazhan guomin jingji de di yige wunian (1953 nian dao 1957 nian) jihua zhixing jieguo de gongbao*; SSB, *Guojia tongjiju guanyu 1958 nian guomin jingji fazhan qingkuang de gongbao*.

101 In 1958, the SSB published a book on the use of calculating machines (SSB, *Jisuanji de susuanfa*).

102 Xue Muqiao, "Di yige wunian jihua qijian woguo tongji gongzuo de chubu jingyan he jinhou renwu," 7. In interviews, retired Renmin University professor Ni Jiaxun (1931–) recalled using an abacus for all calculations when he worked in a county-level statistical unit during the early 1950s.

103 BMA 004-010-00253. The first briefing in this folder is dated 4 April 1956, and the last appears to be from October 1956. In all, there are fifty bulletins, totaling 145 pages.

also sent to the director of the SSB and to the various relevant SSB departments. So, for example, the industry and capital construction briefings were sent to the industrial statistics department, the agriculture briefing was sent to the agriculture statistics department, and the commerce briefing was sent to the department of trade statistics. The second set of recipients included a far greater number of agencies, including various ministry-level bureaus. Four of them, the Bureau of Finance, the Bureau of Taxation, the Beijing branch of the People's Bank, and the Supervision Bureau, received copies of every briefing, regardless of the category. Thus, by the mid-1950s, a fairly robust network of agencies was sharing interim statistical research on the performance of the economy.

Supply

Supplying data was the last of the four principal duties that was the responsibility of the SSB and the statistical apparatus. Although the annual bulletins were the most visible examples of SSB work, they did not represent the bulk of the data that Xue Muqiao had in mind when he put forward his four-part categorization of SSB activities. Indeed, the bulk of the data and research that the SSB supplied was intended not for public consumption but for designated government agencies and the leadership at various levels. Many of the reports and briefings cited in this chapter often carried the stamp "internal issue; not for publication" (内部发行, 不得对外发表; *neibu faxing, bude duiwai fabiao*).[104]

It was the supply of this kind of data that was regarded as absolutely critical for the planning process, and was the final and most important output of the statistical system. Yet, according to Xue Muqiao, its fundamental importance was ignored in favor of more frivolous pursuits. In 1957 Xue noted that "in the past there were some comrades who underestimated the glorious task of supplying data. [Instead,] they particularly emphasized the 'supervisory role' and the 'vertical integration' of statistics." Xue explained that these cadres went to unreasonable lengths to only focus on the supervisory role of statistics. For Xue, this was a gross error, especially when overall statistical capacity remained rudimentary and when the SSB still struggled to produce reliable data in a timely fashion.[105]

104 See, for instance, BMA 133-001-00090: 81. Another formulation included "internal document, outside transmission not permitted" (内部文件, 不得外传; *neibu wenjian, bude waizhuan*) (BMA 004-010-00807: 16).

105 BMA 133-001-00090: 82.

We have already seen some of the problems that plagued the accurate and timely supply of data. Data were supplied at various levels: frequently among bureaus, to various central agencies, and eventually reported up to the SSB headquarters in Beijing.[106] But who used what kind of data was unresolved: The center often only wanted annual data, but at provincial and local levels, monthly and quarterly data were usually more important.[107] The result was duplication, the chaotic issuance of reports, and competing deadlines that resulted in severe delays.[108]

The First Five-Year Plan and Statistics

In most standard (economic) histories, the 1950s are divided into three periods: 1949 to 1952 as the period of recovery, 1953 to 1957 as the period of the First Five-year Plan, and 1958 to 1962 as the period of the Great Leap Forward (GLF).[109] The First Five-year Plan thus occupies a central position in the decade, temporally as well as in terms of its impact on Chinese development. Even though the period began in 1953, it was only after July 1955 that the Five-year Plan was released as a document.[110] Over the years, general scholarly consensus has been that it was less a plan in the strictest sense of the word and more a set of guidelines.[111]

106 See, for instance, BMA 002-020-00670: 12.

107 Xue Muqiao, "Di yige wunian jihua qijian woguo tongji gongzuo de chubu jingyan he jinhou renwu," 9.

108 See, for example, BMA 035-002-00203: 21–27.

109 Needless to say, variations on this theme exist. The original prospective formulation in 1949 called for "three years of recovery, then ten years of development" (Riskin, China's Political Economy, 53). The years from 1949 to 1956 are also frequently referred to as the period of transition, implying that by the end of this period the three major socialist transformations—of agriculture, handicrafts, and the capitalist economy—were largely successfully achieved. The period from 1958 to 1962 would originally have been the period of the Second Five-year Plan, but the Great Leap Forward put such plans to rest. For a survey of scholarship on the Chinese economy during the Mao era, see Perkins, "Research on the Economy of the People's Republic of China."

110 The second part of Li Fuchun's address, delivered at the Second Session of the First National People's Congress in Beijing on 5 and 6 July 1955, presented a full summary of the plan. The Foreign Languages Press produced an official English translation of his speech in October of that year; a full translation of the Five-year Plan was issued in 1956 (China, First Five-year Plan for Development of the National Economy of the People's Republic of China in 1953–1957). For more, including the text of the 1955 Foreign Languages Press publication, see MacFarquhar and Fairbank, eds., The Cambridge History of China, 42–91.

111 The first draft of the plan was rejected by Soviet advisers in the Soviet Union in 1952.

The importance of statistics for planning was stressed in the earliest discussions about building a statistical and planning infrastructure, but the relationship was usually articulated only in general terms. Recall, for instance, Wang Sihua's 1950 speech, wherein he stated that by establishing a system of socialist statistics, China was already on the path toward a planned economy. A somewhat more precise link between planning and statistics can be found in a document entitled "Written Answers to Questions Concerning State Planning in China." This document was prepared several months prior to Li Fuchun's speech in July 1955 and delivered to the Indian Prime Minister, Jawaharlal Nehru (1889–1964) through the office of Vice Premier Chen Yun.[112] According to the document, the plan was drawn up by working out control figures, making a draft plan, and approving it for delivery to lower-level organs. In the first step, the SPC worked out the control figures for the entire nation. Upon approval by the SAC, these figures were sent down to the lower-level organs.[113] At this point, various ministries and local administrations used these figures to draw up their own draft plans and then submitted them to the SAC for approval, with copies also delivered to the SPC. On the basis of these lower-level draft plans, the SPC then revised the national draft plan.[114] The process was recursive and relied on coordination across the various administrative levels and with the statistical organs:

> The State Planning Commission maintains connections with local administrators through the planning commissions of the provinces, municipalities and autonomous regions. In addition to the planning commissions of the

---

The plan that was subsequently approved focused primarily on the state-owned economy and favored industrial production and capital construction; investment in agricultural production constituted an almost negligible proportion of overall investment. Trenchant discussions of the plan can be found in Donnithorne, *China's Economic System*; Li, *Economic Development of Communist China*; Meisner, *Mao's China and After*; and Riskin, *China's Political Economy*, 53–60.

112 The questions were prepared by P.C. Mahalanobis (see Part III), who had requested that Nehru present them to the Chinese on his official visit to the PRC in 1954. Nehru handed the questions to Vice Premier Guo Moruo (1892–1978) on 22 October 1954. The Chinese original of the document (containing the answers to the questions) was translated into English by the Chinese Foreign Ministry and both documents arrived in New Delhi (through the Indian Embassy in Beijing) by the end of April 1955. For more, see NAI EA-FEA-10(48)-FEA/55: 1, 3.

113 Set up in 1952, the SPC came under direct control of the SAC after the 1954 reorganization.

114 NAI EA-FEA-10(48)-FEA/55: 4 (see p. 17 for the Chinese version).

provinces, municipalities and autonomous regions, local planning organs, to be known as "planning and statistical sections" [计划统计科; *Jihua tongji ke*], are set up in the counties.[115] [Original English translation.]

At the lowest levels, planning and statistical functions were thus combined within a single office. At the national level, the SPC's principal partner was the SSB, and both the SSB and the SPC were under the direct control of the SAC. To quote from the original English document: "The state takes the figures of the Statistical Bureau as base for drawing up its plan, and reviews the execution of the plan on the strength of the figures of the Statistical Bureau."[116]

Much like Mao's exhortation to "serve the people," statistical work had to "serve national planning." As a report on Beijing's Third Meeting on Statistical Work explained, such service involved three tasks: the supply of correct data, analysis and research to discover problems, and suggesting solutions to the problems. The statistical numbers so produced were the basis for preparing the plans, inspecting them, deciding on policies, and checking on policy implementation. So, for example, the SSB could check the rate of industrialization and the rate of nationalization, and could formulate a variety of supply plans.[117]

Given the need for close coordination between statistics and planning, what was the reality on the ground? At the level of the central headquarters of each organization, coordination appears to have been the *de facto* norm. This coordination was no doubt facilitated by the fact that the SSB and the SPC were located on different floors of the same building in Beijing. But if we look at the example of statistical and planning work at the city level, things become less clear. According to materials related to Beijing's Third Meeting on Statistical Work in 1954, of the 114 sanctioned participants at the meeting, only 5 were from the planning bureau.[118] The rest were statistical cadres from forty-five units and key factories and mines. At first glance, these numbers indicate that the Beijing Planning Bureau's input in statistical work was limited. But when the total number of cadres at the Beijing Planning Bureau is taken into account (forty-one), we find that a significant number of the cadres at the Beijing Planning Bureau was actually in attendance (roughly 12 percent; or five out of forty-one). In order to further investigate the relationship between the two

115 NAI EA-FEA-10(48)-FEA/55: 4 (see p. 17 for the Chinese version).

116 NAI EA-FEA-10(48)-FEA/55: 4 (see p. 17 for the Chinese version).

117 BMA 004-010-00807: 81.

118 BMA 004-010-00807: 9. The postmeeting report stated that in reality there were as many as 148 participants and over 60 observers. See BMA 004-010-00807: 80.

bureaus, let us consider the 1953 annual work summaries of the Beijing Planning Bureau and the Beijing Statistics Bureaus.

Reflections on Planning and Statistics, 1952–1958

A summary report of statistical work carried out in 1953 in Beijing was prepared by the Beijing Statistics Bureau in April 1954.[119] The report stated that there were 3,829 full-time and part-time cadres involved in statistical work in Beijing.[120] One of the report's five sections (section 3) was devoted to a discussion of the bureau's contribution to planning work.

The section on planning began by observing that the bureau had provided statistical data to the leadership to assist in drawing up plans and deciding upon policies. For instance, in July 1953, under the leadership of the Finance Committee, the bureau compiled and consolidated annual reports for every year from 1949 to 1952 to aid in long-term planning. In the few cases where data were unavailable, it provided estimates. These materials were then crosschecked by the relevant ministries and organs, and printed in booklet form. In order to supply the data needed for drawing up the plan for 1954, the bureau also provided forecast figures for 1953. Such data were to assist in policy making and in carrying out key tasks. For example, in order to research the relative magnitudes of the private and public economies, the bureau provided quarterly data on the share of the commercial economy. In similar fashion, Beijing's population was verified so that the state monopoly for the purchase and marketing of grain could be effectively implemented. Data on average wages and labor productivity were also provided to facilitate research on the problem of wages.[121]

Even though the bureau felt that its various offerings basically met the planning requirements, it acknowledged that shortcomings remained. For starters, it admitted that it was not good at utilizing statistical numbers from all possible sources, ignoring in particular, statistical data from the administrative and professional departments. Even in those cases where coverage of all the numbers was achieved, the data were often not fully utilized.[122] A second concern was that specific needs were not addressed directly. Problems were discovered, but

---

119 BMA 004-008-00291: 9–22. For earlier drafts of this report, see BMA 004-011-00034: 13–22, 24–42.

120 BMA 004-008-00291: 16.

121 BMA 004-008-00291: 13.

122 BMA 004-008-00291: 14.

the leadership was informed only once the data were consolidated. As a result, even though the data provided to the leadership were plentiful, they played a limited role in decision making.[123]

What did the Beijing Planning Bureau's annual summary of work for 1953 say about statistics? Drafted in January 1954, the annual summary noted that the Beijing Planning Bureau was officially set up on 21 October 1953 and consisted of more than forty-one cadres distributed across departments responsible for planning in industry, agriculture, infrastructure, trade, health and education, and so forth.[124] The bureau received its instructions from the SPC, the Beijing Party Committee, the Beijing Economics and Finance Committee, and the Beijing City Government. Based on data obtained from the Beijing Statistics Bureau, the tasks of the Beijing Planning Bureau were to examine, consolidate, and balance long-term, annual, and seasonal plans for local industry, agriculture, trade cooperatives, capital construction, health and education, and wages.[125] The report went on to note that planning work had long suffered for three reasons. Leaders were unable to focus on the organization of planning because they were busy with the state purchasing monopoly of food grains. The planners also encountered problems with the quality of index numbers and often did not receive the revised numbers on time. Finally, they complained about the inadequate number of cadres and their poor overall training.[126] Neither statistics nor statistical work was directly mentioned, and yet two of the three problems were essentially statistical in nature.

It was only later in the year that the SPC directly addressed the issue of planning and statistics at the national level. "Regarding Uniform Provisions for a Number of Problems [Encountered] in Planning and Statistical Work" was a 24 August 1954 six-page document that directly addressed the issue of planning and statistics at the national level.[127] The document noted that problems existed both at a systemic level and at the level of coordination. Coordination between planning units and statistical units, in particular, was frequently inconsistent. Such inconsistencies, in turn, affected the calculation and meaning of specific indicators. Furthermore, subordinate planning and statistics

---

123 BMA 004-008-00291: 14.

124 BMA 004-011-00034: 1–6.

125 BMA 004-011-00034: 2.

126 BMA 004-011-00034: 3.

127 BMA 004-011-00255: 1-7. The regulations were sent to all central committees, departments, and institutes, and to the statistics and planning bureaus of every province and municipality. Copies were also delivered to all units of the SPC and the SSB in Beijing.

departments often did not strictly enforce the joint regulations of the SPC and SSB, or they often supplemented these regulations with their own contradictory regulations, fueling duplication and chaos. Finally, a single unit frequently possessed multiple sets of planning numbers and did not clearly specify which set was to form the basis for checking on plan implementation. As a consequence, there was no coordination between statistics and planning at any given level.[128]

In order to solve the problems of coordination, the SPC presented five proposals. First, it insisted on greater cooperation and understanding between the planning and statistics offices. At every level of administration, the statistics office had to be in compliance with the corresponding planning office. It was the statistics office's responsibility to provide, examine, and approve all of the basic data required for planning. This structure was to be replicated at all levels, from the province down to the county. In order for this to work in practice, planning offices were required to request that the statistical data be submitted well in advance of their use, so that statistical activities could be properly planned, and the data could be delivered in a timely fashion.[129]

The second provision focused on indicators, stressing that their meaning and the methods used to calculate and classify them should be in agreement across all statistics and planning work. Without such commensurability, it would be impossible to produce any meaningful comparisons. Where planning guidelines already existed, they were to be used to rationalize the indicators. Similarly, if only statistical guidelines existed, then they were to serve as the basis for the rationalization. If neither statistical nor planning guidelines were available, then representatives from both the statistical and planning offices were required to work together and devise mutually agreeable guidelines. In cases of special conditions or problems, the opinions of both the SSB and the SPC were to be sought. Under no circumstances were local officials to act before obtaining approval from the SPC and the SSB.[130]

Among the key tasks for statistical organs was checking the rate of plan implementation. Accordingly, the third provision stressed the need for statistics reporting units and summary (comprehensive) units to be in compliance with the planning units. In particular, the planning units that fell within the national economic plan were required to use the SSB's periodic reports and survey estimates. Any changes in the relationship among planning units

128 BMA 004-011-00255: 2.
129 BMA 004-011-00255: 3.
130 BMA 004-011-00255: 3–4.

or in economic type could only be carried out in consultation with both the SPC and the SSB, and all catalogs used by planning units had to be copied to statistics offices.[131]

The fourth provision addressed methods of plan implementation. In all cases, nationally approved numbers were to be used. So, for instance, when checking the implementation of a particular annual plan, the national annual plan was to serve as the basis for comparison. Similarly, when checking quarterly plans, quarterly numbers were to be drawn from the national annual plan and used as the basis for comparison. If such numbers did not exist, corresponding quarterly figures from the next-higher level were to be used. In the case of monthly plans for grassroots industries, the quarterly plan of the superior administrative unit broken into monthly estimates was to be used. As with the earlier provisions, any major changes or corrections could only be made after obtaining approval. In the absence of any approval, the units were to continue using the older (incorrect) numbers and to copy all results to the statistics office.[132]

The final provision reiterated the relative administrative ranking of planning and statistical units. At every level, planning was to provide leadership to statistical work. Cadres from both units were required to meet periodically to discuss statistical work, so that day-to-day work would be more closely coordinated. In addition, cadres from each unit were required to attend each other's work meetings and, when necessary, to jointly report and discuss their work with the central government.[133]

Reading against the grain, these five provisions and the 1953 reports from Beijing suggest that in the first year of the First Five-year Plan, communications among planning and statistical activities were not only limited but poor. Planning and statistics offices frequently operated on an ad hoc basis, calculating their own indicators without appropriate approval, thereby adding to the cacophony of data and analyses. The relatively skewed number of personnel in Beijing, where a little more than forty-one planning cadres had to coordinate their work with nearly 4,000 statistical cadres, is also suggestive of how difficult it must have been for the planning organs to provide leadership and oversight of the statistical organs. Oversight by the planning organs also meant that the statistical organs were under two different authorities—the people's committees and the planning committees. In many instances, especially at the

131 BMA 004-011-00255: 4–5.
132 BMA 004-011-00255: 5–6.
133 BMA 004-011-00255: 6.

local levels, there may have existed significant overlap in terms of personnel. But administratively they were different and therefore could easily be the source of contradictory requests as far as the statistical units were concerned. In spite of these difficulties, the SSB did produce annual reports on economic development and plan implementation for every year during the period of the First Five-year Plan.

Even though the problems plaguing statistics and planning were accurately assessed by 1954, their resolution remained elusive during the following years. A notice issued jointly by the SPC and the SSB in April 1958 repeated many items from the 1954 provisions.[134] At the same time, developments during the preceding four years also generated new challenges. The 1958 notice remarked that statistical work had to keep pace with the continually expanding scope (口径; *koujing*, lit. aperture) of the annual plans. As a result, statistical work surpassed the scope of the First Five-year Plan. For example, in the case of investment in infrastructure, the First Five-year Plan did not include investment in the military commission, in the bureau for the management of government offices, or in the administrative departments. However, such investments were all part of the individual annual plans by the end of the First Five-year Plan period.

Continuing differences in the classification methods used by the statistical and planning organs also resulted in inconsistencies. For example, statistics on the sown area for tobacco and for different kinds of hemp and flax (大麻; *da ma*, cannabis hemp), velvet leaf (青麻; *qing ma*), and kenaf (羊麻; *yang ma*) was included in the First Five-year Plan. However, in regular statistical work, which was more attuned to the requirements of the annual plans, they were listed within the sown area for technical crops. In another example, the First Five-year Plan did not include tubers and catering in its figures for retail sales of food commodities. However, these were included in the annual statistical figures. The indicator for food did not include data for soybeans; and indicators for infrastructure investment in agriculture, forestry, or water management did not include investment data from the forestry industry.[135]

Much as in 1954, the SSB and SPC recommended a slew of measures to combat these problems. In those instances where the scope of statistics exceeded that of planning, they recommended that the results be publicly announced, without correcting for the differences. This was acceptable when the

134 BMA 005-001-000298: 1–6.
135 BMA 005-001-000298: 3.

expansion in scope was more-or-less offset by the reduction in prices or other factors, such as in the case of infrastructure investment.[136] In those cases where the scope of planning exceeded that of the statistics, the statistical data had to be adjusted by making appropriate estimations. On the issue of inconsistencies in classification, the SPC and the SSB called for planning categories to take precedence over statistical categories. The notice recognized that changes in the scope of the plan were sometimes made during the implementation process through a variety of actions, such as increasing or decreasing the number of construction units. Under such circumstances, instead of constantly revising the plan, the notice asked that changes be carefully recorded so that once the plan was completed, the figures could be compared to the figures in the original plan. Furthermore, because the nature of the economy underwent substantial changes during the plan period, when analyzing plan completion rates, it was best to use annual statistical data, since planning data were not commensurable for different years.[137] It was with such considerations in mind that discussions about the Second Five-year Plan were undertaken.[138]

Conclusion

In the end, for all the valorization of an extensive, complete, and objective statistical system, Xue Muqiao admitted in September 1957 that in certain important situations, the timeliness of the data outranked their accuracy:

> Even though statistical data should possess a high degree of accuracy, it is not necessary that all data must be highly accurate [all the time]. In order for the leading authorities to understand the situation, research questions, and decide on policies, they frequently need reference data on a timelier basis. Such data need not possess a high degree of accuracy or be comprehensive, but it must be supplied in a timely fashion.[139]

Implicit in his statement was the admission that in most cases data were neither timely nor accurate. The archival materials that comprise the principal source base for this chapter suggest why such a bleak picture of statistical work

---

136 BMA 005-001-000298: 3.

137 BMA 005-001-000298: 3–4.

138 For discussions by the Fujian, Xi'an, and Inner Mongolia Statistics Bureaus, among others, see BMA 133-001-00091: 43-45, 49, 63–64.

139 Xue Muqiao, "Di yige wunian jihua qijian woguo tongji gongzuo de chubu jingyan he jinhou renwu," 13, 10.

during the 1950s is justified. Although they demonstrate a tremendous degree of engagement and professionalization, they also clearly highlight the struggles involved in manifesting an idealized system centered on exhaustive enumeration via a periodic reporting system.

This was not, however, the way statistical work was presented in more public materials. To publicize their work and to keep morale high, the SSB published an entire range of books and pamphlets. Typically composed of essays written by cadres in different local, regional, and provincial statistics offices, these works carried titles such as *Jiceng nongye tongji gongzuo jingyan huibian* (Compilation of experiences of grassroots agricultural statistics work), *Woguo gongye tongji gongzuo de jingyan* (The experience of industrial statistical work in our country), and *Zenyang zuo gongye tongji fenxi gongzuo* (How to carry out analysis of industrial statistics). Even though such works extolled the successful experience of selected cadres and statistics offices, read against the grain they only serve to sharpen the sense we already have of a system bogged down by the weight of its own expectations. At the same time, they also highlight the fact that statistical activity was as much about capacity building as it was about claiming legitimacy; that is, creating the appearance of capacity and ensuring that the Communist revolution was regarded as a success. This is precisely what the annual bulletins cited above sought to achieve. Publication of *Weida de shinian* (Ten great years) in 1959 was perhaps the ultimate legitimizing weapon in this regard.

A critical element in both capacity building and claiming legitimacy was the training of adequate numbers of statistical cadres.[140] It is to that topic that we turn to in the next chapter.

---

140 Xue Muqiao, "Di yige wunian jihua qijian woguo tongji gongzuo de chubu jingyan he jinhou renwu," 5.

PLATE 1. Wang Sihua. (Author's personal collection.)

PLATE 2. Photo marking the completion of Soviet expert G. M. Maximov's term in China, August 1957; with: Sun Yefang (2nd from L), N. P. Semidevkin (4th from L), Xue Muqiao (center), Maximov (4th from R), Jia Qiyun (2nd from right), and Wang Sihua (1st from R). (Author's personal collection.)

PLATE 3. Jin Guobao. (Author's personal collection.)

PLATE 4. Xue Muqiao. (Author's personal collection.)

PLATE 5. P.C. Mahalanobis. (Courtesy Indian Statistical Institute, Kolkata.)

PLATE 6. Mahalanobis and Zhou Enlai at the Indian Statistical Institute, Calcutta, 9 December 1956. (Courtesy Indian Statistical Institute, Kolkata.)

PLATE 7. Zhou Enlai hosts Mahalanobis and Lahiri over dinner, 9 July 1957; with R.K. Nehru (2nd from L), D.B. Lahiri (3rd from L), Deng Yingchao (4th from L), Mahalanobis, Mrs. Mahalanobis, Zhou Enlai, Mrs. Nehru, Xue Muqiao, and Wang Sihua. (Author's personal collection.)

PLATE 8. Mahalanobis welcomes Wu Hui and Gong Jianyao to the Indian Statistical Institute, Calcutta, January 1958. (Courtesy Indian Statistical Institute, Kolkata.)

PLATE 9. Jia Qiyun. (Author's personal collection.)

6

To "Ardently Love Our Statistical Work"

STATE (IN)CAPACITY, PROFESSIONALIZATION, AND THEIR DISCONTENTS

"Quite honestly, Minister, I want a job where I don't spend endless hours circulating information that isn't relevant about subjects that don't matter to people who aren't interested."[1]

Comrade Li Ming Dismisses Statistics

In July 1955, the *Tongji gongzuo tongxun* (*TJGZTX*), the principal journal of statistical work in the PRC, published a short letter from a youthful statistical worker.[2] Just twenty years old, Li Ming noted that he had been engaged in statistical work for three years. Prior to beginning work he had received a year and a half of statistical training at a statistics and accounting vocational school. When he began work, Li told the editors of the journal, he had been full of enthusiasm. He hoped to wholeheartedly devote all his abilities to statistical work and thereby serve the people. To that end, he spent an additional three months during the latter half of 1954 at the cadre school of the Central Fuel Industry Ministry to supplement his training. Such training and experience should have allowed him to be successful in his work. Yet, as he lamented, this was not the case. Quite to the contrary, Li Ming discovered that the more he

1 Lynn and Jay, eds., *The Complete Yes Minister*, 371.
2 Li Ming, ed., "Li Ming tongzhi weishenme buyuan zuo tongji gongzuo?" The letter was republished in SSB, *Re'ai women de tongji gongzuo*, 8–9 (hereafter *RWTG*).

worked, the more he lost faith in statistics and the less he knew about his future prospects. Having already worked for three years, he could not deny that he had a basic grasp of statistical work. But was this not merely the addition and subtraction of numbers or the calculation of percentages (加加减减, 算算 百分数; *jiajia, jianjian, suansuan baifenshu*)? After all, he did not directly participate in production work and, as such, he was reliant on information provided by others. Even if he visited the factory floor, there was much about the production process that he did not understand. Such ignorance meant that the blind application of statistical methods and tools was not the least bit useful and instead was a waste of time. So were born his doubts regarding the usefulness of statistical work; doubts so acute that he had reached a point where he was unwilling to do any work.

Statistics' perceived worthlessness made unambiguous and his own lack of desire powerfully expressed, Li Ming proceeded to note that his frustration was all the greater given his youth. The better part of his life lay ahead of him, and he still yearned to make practical and real contributions to the motherland. Statistical work, clearly, did not qualify. And so, he found himself in a dilemma. On the one hand, he was increasingly unwilling to participate in statistical work; on the other, he recognized the resources the nation had already devoted to his statistical training. How could he now let the nation down by seeking other work? Caught between the Scylla of patriotic desire and the Charybdis of patriotic duty, Li Ming implored the editors of the journal to help him find a solution.

In publishing Li Ming's letter, a solution is precisely what the editorial team at the journal hoped to unearth. In a short preface to the letter, the editors explained that the kinds of doubts and general attitudinal problems that Li Ming had encountered were present among a not insignificant number of statistical workers.[3] Accordingly, they invited readers to respond with their reactions and suggestions, thus beginning a national debate on statistical work. Responses were published over the following several months. The discussion was brought to an end in the December issue of the journal, which included a brief overview of unpublished letters, as well as an essay summarizing the discussion and its lessons written by Jia Qiyun (1914–2004), a deputy director of

3 For instance, according to a 1954 survey, of the thirty-seven statistical workers at a petroleum factory in the city of Fushun (in Northeast China), 40 percent did not pay attention to their work. In another example, statements delivered at a statistical meeting of the Kaifeng (Henan province) group noted that as many as 60 percent of the workers neglected statistical work. *RWTG*, 1.

the SSB. In the following year, the SSB compiled what it deemed were the best responses in a small booklet for wider dissemination. To those responses it added introductory and summary essays that provided additional didactic content regarding the value and revolutionary nature of statistical work. For the booklet's title, the SSB appropriated the title of Deputy Director Jia Qiyun's article in the December issue: *"Re'ai women de tongji gongzuo"* (热爱我们的统计工作; Ardently love our statistical work; *RWTG*). Published in August 1956, the booklet had a publication run of some 5,500 copies.[4]

It is more than likely Li Ming did not exist. He was probably a composite character created to embody the real concerns and problems confronting Chinese planners and statisticians.[5] Indeed, the putative Li Ming and the campaign his letter precipitated lay bare a crucial anxiety during much of the early PRC—the creation and maintenance of the ideal (statistical) worker. Several forces, each pulling in a different direction, made this an especially challenging task. Whereas some of these forces were part of the general problem of bureaucratization following revolution, and have been addressed in earlier scholarship,[6] many others were specific to the nature of statistical work and to the role of statistics in the economy and what it "produced." It is Li Ming's notable lack of reverence for the power and positivism of numbers that sets up the basic contradiction. In Part I we saw that the early 1950s was a time when a new theoretical and methodological paradigm for statistics and statistical work was established. But what was one to do when the common worker refused even to acknowledge the transformative power and potential of statis-

---

4 *RWTG.*

5 Such practices are well known in PRC history. The most notorious example is, of course, Lei Feng (1940–1962). A young People's Liberation Army (PLA) soldier, Lei Feng was at the center of a 1963 PLA-orchestrated campaign under the leadership of Lin Biao (1907–1971) that sought to "emphasize the basic values of service to the party." As explained by Spence, *The Search for Modern China*, 597: "The posthumously discovered *Diary of Lei Feng* emphasized again and again the soldier's undying love for the revolution, for his country, and for his comrades, as well as his unswerving devotion to Chairman Mao. The fact that the 'diary' was fictitious, concocted by the PLA propaganda writers, should not conceal its basic significance, which was to launch an attack against the lack of revolutionary fervor displayed by many intellectuals and writers in the People's Republic." A possible inspiration for Lei Feng was Alexsei Stakhanov (1906–1977), whose 1935 "achievements" in coal mining were part of the eponymous Stakhanovite campaign to spur production in the Soviet Union.

6 See, for instance, Lee, *From Revolutionary Cadres to Party Technocrats*; Orleans, *Professional Manpower and Education*; Schurmann, *Ideology and Organization*; and Vogel, "From Revolutionary to Semi-Bureaucrat."

tics, far less any putative resolutions to definitional and methodological distinctions? Indeed, on the ground, the concerns were far more basic: The creation of professional cadres and an attendant infrastructure, ensuring an adequate number of such cadres, and maintaining their morale, all while keeping in mind the central and fundamental role that statistics and statistical workers were supposed to play in the economy. It is with almost metonymic, and perhaps entirely predictable, precision that Li Ming's minibiography and complaints mirror these concerns.

This chapter, then, is about state attempts to build statistical capacity; it is also about how the state dealt with incapacity. My thinking about capacity draws upon Michael Mann's formulation of infrastructural power. In his seminal 1984 essay, Mann characterized infrastructural power, in contradistinction to despotic power, as "the capacity of the state actually to penetrate civil society, and to implement logistically political decisions throughout the realm."[7] Statistics is a powerful example of infrastructural power. Such power, Mann argued, is a characteristic of modern industrial societies. Accordingly, I investigate how the PRC state went about establishing infrastructural power in the realm of statistics, something that the Republican state arguably both lacked and suffered from. In a more recent working paper, Francis Fukuyama, building on Mann's ideas, has stressed the need to better understand such capacity as part of a larger definition of governance, regardless of the nature of the government in power.[8] According to Fukuyama, one key marker of this capacity is the level of education and professionalization among government officials.[9] Thus, if the previous chapter focused on the nature of statistical work, this chapter shifts the spotlight to the creation and maintenance of cadres capable of doing such work. Our opening story, the ensuing national debate

7 Mann, "The Autonomous Power of the State," 189. Mann focused primarily on capitalist countries, observing that such infrastructural power may have begun by bolstering despotism but was eventually also used by actors outside the state, including members of civil society and powerful capitalist corporations. However, my interest here is not to investigate the checks on power that these processes might generate. Regardless of the system of governance, the category of infrastructural capacity has great purchase for China in the 1950s (and indeed much of earlier Chinese history as well).

8 Fukuyama, "What Is Governance?" Referring to the field of political science and policy studies, Fukuyama noted in his introduction: "Everyone is interested in studying political institutions that limit or check power—democratic accountability and rule of law—but very few people pay attention to the institution that accumulates and uses power, the state."

9 Ibid., 7.

in *TJGZTX*, and the eventual publication of *RWTG* all hint at how massively challenging this task was.

A variety of stratagems—training, supplementary training, self-study, motivation, and reward—were employed to tackle these challenges. They began with training in preparation for employment and, in theory, were continually mobilized throughout one's career. This chapter traces this training and motivation work, using Li Ming as a foil wherever possible.[10] Beginning with his claims of youthfulness and education, we first explore the kinds of statistical training offered by the state. For a select few, such training may have been available at the university level, but the vast majority of statistical workers was trained in technical schools through training courses (训练班; *xunlian ban*). But these were inadequate, and by the middle of the decade the SSB observed that it was unable to train enough statisticians through such means. The suggested solution was self-study while on the job. Self-study, in turn, casts a firm and unwavering spotlight on morale and its maintenance, which are examined in the final third of this chapter. Starting with an exploration of the role of *TJGZTX*, which straddled the line between training and motivation and served as a forum for both activities, I proceed to a sustained discussion of *RWTG*, dwelling on its production and content. I conclude by observing that if Li Ming was at the heart of the campaign, serving as a model subject in need of reformation, then by 1956 other more locally organized campaigns began to promote local activist model statistical workers as subjects worthy of emulation.

Red and Expert: The Need for Revolutionary Professionals

Prior to the 1949 victory, the CCP's experience in governance was restricted to small pockets of territory and tiny populations in Jiangxi, the Shan-Gan-Ning border region, and wartime base areas such as Yan'an.[11] The transition to greater political control, starting in 1948 in the Northeast and spreading elsewhere over time, was hamstrung by a severe shortage of trained and educated personnel.[12] Such personnel were needed to effectively oversee the various

---

10 For a general discussion of motivation and work incentives, with a focus on the late Mao era, see Riskin, "Maoism and Motivation."

11 In some of these areas, statistical work was undertaken as early as 1941. For more, see Renmin University, Department of Statistics, *Tongjixue jiangyi* (*chugao*), Vol. 6.

12 As Ezra Vogel observed in "From Revolutionary to Semi-Bureaucrat," 40: "In China, with a population larger than Europe and Africa combined, a vast majority of the population in 1949 had no school training, did not speak the major dialect, and was not part of a national com-

and expanding organs of the state and to help govern the newly established republic. Given that the revolution and civil war era "conception of the cadre as a combat leader was not ideally suited to the tasks of civil administration that faced the Chinese Communists after 1949," new cadres were needed.[13] For the most part, recruits were drawn from three strata of society:

> First, the "positivists" who gushed up from the movements; most of these were working-class people with more "virtue" than "ability"; we know that many workers were rapidly promoted to supervisory positions in factories, but probably only a few of them were given complex administrative positions. Second, the students, undoubtedly a much more promising source of recruitment; most of them had been sympathetic to the Communists, even though they were usually not involved in the illegal Party organizations of the cities; these were the "new intellectuals." Third, the "old intelligentsia": the new regime preferred young to old intellectuals, but, as An Tzu-wen [An Ziwen] indicated, the older ones were needed as well.[14]

Accordingly, by 1953 the number of civilian cadres had increased fourfold: from 720,000 in 1949, to 2,750,000 in September 1952, and to 3,310,000 in the following year.[15]

Although discussions linking "virtue" (德; *de*) and "ability" (才; *cai*) have a long history in Chinese political and moral philosophy, they appear to have been first deployed in a Communist revolutionary context by Mao in a 1938 speech.[16] In its later expression as Red and Expert, the dichotomy and relative

---

munication network." In anticipation of this problem, which would only grow in magnitude, at a September 1948 meeting in Xibaipo the Politburo articulated the need to train cadres. See Gao, *The Communist Takeover of Hangzhou*, 16; and Lee, *From Revolutionary Cadres to Party Technocrats*, 14, 45–46.

13 Schurmann, *Ideology and Organization*, 164.

14 Ibid., 168; see also Vogel, "From Revolutionary to Semi-Bureaucrat," 54–59, for a discussion of the major groups of cadres based on their origins. On the general question of how the PRC state dealt with intellectuals in the early 1950s, see U, "The Making of *Zhishifenzi*."

15 Schurmann, *Ideology and Organization*, 168; and Vogel, "From Revolutionary to Semi-Bureaucrat," 40, with each citing different works by An Ziwen.

16 Schurmann *Ideology and Organization*, 165. Schurmann observes that Mao may have been inspired by a speech by Stalin. Possibly the earliest mention of the virtue-ability dyad can be found in Chinese antiquity in the *Book of Rites*, where it appears early in the "Liyun" (禮運; Conveyance of rites) chapter in a discussion on the nature of an ideal society—the Great Unity (大同; *datong*): "When the Grand course was pursued, a public and common spirit ruled all under the sky; *they chose men of talents, virtue, and ability*; their words were sincere, and what

dominance of one over the other at any given point in time has been an influ-ential heuristic device for understanding events and policies in China since the 1950s.[17] But we would do well to remember that the ideal cadre was to be Red and Expert in equal measure.[18] He was to have the technical and organi-zational skills to promote the work done in his unit while at the same time also be brimming with ideological purity and revolutionary fervor so that any counter-revolutionary elements could not regain a foothold. Mao would note twenty years after his 1938 speech: "There is no doubt that politics and econ-omy, and politics and technology should be united. It has been so in the past and will be so forever. This is what red and expert mean."[19] The Red Expert thus drew his authority both from his expertise and from his ideological cor-rectness. To the extent that experience was important, it was the cadre's per-sonal experience garnered in the present. The only significant past experience of value in this formulation was the advanced experience of the Soviet Union. In this way, the Republican-era dichotomy between experience and expertise as two clashing sources of authority was neatly sidestepped.[20] The PRC aimed to produce a new kind of (professional) cadre.[21]

Students, and young people more generally, were quickly identified as the target group for achieving the ideal balance of Red and Expert. The "positiv-ists," because they were often too old to train, and the "old intelligentsia," who

---

they cultivated was harmony (Emphasis added)". See http://ctext.org/liji/li-yun, accessed 14 December 2018, which uses the translation by James Legge.

17 For a more recent instance, see Andreas, *Rise of the Red Engineers.*

18 Meisner, *Mao's China and After.*

19 Mao continued: "Political workers must have some knowledge of business. It may be difficult for them to have a lot, but it may not do for them to have only a little. They must have some. To have no practical knowledge is to be pseudo-red, empty-headedly political. Politics and technology must be combined together. In agriculture, this means carrying out experi-ments; in industries, understanding advanced models, trying out new techniques, and produc-ing new goods" (Mao Tse-tung, "Red and Expert").

20 This distinction has been illustrated particularly well in the field of medicine, where the experience of traditional Chinese medicine was pitted against the expertise of modern medical science, setting up two contrasting sources of authority (Asen, *Death in Beijing*; Lei, "How Did Chinese Medicine Become Experiential?; and Lei, *Neither Donkey Nor Horse*).

21 On an early discussion of the professional as possessor of authority, see Parsons, "The Professions and Social Structure." Recent work on questions of professionalization has called for the need to move beyond Talcott Parsons's now "traditional" definition and focus instead on precisely the issues of expertise, skill, technique, judgment, and experience (Gordin, "Statis-tique et revolution en Russie," 805; and Mackenzie, *An Engine, Not a Camera,* 11).

were needed in the immediate short term but would always be politically and ideologically suspect, were both deemed less suitable. Accordingly, the training and ideological indoctrination of the young was carried out at all possible levels, keeping in mind both expediency and long-term goals.[22] On a national scale, "of the 2,750,000 cadres in China in mid-1952, 66,000 [2.4 percent] had graduated from regular universities, 100,000 [3.6 percent] had been processed at 'people's revolutionary universities,' but over 1,100,000 [40 percent] had been trained in rotating training courses."[23] Furthermore, this ideal of Red and Expert was applied not only to cadres who were, after all, defined by their leadership roles and by specific minimum ranks but to all young skilled and semiskilled workers needed by the state.[24]

University Education and Training:
The Case of Renmin University

Discussions concerning the setting up of educational institutions to train cadres picked up pace after the official declaration of the PRC in October 1949.[25] Surveys by visiting Soviet advisers had already investigated the nature and content of education at various institutions across the country. Eventually, it was decided that most major pre-existing universities would not be disturbed.[26] Instead, a new university, Renmin University, would be established that would serve as a national model. Dubbed the People's University, its

22 On the challenges faced by Wang Sihua in 1950 to train adequate numbers of statistical cadres in the Northeast, see Wang Sihua, *Wang Sihua tongji lunwenji*, 9–10.

23 Vogel, "From Revolutionary to Semi-Bureaucrat," 45, citing An Ziwen.

24 It bears pointing out that the words *de* (德) and *cai* (才) were never mobilized, though they were very much the implicit categories at work in most of the materials I consult in this chapter.

25 For a summary of pre-1949 Communist education and cadre training, see Stiffler, "Building Socialism at Chinese People's University," 89–102; and Lindsay, *Notes on Educational Problems*. Chen, *Creating the "New Man,"* 61, provides a list of the key schools set up in Yan'an (such as the "Anti-Japanese Military and Political University, the Lu Xun Academy of the Arts, the Northwest Public School, the Central Party School, the Academy of Marxism-Leninism, the Women's University, Yan'an University . . .").

26 In Beijing, for instance, Beijing and Tsinghua universities were allowed to continue operations without interference (Stiffler, "Building Socialism at Chinese People's University," 57). The policy of noninterference would be set aside in 1952 when a major reorganization of schools and departments was undertaken and included the shutting down of all private universities. In Beijing, Yenching University was closed, and both Beijing and Tsinghua universities were

principal focus was to train students in "economic-managerial specialties that were the regime's highest priorities in 1949–1950."[27] The university was inaugurated on 3 October 1950.

In plans drawn up in November and December of 1949 it had been proposed that the university would offer two types of courses. The Regular-Course Division (本科; *benke*) would involve two to four years of study in one of eight departments: Economics, Economic Planning, Finance-Credit, Trade, Cooperatives, Factory Management, Law, and Diplomacy. Short-term training classes envisaged as six-month courses, offered by the same departments, emphasized economic management and state administration, with the idea of producing trained cadres who could be put to work straightaway. In the December plan, these short courses, previously called "training classes," were renamed "specialized advanced classes" (专修班; *zhuanxiu ban*).[28] As for who would attend these classes, it was decided that "students of the 'young intellectual' category should be eighteen to thirty years old and graduates of senior middle schools or 'young intellectual' CCP cadres who had graduated from junior middle school or the equivalent and had at least three years of experience in CCP revolutionary work or specialized organs. These students would be enrolled in the Regular-Course Division." In contrast, the Short-Courses would enroll cadres with educations equivalent to lower middle-school, who had more than three years' experience working in "liberated

---

brought under greater state control. See Hayhoe, *China's Universities* (in particular, ch. 3: The Socialist Story, 1949–1978); and Stiffler, "Building Socialism at Chinese People's University."

27 Stiffler, "Building Socialism at Chinese People's University," 22. Later, Stiffler elaborates: "In directing that a substantially revised plan be drafted, the CCP leadership reaffirmed Liu Shaoqi's clearly stated priority for the university: Renda would train Chinese cadres in economic management and state administration. It would not emphasize the training of cadres in Marxist-Leninist ideology, as in the Yan'an and Civil War–era cadre-training schools, and as the Soviet advisors recommended" (64). "By 1952, however, Renda's priorities would shift back in the direction of the Marxist-Leninist ideological training that the Soviet experts had recommended" (83). On a possible precursor in the Northeast (东北人民大学; *Dongbei renmin daxue*), see Wang Sihua, *Wang Sihua tongji lunwenji*, 9–10.

28 Stiffler, "Building Socialism at Chinese People's University," 66. The number of Soviet advisers was 50 (down from 185). "The CCP leadership decided to put in first place the six-month Short-Course Division, which emphasized economic management and state administration, while cutting the two- to four-year Regular-Course Division. This represented a clear victory for Liu Shaoqi's pragmatic, short-term, technocratic priorities, and for the regime's goal of rapid economic recovery."

TABLE 6.1. Renmin (People's) University Statistics Graduates/Enrollees

	1951	1952	1953	1954	1955	1956	1957	1958	1959	1960	Total
专修生 Short-term students	71	145	493	172	136	0	135	99	80	51	1382
研究生 Research students	0	13	5	0	25	31	0	2	0	0	76
本科生 Regular students	0	0	23	109	0	192	99	104	82	135	744
Total	71	158	521	281	161	223	234	205	162	186	2202

Source: Based on annual rosters from the Statistics Department alumni publication. For details, see Renmin University of China, School of Statistics, *Xiaoyou jiniance,* 1–7.

areas" and were between twenty and forty-five years in age, as well as retrained old intellectuals.[29]

While not part of the November list of disciplines, statistics was added by the time of the revised December plan.[30] At this early stage, statistics was offered only as a short-term course, and the November plan called for 300 students of statistics out of a total planned enrollment of 3,800 students. Of the planned 2,700 Regular-Course seats, not a single one was in statistics. In the December plan, which involved a general downsizing, the corresponding numbers were 200 out of 3,000; and zero out of 1,400.[31] The number of dedicated teachers was also downsized, from 5 out of 185 in the November plan to 2 out of 50 in the December plan.[32] It was from such humble origins that Renmin University became the center of research and training in statistics over the course of the ensuing decade.

By June 1952, when a formal Department of Statistics was established, the university had become the central node for theoretical and pedagogical research in statistical training. As Table 6.1 shows, during the remainder of the decade at least 2,200 students received some form of training in statistics at the university.[33]

29 Stiffler, "Building Socialism at Chinese People's University," 145.

30 Pedagogy was also added at this time.

31 Stiffler, "Building Socialism at Chinese People's University," 70.

32 Ibid., 72. Stiffler also notes that early curricular plans had an extreme scarcity of specialized courses; the curricula for several different courses were actually similar (79).

33 This estimate and the individual breakdowns in Table 6.1 are based on annual rosters of

It is unclear whether the targets set for the first batch of students in November and December 1949 were actually met. Data from *Xiaoyou jiniance* (校友纪念册; Alumni Record) indicate that only students attending "advanced specialized classes" graduated in the first two years. The first class of graduates with the equivalent of an undergraduate degree was produced in 1953.[34] This makes sense if, per the plans, Regular Courses were to last two to four years and students began instruction in 1950. Even so, only 23 undergraduates were produced that year, less than 5 percent of the total (521) students enrolled in the statistics program. The year also witnessed the largest graduating class, which coincided with the start of the First Five-year Plan (1953–1957) and a push toward greater planning and statistical work.[35] After that peak, the number trained in any given year hovered around 200, with the share of undergraduates gradually rising to constitute the majority by 1960. The total number of graduate students or researchers (研究生; *yanjiusheng*) was just seventy-six for the entire period, with the majority receiving their degree in 1955 and 1956. For our purposes here, the most significant and suggestive result is that, even at a place like Renmin University, the focus was on preparing statistical workers via short-term courses: Over the entire decade 63 percent (1,382 out of 2,202) of all students were trained in advanced specialized classes.

In addition to Renmin University, departments and schools were also set up in a number of other places during these early years. Table 6.2 provides a basic chronology of some of the major statistical schools or departments that were established during the 1950s.[36] Some of these schools and departments offered Regular Courses, but most also provided short-term training. In many instances, these were pre-existing institutions that were reorganized and redesignated as special schools after 1949. Even if we assume that these schools

---

graduates produced by Renmin University. I do not know if these lists are exhaustive, so the likelihood remains that additional students were trained.

34 Throughout the entire decade, statisticians could not attain a doctorate or associate doctorate degree, nor did they gain professional designations typically granted to engineers (SSB, *Tantan tongji ganbu de zixue*, 1).

35 After 1955, the urgent demand for new cadres appeared to stabilize (Vogel, "From Revolutionary to Semi-Bureaucrat," 46), and this seems to fit with the data above, with the number of short-term students tapering off after 1955.

36 Renmin University of China, School of Statistics, *Xiaoyou jiniance* (the 2007 issue observes that the Department of Statistics at Renmin was the first to be set up under the aegis of the new economics of the PRC). I am not sure how to evaluate this claim against the information in Table 6.2.

TABLE 6.2. Establishment of Dedicated Statistical Schools or Departments

Date	Name of Bureau/Activity	Province/Region
About 1950	Xiamen University Statistics Department	Fujian
August 1951	Northeast Planning Statistics Institute	Northeast
June 1952	Renmin University Statistics Department	Beijing
29 November 1952	Northeast Finance and Economics Institute	Northeast (Changchun?)
July 1954	Chongqing School of Statistics	Chongqing
August 1954	Xi'an School of Statistics	Xi'an
June 1955	Shanghai School of Statistics	Shanghai
27 March 1958	Control over the Shanghai, Chongqing, and Changchun Statistical Schools is transferred to the respective provinces	Shanghai, Chongqing, Changchun
27 March 1958	Control over the Nanchang and Xi'an Statistical Schools is transferred to the respective provinces	Nanchang; Xi'an

Source: DSJ.

trained a similar number of students as Renmin University, we would arrive at approximately 20,000 students trained over the course of the decade. Yet, by 1956 the number of statistical workers was claimed to be as high as 200,000.[37] Clearly most of the training occurred not at major universities but elsewhere. As Li Fuchun explained in 1951, "[We] cannot train advanced statisticians only at the university level; rather . . . every province, municipality, [and] large region, should train grassroots statisticians."[38] One major site for such training was the vocational school (专科学校; *zhuanke xuexiao*).

Vocational/Technical Schools

Li Ming did not have the good fortune to be trained at Renmin University or at any of the dedicated schools of statistics established in the early 1950s that are identified in Table 6.2. In his letter, he denies us the name of his training center, noting only that it was a vocational/technical school. In this experience, he was not alone. Expediency and overwhelming demand meant that most of the tens of thousands of statistical workers were trained at vocational

37 *RWTG*, 3; elsewhere in *RWTG*, a figure of 190,000 is cited. See *RWTG*, 10. Huang, "The Statistical Agency in China's Bureaucratic System," notes a similar figure.

38 *DSJ*, 5.

or technical schools. Such schools, often established anew or created by the takeover of pre-existing institutions, dotted the entire country and came to supply an entire generation of state employees, including, of course, statistical workers. This section uses the example of Beijing Finance and Economics School (cited hereafter as BFES) to provide a sense of what such schooling may have been like.

Established on 31 March 1951, BFES traced its roots to Beijing Business and Vocational High School, which was established in 1929.[39] Originally located in crowded Weiying Hutong, an alleyway inside the walled city between the Xuanwu and Heping gates, by the time of its conversion the school had moved to more spacious environs south of Qianmen Gate and could accommodate 900 students in 18 classrooms and 180 dorm rooms. Facilities included a library and rooms for various other activities. By the end of the year, total enrollment stood at 575 students, divided among Regular and Intensive classes. During those initial months, the school also graduated 179 students in the Regular Course stream and 98 students through short-term courses in accounting and statistics.[40]

Four courses of study were offered by BFES during its initial years: Regular Classes, Intensive Classes, and two types of Short-term Classes. Each course of study was designed with different goals and a different target audience. Table 6.3 summarizes these differences. The Regular Course was designed as a three-year program for young teenagers with the equivalent of a middle-school education. The school offered more than sixteen Regular Courses (more than the required twelve) but could not meet the stipulated teacher ratio of 2.5 to 3 teachers per course.[41] Graduates were subsequently assigned to work in various bureaus of the central and city governments as mid-level cadres. A report written in November 1951 observes that although they were enthusiastic about work and willing to take on responsibility, their skill levels were inferior to the average university-trained economics graduate.[42] The content of the Regular Courses was often modified to meet the requirements of the relevant city and central government bureaus, but, in general, three types of majors were offered: Accounting, Statistics, and Factory Management. Accounting majors focused on cost accounting, government accounting, banking, and other specialized areas. Instruction in Statistics included advanced

39 BMA 004-010-00008: 7–9.
40 BMA 004-010-00008: 3–4.
41 BMA 004-010-00008: 3–4.
42 BMA 004-010-00008: 7–9.

TABLE 6.3. Beijing Finance and Economics School (BFES): Early Characteristics (1951–52)

	Regular classes	Intensive classes	Short-term classes: Accounting and taxes	Short-term: Statistics
Educational Qualifications	Middle-school equivalent	High school equivalent	Middle school equivalent	Middle school equivalent
Qualities	Correct thought, healthy, willingness to work	Correct thought, healthy		
Age	15–20	18–28 unemployed intellectuals	20–40; mostly unemployed intellectuals	18–25; mostly unemployed intellectuals
Course of Study	3 years; receive proper education, master modern finance and economics	3 months; per pressing need, receive basic scientific and economic/financial skills	3 months; per pressing requirements, specialized training in relevant skills	6 months; per pressing requirements, specialized training in relevant skills
No. of classes	16	1	1	1
No. of students	543	32	49	49
After graduation	Mid-level cadres	Participate in city government work (likely low level?)		
Remarks			Discontinued after July 1951	Discontinued after July 1951

Source: Based on information in BMA 004-010-00008: 2, 1–9.

statistics, industrial statistics, and other related specialized fields. Factory Management comprised courses in industrial economics and management and labor organization. Each major also included basic classes in math, introductory statistics, and economic geography.[43]

In contrast to the Regular Classes, the Intensive and Short-term Classes were designed to meet immediate and urgent staff shortages, usually in the city government. Such training normally lasted only three months and dealt with the highly specific skills required to carry out specific tasks. Students were generally older, and in many cases had complicated backgrounds because of their identification as unemployed intellectuals.[44] For these students, political study was especially vital to ensure their loyalty to the revolution. BFES reports lamented, however, that this was not carried out in an organized manner, and as a result the students' subsequent work performance suffered. The two short-term training classes in Accounting/Taxes and Statistics appear not to have been permanent course offerings, since they were discontinued after July 1951.

Although Li Ming's own description of his education in statistics does not exactly fit any of the forms of training offered by BFES during its first several years, it was probably closer to a truncated version of the three-year program described above. This may have been as much an outcome of differing courses of study across the country as it was of expediency, which may have led to a premature work assignment. It is likely, however, that the nature of study was a close approximation of the situation at BFES. The standard teaching method involved the use of the Five-Stage Teaching Method: planning the class; checking previously assigned homework; teaching the lesson; consolidating (revising?) the lesson; and, finally, assigning homework for the next class.[45] Study materials involved a variety of textbooks designed for specific courses. In many instances, some of the Renmin University texts discussed in Part I were the principal study materials.[46] Just as significant was study outside of the classroom, which typically involved three activities: listening to monthly

43 BMA 004-010-00008: 5.

44 On the categorization and treatment of intellectuals in the early 1950s, see U, "The Making of Zhishifenzi."

45 These were a likely adaptation of Johann Friedrich Herbart's (1776–1841) pedagogical theories, which were influential by the turn of the nineteenth century. In a small minority of instances, the three-stage method was also used (BMA 004-010-00008: 6, 7–9).

46 At this early stage, books on mathematical statistics appear to also have been included.

current affairs broadcasts; reading newspapers; and reading classwork-related materials. In addition, various forms of teaching, support, and guidance were provided both inside and outside the class.[47] We will return to these forms of training in a later section.

On occasion, BFES facilities were borrowed by other agencies and companies to carry out additional training. This was especially sensible in the early years when the emphasis on practicality and expediency meant using all available resources to solve pressing concerns and because the school had space to spare. The November report on the state of affairs at BFES noted with some satisfaction that after the move to its new location, the school's infrastructure was more than adequate. In fact, in November 1951 only 575 students were registered even though there was space for 900; similarly, only 521 students lived in dorms that could accommodate as many as 900.[48] As various agencies and companies sought to adapt to the changing economic goals put forward by the state, they were able, on their own initiative, to take advantage of this extra space to meet their short-term requirements.

One such instance is that of Beijing Public Enterprise Company and its attempts to deal with personnel shortages. In June 1951, the manager and deputy manager of the company approached the City Government and the City Finance and Economics Committee with a request that they be able to use BFES facilities and arrange short-term statistics and accounting courses during the summer. They explained that the company was experiencing a major expansion but the lack of statistical workers and accountants was gravely impeding its work. They estimated that the number of university graduates already assigned to them for that summer was insufficient to meet its ever-mounting needs. To solve this widening disparity, the company planned a three-month short-term course to develop statistical and accounting talent. To that end, the manager and deputy manager requested they be permitted to borrow empty classrooms at BFES to carry out this training.[49] The request was subsequently approved, and in August the company reported that classes had commenced the previous month.[50]

By 1954, such ad hoc arrangements appear to have become less common. This is most obvious in the changed nature of reports and planning documents

47 BMA 004-010-00008: 6.
48 BMA 004-010-00008: 3–4.
49 BMA 004-007-00065: 1–2.
50 BMA 004-007-00065: 6–7.

from 1953 and 1954.[51] Comparisons with a report from 1951 indicate differences in both form and content. The nine-page 1951 report was handwritten, concise, and focused primarily on an empirical discussion of classrooms, types of classes, and teaching style. Everything about it suggested immediacy, urgency, and improvisation. In contrast, the planning documents from 1953 and 1954 ran into several tens of pages, were typed or printed, and adopted formulaic language and standard outlines. Planned action, via regulation and the setting of specific goals with specific deadlines, was the stated ideal. By 1953, certain basic themes were already evident. The first and foremost was elevation of the Soviet Union to the status of role model. As a planning document noted, "the teachers, after having studied and analyzed Soviet syllabi and teaching materials, have realized their superiority and the concordant importance of studying them intensely."[52] A 1954 planning document was even more explicit in discussing past achievements and future plans, noting that an educational system was being built on the foundation of Marxism-Leninism.[53] The construction of a socialist society required relying just as much on the advanced experience of the Soviet Union as on the devotion of students and teachers. Accordingly, the report noted that the level of planning had been raised over the past several years. The political and professional standards for teachers had also been raised. New testing methods were put in place, and specialized syllabi and course materials were prepared.

By this time, the school's brief had also become narrower. From a general vocational school for the training of cadres, it had been transformed into a dedicated school focusing on Industrial Management to train mid-level vocational/technical cadres (中等专业干部; *zhongdeng zhuanye ganbu*). Statistical training remained a core component of the school's activities, though the scope of the courses was reduced to reflect the school's new character. In 1953, statistical offerings had included: General Principles of Statistics, Industrial Statistics, Capital Construction Statistics, and Social and Economic Statistics.[54] By 1954, the latter two were jettisoned, leaving two qualitative forms of classes, one focusing on general principles and the other on specialized vocational statistical skills.[55]

51 BMA 004-001-00096 and BMA 004-001-00103.

52 BMA 004-001-00096: 3. The quoting of teachers was becoming an increasingly common strategy.

53 BMA 004-001-00103:10–25.

54 BMA 004-001-00096: 2–3.

55 BMA 004-001-00103: 12.

Supplemental Training

A further form of training was supplemental, often carried out by taking a break from one's job. Li Ming claimed to have received three months of supplementary training. Such training likely became more common as the nature and amount of work increased during the First Five-year Plan period. For instance, in February 1954, a joint notification, concerning the transfer of already employed cadres for six months of supplemental training, was delivered from the Beijing Statistics Bureau and the Bureau of Personnel Administration to eight city bureaus.[56] The notification stated that BFES, in order to co-operate with the needs of the large-scale economic construction going on throughout the country, would train on-the-job (在职; *zaizhi*) statistical cadres in a planned way and thereby raise the level of their professional skills. The notification lamented that the city's statistical cadres were lacking key personnel (lit: lacking backbone; 缺乏骨干; *quefa gugan*) and this problem had to be addressed. After an investigation, it was resolved that additional training was especially needed for cadres involved in capital construction statistics. As a result, the majority of seats, thirty-five out of sixty-one, were reserved for the Construction and Engineering Bureau. The remaining twenty-six seats were divided across seven other bureaus:

Local Industry Bureau	8
Public Works Bureau	5
Sanitation Engineering Bureau	5
Public Health Bureau	1
Public Services Bureau	3
Education Bureau	1
Housing Administration	3
Total	26

By 1957, a two-tier structure had emerged for such supplemental training classes. In the first tier, classes were organized by schools that were directly overseen by the SSB. These were normally attached to universities or to dedicated statistics schools (see Table 6.2). Every year, the SSB approached provincial statistics bureaus with a request to send students to these schools. Candidates had to have the rank of section chief or higher and had to be vetted

56 BMA 004-010-00812.

by the SSB before they could begin their studies. In addition, some schools insisted on administering admission tests. The training usually lasted for one year. In the second tier, training classes were set up by provincial- and city-level statistics bureaus, on occasion in concert with the local planning committees. BFES was likely one such school. The student body consisted of cadres already employed as statistical workers in the local district/area. The training period was usually in the three-to-six–month range. In 1956, approximately 5,000 students were attending such classes in schools in fourteen provinces (including Shandong, Zhejiang, Heilongjiang, Henan, and Hunan).[57] In addition, a similar two-tier system of cadre schools also provided training at the central and provincial levels. Employees of the central government were sent to cadre schools attached to various central ministries, whereas provincial employees attended provincial cadre schools. Statistical training in these schools was organized in an ad hoc manner. Most likely, substantial variation existed with regard to students, curricula, and duration of study.[58]

Two other methods were also employed to provide supplemental training: night schools and correspondence courses. In 1957, night classes were offered by the Shanghai, South-Central, Northeast, and Sichuan Academies of Finance and Economics. Correspondence courses were offered by the Correspondence Division of Renmin University directly from Beijing. The division also established coaching stations (辅导站; *fudaozhan*) in nearby cities, such as Tianjin, Taiyuan, and Baoding. Each of these schools targeted already employed statistical workers who possessed the equivalent of a senior-middle-school education. Even so, they were usually required to pass an admissions test and the nature and content of the training was likely modeled after the Regular Classes.[59]

The formal and supplemental training described here, in addition to the university-level education described in the earlier section, comprised the central bulwark of the state as it sought to create basic human resource infrastructural capacity in statistical work. Li Ming was a recipient of such a mix of training. His initial training was at a vocational school, which in 1954 was supplemented by a short-term course. Although I have found no aggregate data that indicate how many cadres were trained through these methods, we are now in a position to make some very general estimates.

---

57   SSB, *Tantan tongji ganbu de zixue*, 76.
58   Ibid., 76–77.
59   Ibid., 75-76.

Confronting Incapacity—Personnel Shortages and the Advocacy of Self-Study

Let us begin with some simple arithmetic. By 1955, the SSB claimed a workforce of 200,000 cadres. In an earlier section, I estimated that the number of graduates trained in statistics at the university level was unlikely to have exceeded 20,000 during the entire decade of the 1950s. This gives us an estimate of about 10,000 graduates by mid-decade. It is somewhat more difficult to estimate the numbers trained at vocational schools. By the mid-1950s there were such schools in at least fourteen provinces. In 1951 BFES claimed a little less than 600 students enrolled in three-year courses. This would mean three full cycles of students by 1955. However, we do not know how many of these were trained for statistical work. For the purposes of estimation, let us assume other schools in the fourteen provinces were operating in a similar fashion. In addition, we know that in 1956 about 5,000 students were attending three-to-six-month–long supplemental courses in these schools. This gives us an estimate for 1956 that ranges from 10,000 to 20,000. Assuming the same rate, this would give us a range of 50,000 to 100,000 students trained via this method from 1951 to 1955. However, we should remember that supplemental education implies that these cadres had already received basic training in statistics, either at vocational schools or on the job, or they were holdovers from the pre-1949 regime. Like Li Ming, many should not be counted twice in our attempt to estimate the total numbers trained. Statisticians were also trained at other venues, notably cadre schools and schools run by various ministries. Accepting that this method offers us an extremely imprecise estimate, we can still reasonably claim that it would be a stretch to make the numbers add up to 200,000.[60]

So what can we conclude? The state was unable to train enough personnel through traditional schooling. The training of cadres, where it existed, consisted primarily of short-term courses, after which cadres were assigned to work in the hope that they would learn on the job. One response to this incapacity was advocacy of "self-study."[61] In fact, in 1957 the Education Department

60 A regional example of such incapacity is the case of statistical work in Beijing in 1953. Of the 3,829 full-time and part-time statistical workers spread across the statistics bureau and various other industries, government departments, and related organs, only 618 were at the rank of scientific staff or above. Furthermore, only 1,291 of the 3,829 had studied any statistics at all. See BMA 004-008-00291: 16.

61 Another way to understand the need for self-study can be appreciated from a general sense of the statisticians' brief: Personnel in Beijing and the provincial bureaus had to oversee

of the SSB published a volume entitled *Tantan tongji ganbu de zixue* (Discussions [concerning] self-study by statistical cadres), in which it went so far as to substantially abnegate sole responsibility for training a sufficient number of personnel:

> But, we must realize one thing: no matter whether in recent times or in the future, the country cannot nor must it establish statistical schools and training courses on a large scale and get students to attend them. Even night school and correspondence classes can only realistically be organized in large cities and in limited numbers. Therefore, the education of statistical cadres must rely on one method—self-study while on the job [在职自学; *zaizhi zixue*].[62]

Besides addressing the key limitation that not enough personnel were being trained, self-study also offered a solution to another common complaint. In his letter Li Ming had expressed dissatisfaction that in spite of his training there was much about the production process that he did not understand. He worried that the blind application of statistical method was not in the least bit productive. One of the published responses to his letter began by confirming this precise problem: In his October 1955 letter, university graduate Li Yuren acknowledged his dismay at being unable to apply to statistical work the theory he had learnt in school. Only by learning on the job was he able to bridge this divide.[63] Publication of *Tantan tongji ganbu de zixue* was aimed at addressing both these issues.

In justifying the need for self-study, the volume offered some preliminary clarifications. The first had to do with the goal of their study: to unceasingly raise the standard of statistical work, to enrich the content of statistical science, and to bring together statistical theory and practice, thereby contributing to nation-building. In other words, they were to transform themselves into statistical experts. This kind of expertise was particularly important; empty-headed theoreticians (空头理论家; *kongtou lilunjia*), whose study was not functional and was divorced from actual requirements, were expressly undesirable. The second clarification addressed the question of whether dedicated

---

the work of 2,200 county offices, which in turn oversaw village cooperatives responsible for 750,000 villages. Add to this the thousands of industrial units and ministry-level units. Suddenly, 200,000 personnel appear to be insufficient!

62 SSB, *Tantan tongji ganbu de zixue*, 3.

63 Li Yuren, "Wo dui tongji gongzuo de renshi," 53 (also published in *RWTG*, 33–35).

study was necessary. The volume explained that although statistical work naturally involved addition and subtraction, it definitely was not limited only to such activities. Nor was statistics merely a game of numbers. Rather, statistics possessed abundant political and economic content because it indicated the results of the labor of China's hundreds of millions of people. Study would ensure the avoidance of errors and promote accurate analysis, and thereby help in the deduction of the intrinsic laws of all objective things. In particular, knowledge of economics and business, political theory, and statistical theory were all critical to becoming statistical experts. But the volume went on to lament that in recent times the level of statistical workers was lacking. The only way to solve this contradiction was for workers to self-study. Admittedly, this was no easy task—time was limited and guidance often absent. And yet, the volume exhorted workers to persevere and carry out long-term self-study.[64]

Guidelines for self-study were divided into two sections: those dealing with elementary or foundational knowledge common to all statistical workers and those specialized materials and guidelines for subdiscipline-specific statistics (industrial statistics, agricultural statistics, etc.).[65] Undergirding both was practical professional work, which consisted of familiarizing cadres with the two most important methods of statistical data collection: the periodic reporting system and survey plans (调查方案; *diaocha fang'an*).[66] General statistics-related classes were divided into three categories.[67] The first was the study of the basic problems in statistical theory, in which the goals and methods of statistics, the collection of data, and their ordering and analysis were introduced. Concepts included mass surveys, summarization, stratification, averages, trends, and index numbers.[68] The next category consisted of

---

64 SSB, *Tantan tongji ganbu de zixue*, 2–3.

65 Ibid., 4.

66 Ibid., 5–6.

67 Ibid., 7.

68 Recommended study materials included: Xu Qian et al., *Tongji lilun yiban wenti jianghua*; SSB, *Tongji jiangzuo jiangyi: Di 1–11 jiang*; Renmin University of China, Department of Statistics, *Tongjixue yuanli jiangyi*; of these, the first, which was popular and easy to understand, was intended for those who had just begun the study of statistics and had low educational levels. The contents of the second and third categories were relatively more detailed and suited for cadres with higher levels of education (possibly high school or some college). For those who wanted to study further, they could consult: *Tongji lilun di 1–10 zhang*, which was a 1957 SSB translation of a Soviet Central Statistical Administration edited volume; Strumilin [斯特鲁米林], *Tongjixue*; and Kozlov (?) et al., *Tongji yiban lilun wenti*. See SSB, *Tantan tongji ganbu de zixue*, 9–10.

branch-specific statistics, for example, capital construction statistics focusing on those phenomena related to construction work and their analysis.[69] Finally, there was the study of statistical economics, which involved the collection of data related to economic activity (indices), their content, calculation, and analysis. The objects of such analysis were central economic issues: national income, balance sheets, the organization of economic work, and so forth.[70] In addition, two related subjects that workers were also encouraged to study were accounting and analysis of economic activity. Finally, there were materials on Marxism-Leninism and related treatises on economics and statistics.[71]

The organization of self-study was calibrated using a fourfold categorization based on the education and experience of the individual cadres.[72] For those cadres who did not possess a junior-middle-school education, the first task was to attain the equivalent of that level and to familiarize themselves with the basic professional knowledge required for their jobs. At the next level were cadres who possessed education equal to or above the junior-middle-school level but who had not systematically studied political and statistical theory. The goal was to convert them into statistical cadres who were able to work independently. Cadres with several years of statistical-work experience comprised the third group.[73] In the final category were cadres who had attained a university degree or had some university-level education. Their main tasks were to further research political economy and branch-specific economics, and to master economic statistics and branch-specific statistics. In so doing, they would become statistical experts, capable of producing reports containing statistical analyses and able to scien-

---

69 Much of the remainder of SSB, *Tantan tongji ganbu de zixue*, is dedicated to a discussion of such branch-specific study (more on this below).

70 Recommended study materials included: SSB, *Tongji jiangzuo jiangyi*; Xu Qian et al., eds., *Jingji tongjixue jiangyi*; Bideluofu, *Jingji tongjixue jiaocheng*; and Soviet Central Statistical Administration, ed., *Tongji lilun, di 11–26 zhang*. The last of these, which could serve as a principal reference book, was described as a good combination of statistical economics and basic problems in statistical theory; its sections on statistical theory were relatively comprehensive. For details, see SSB, *Tantan tongji ganbu de zixue*, 9–10.

71 Study materials included documents related to the Eighth Party Congress and other important meetings of the party and government; articles from *RMRB* and journals such as *TJGZ* and *Jihua jingji* (计划经济); internal departmental newsletters (内部刊物); SSB, *Tongji gongzuo zhongyao wenjian huibian*, 3 vols.; and Anon., *Jisuan jishu jiangyi*.

72 SSB, *Tantan tongji ganbu de zixue*, 11–13.

73 They were not, however, to neglect the "Party's Policies and Knowledge of Professional Duties" (党的方针政策和实际业务知识), since there was often new material to keep up with; see SSB, *Tantan tongji ganbu de zixue*, 12.

tifically and practically address complicated statistical and methodological questions.

Once the general foundational content was clarified, statistical workers were directed to the main content of *Tantan tongji ganbu de zixue*. Detailed descriptions of pedagogical materials and study methods were presented in seven branch-specific chapters, their sequence indicating their relative importance: Industrial Statistics; Agricultural Statistics; Capital Construction Statistics; Commercial Statistics; Material Supply; Culture, Education, Health, and Hygiene; and Surveys of Family Income and Expenditures.

Finally, instructions were provided on how workers might access the various study materials listed in *Tantan tongji ganbu de zixue*. All books that had been made public could be purchased at the local Xinhua Bookstore. If a particular book was not available locally, then it could be ordered from the Beijing Mail-Order Bookstore (邮购书店; *yougou shudian*). Since the volume only listed a small proportion of the actual number of books available on economics and statistics, cadres were encouraged to browse catalogs and purchase other relevant books. As for those materials published or compiled and printed internally, somewhat different procedures were outlined. Materials published by Renmin University Press or by Statistical Press, such as the different volumes of the compendia *Tongji gongzuo zhongyao wenjian huibian* (Compilation of Important Documents on Statistical Work), could be purchased from the relevant presses with a letter of introduction from the cadre's home bureau. Statistics and planning forms and their rules, survey plans, compilations of answered problems, as well as materials such as documents from the previous year's activist statistical workers' meeting, could usually be obtained from the workers' own units. Failing that, they could be sourced from a higher administrative statistical agency.[74]

It is not hard to imagine that this kind of self-study was largely a fiction, a utopia. The program of study outlined in *Tantan tongji ganbu de zixue* is sufficient to intimidate a full-time student, let alone a worker with a full-time job. I have not found any conclusive evidence that its adoption and use as a manual achieved any degree of popularity among statistical workers. And yet, such an assessment of *Tantan tongji ganbu de zixue* remains important for it points to another basic incapacity. Self-study was a road to self-improvement, a pathway to greater contribution to the nation. But such self-improvement could only be actualized if morale was high. Only when morale was high would one expect to see cadres willing to dedicate their energies to advancing their own

74 SSB, *Tantan tongji ganbu de zixue*, 75.

training and education. Let us, therefore, return once again to Li Ming and his frustrations.

Responding to Li Ming

Since its first issue in 1953, *TJGZTX* [75] served the dual role of a professional journal and didactic venue.[76] By 1955, each monthly issue contained approximately twenty-five articles that spanned such topics as methodology, statistical data on specific questions or economic activities, reports of annual and local meetings concerning statistical work, translations of works by Soviet statisticians, reports from specific work units on recent achievements, announcements declaring the start of a new campaign, and descriptions or transcripts of major speeches that touched upon statistical work. By disseminating experience and information about statistical activities to various parts of the country, many of these essays served a pedagogical purpose. Every issue also carried reports that touted achievements in statistical work or essays that wished to share experiences about a particular kind of recently completed activity. Included also were essays by enthusiasts/activists (积极分子; *jiji fenzi*), meant to serve as encouragement to fellow workers. On occasion, one could also find articles publicly shaming inferior work by a particular statistical worker or unit.[77]

By publishing Li Ming's letter, however, the editors of *TJGZTX* were adopting a new strategy. In what appears to be the first instance of its kind, the journal was used as a platform for a more open-ended debate on the very nature and purpose of statistical work. Instead of specific discussions on method or exhortations to execute a specific kind of statistical work, the editors (and the SSB as a whole) hoped to demonstrate and disseminate a mass "correct" response to several key questions.[78] The first of these centered on the claim

75 In 1957, the journal was renamed *Tongji gongzuo* (*TJGZ*).

76 Related journals fulfilled similar roles; see, for instance: *Xin jianshe* (New construction) and *Jihua yu tongji* (Planning and statistics), among others.

77 See, for instance, the shaming of a unit leader for not filing statistical reports in Hubei Province People's Committee, "Hubeisheng renmin weiyuanhui fachu tongzhi," 48.

78 As it turned out, Li Ming was not the only one to cast doubts on statistics. The publication of his essay created space for other voices to echo his concerns. An article in *RWTG*, authored by the Editorial Office of *TJGZTX*, noted that such voices were a small minority, but then went on to name some of them: Liu Guang, Zhi Ge, and an unnamed reader from a northern Beijing suburb. The penultimate essay in the volume again made reference to some names (*RWTG*, 23–24, 63). Much like Li Ming, it is unclear whether Li Guang and Zhi Ge were actual people or useful effigies for the rebuttal of specific perspectives.

that statistical work was mere arithmetic jugglery—adding, subtracting, calculating percentages, and so on—and, therefore, utterly boring.[79] Was such an understanding correct, the editors asked? How should one arrive at an accurate understanding (or appreciation) of statistical work? The second question dealt with the issue of usefulness. The editors invited readers to ascertain whether Li Ming's verdict on the uselessness of statistics was correct. They encouraged readers to consider the role and status of statistics in the ongoing national construction. How could the full potential of statistics be brought into play? The final question addressed Li Ming's fears of a lack of advancement and the poor future prospects of statistical workers in general. Li Ming had lost faith in statistics, they acknowledged. But was this justified? What, after all, were the prospects for statistical workers; could they rise through the ranks?[80] In October 1955, two months into the debate, the editors of the journal decided to collapse the three questions into one central question: Why [are we] unwilling to do statistical work?[81]

Over 600 responses were received by the journal in the four months from August to November.[82] A representative handful was published each month: five in August, four in September, five in October, and three in November. The December issue brought the debate to a close and carried two articles. The first was a general overview of the debate, summarizing content from the numerous letters that had been received but not published. The second was the December article by Deputy Director Jia Qiyun of the SSB. The eighteen responses as well as the summary essays were subsequently published as *RWTG*. Before an analysis of the debate can be offered it is worth probing the logic of switching to a single question in October—the unwillingness to do statistical work. For that, we should consider the peculiar status of statistical work itself.

The (Peculiar) Status of Statistics

For a whole generation that came of age in the late 1940s and early 1950s, the gun-toting, landlord-slaying peasant revolutionary was a heroic ideal.[83] Even

79 枯燥无味; *kuzao wuwei*; lit. dry and tasteless.

80 Li Ming, ed., "Li Ming tongzhi weishenme buyuan zuo tongji gongzuo?" (the questions were republished in RWTG, 7).

81 RWTG, 64.

82 Ibid., 55.

83 Schmalzer, "Youth and the 'Great Revolutionary Movement' of Scientific Experiment," 170–171) has discussed how science and the role of the scientist provided a similar kind of heroic (and romantic) ideal for many rural youth during the Cultural Revolution.

if many did not directly participate in the revolution, sympathy for the Communists was widespread. Accordingly, one of the principal challenges for the CCP once it came to power in 1949 was to successfully channel this popular fervor into the task of national construction—the building of New China. The broad economic strategy adopted to achieve this goal was industrialization. Any surplus produced in the economy was to be invested in industrial production. Agriculture, the dominant sector of the economy, was marginalized to the extent that its primary purpose was to support the ongoing industrialization. This bias in favor of industrial production, a matter of economic policy, in turn dictated certain social policies. The emphasis on productivity and rapid industrialization at the expense of agriculture also went hand in hand with a shift in focus from the countryside to the city. For a rural, peasant-based, and largely peasant-led revolution, urban China had become the new field of action and achievement.[84] It is not surprising that concerns with the unregulated movement of people, especially from the rural countryside into the cities, began to grow and were often justified based on a need for stability and the desire to ensure the best conditions for production.

One of the results of this shift was the birth of a new type of hero who engaged in production for the country. This applied to agricultural and industrial work in equal measure, but the emphasis on industry meant that the most heroic work was done by those working in factories, producing the raw materials and finished goods that would build the New China. This had implications for the valuation of administrative and service-based work. The "new hero was not the administrative cadre but the engineer."[85] For skill-based technical professions, such as statistics, the repercussions were dire.[86] As Wang Xiangyuan explained in an essay in *TJGZTX* that was reproduced in *RWTG*:

> Right now, there are many comrades who are unwilling to do nonproductive technical work [非生产技术性; *fei shengchan jishuxing*], but are willing to do productive technical work. This kind of thinking is

84 Land reform, of course, was rural, but the major thrust now turned toward the city. Old and new works have focused on this transition from a variety of perspectives: Brown, *City Versus Countryside*; Gao, *The Communist Takeover of Hangzhou*; Brown and Pickowicz, eds., *Dilemmas of Victory* (on various regions in China); and Vogel, *Canton under Communism* (on Canton).

85 Vogel, "From Revolutionary to Semi-Bureaucrat," 48.

86 An interesting contrast to the PRC is India, where statistics in the early post-independence years was highly valued and its practitioners were very powerful. See, for instance, Menon, "Planned Democracy."

present not only among statistical workers but also among those who work in accounting, note-taking, official correspondence, typing, and so forth.[87]

Statistics, along with the other professions that Wang listed, did not enjoy the common man's respect because it did not produce tangible goods. This judgment is reflected in a table in *Weida de shinian*, which lists the relative annual labor distribution between productive work (生产部门; *shengchan bumen*) and nonproductive work (非生产部门; *fei shengchan bumen*) annually from 1949 to 1958.[88] Entitled "Rapid Increase in Percentage of Workers and Other Employees in Productive Work," the table shows a clear upward trend in the share of productive work: climbing from 65 percent in 1949 to 85.1 percent in 1958; the share of nonproductive work declined correspondingly from 35 percent to 14.9 percent during the same period.[89]

Even among "nonproductive technical" professions, statistics stood out. A science that straddled the administrative and the productive, statistics required skilled workers capable of collecting and analyzing vast amounts of data so that tangible production could be better understood, organized, and improved. Its close cousin, accounting, had similar characteristics but was different in significant respects.[90] Unlike statistics, accounting was not part of a national organization linking the center to the village and to work units. Accountants were usually restricted to a work unit, village, cooperative, factory, and so on. Statisticians, in contrast, were part of a national organization, and statistics was fundamental to the rhetoric (and reality) of the socialist state, whose logic dictated that the centrality of planning was not merely economic theory but also a philosophical and practical necessity, predicating all other

---

87 Wang Xiangyuan, "Wo yiding ba gongzuo zuode genghao," in *RWTG*, 38.

88 On the distinction between productive and nonproductive work and its roots in Smithian and later Marxian analysis, see Smith-Peter, "Defining the Russian People," 50. On broader discussions on the topic and how it has remained relevant in politics, see also Gough, "Marx's Theory of Productive and Unproductive Labour"; and Howell, "Once Again on Productive and Unproductive Labour." On how this distinction between labor that produced goods versus labor that produced services underpinned the 1952 adoption of a Material Product System (MPS) of national accounting in China, see Xu, "The Establishment, Reform, and Development of China's System of National Accounts."

89 SSB, Weida de shinian, 164 (also see *Ten Great Years*, 185). The last year in the table, 1957–1958, accounts for a significant share of the change: a swing of about 12.2 percent.

90 See SSB, *Tantan tongji ganbu de zixue*, 7–8, where the focus is on this distinction; and Ezhov, *Soviet Statistics*, 40–41.

activities. The collection and tracking of all information—productive and non-productive—so that informed and appropriate decisions regarding planning and the allocation of men and material could take place were the very corner-stone upon which everything else depended. But making the argument theoretically and in the halls of power was one thing, getting the average statistical worker to do the work was quite another.

To address this problem, two related strategies can be located; one a broad appeal to logic and the other an emotional appeal to the romance of production. Consider the broad appeal first. The logical solution was to claim that all kinds of activity were important and revolutionary—this, after all, was the view put forward in all textbooks and self-study books, either explicitly or implicitly. As one statistical worker explained in his response to Li Ming:

> I have come to realize [through my work] that all kinds of revolutionary work are important. There are no distinctions such as "important and un-important," "small and large," "valuable and worthless"; rather, every profession produces its own leading authority [行行出状元; *hang hang chu zhuangyuan*]. All one needs to do is to work hard, and no matter the nature of the work, success will follow.... [91]

The valorization of productive work over nonproductive work was also attacked. Another interlocutor observed that "although statistical workers did not directly participate in production and, therefore, did not create value, their effect within the larger process of socialist development could not easily be calculated using such a metric."[92]

The most cogent and sustained exposition of this line of reasoning was provided by Deputy Director Jia Qiyun, who placed statistics at the very heart of revolutionary work and the socialist enterprise. In his December essay he wrote that he wished to explain why every type of work was an important part of national development and that if one worked hard and with a focused mind, then success was guaranteed. Jia proceeded to explain that even though heavy industry was the basis for building a socialist society, and working in that sector of the economy was no doubt an honor, other sectors of the economy were just as important. Indeed, without development in other sectors of the economy, such as light industry, agriculture, transportation, commerce, and so on, developing heavy industry was impossible. Therefore, each type of work was equally honorable, equally glorious. Furthermore, it was imperative to recog-

91 *RWTG*, 37. For another instance, see *RWTG*, 42.
92 *RWTG*, 57.

nize that none of these important activities could proceed unless they were supported by administration, a justice system, diplomacy, national defense, education, and health. Regardless of whether it was productive or nonproductive, each type of work was of equal importance and as honorable and full of glory as any other. And underlying all these activities was planning, for which statistics and numbers were absolutely critical. Statistics was thus the basis for everything else: Whatever the job, if statistical work was set aside, then it was tantamount to setting aside any prospects for progress.[93]

Cold, calculated logic, however, also found a worthy partner in powerful imagery and rhetorical flourish. If one was unmoved by the seemingly irrefutable logic of Jia Qiyun's argument, as indeed many may have been, then Feng Jixi's formulation may have held greater appeal. Feng, another interlocutor in the Li Ming debate, began his essay with the following:

> Every time I complete a statistical table/chart, my happiness is like that of a peasant on his field catching sight of a golden ear of wheat, my excitement like that of a steelworker observing molten steel emerging from a Martin furnace,[94] [and] my elation like that of an artist completing a beautiful painting.[95]

The sleight of hand was obvious. In one sweeping sentence, Feng equated statistical work with work in agriculture, industry, and the arts. The key lay in equating a statistical table (and the report it would invariably be a part of) with the tangible goods produced in other fields. Thus, the table was the tangible, material product that a statistician contributed to the economy. In the previous chapter we already encountered the kinds of problems this valorization of the table-as-product generated, in particular the oft-repeated and much-lamented phenomenon of the excess issuance of reports.

---

93 *RWTG*, 67–68. In a subsequent passage, Jia Qiyun similarly dismissed the notion that statistical work was subordinate to other kinds of work.

94 马丁炉; *mading lu*; a likely reference to the mid-nineteenth century Siemens-Martins process for mass producing steel from molten iron. The process used an open-hearth furnace and was developed shortly after the Bessemer process. These processes were complementary—one produced steel faster, while the other gave greater control over the composition of the steel—and remained in use until the middle of the twentieth century, when they were largely replaced by basic oxygen steelmaking. China's last open-hearth furnace reportedly ceased operations at Baotou Iron and Steel Company in Inner Mongolia on 28 December 2001. See "China's Last Open-Hearth Furnace Shuts Down." *People's Daily Online*, 29 December 2001. Accessed 7 September 2017, http://en.people.cn/200112/29/eng20011229_87700.shtml

95 *RWTG*, 10.

Feng's was not the only case in which Li Ming's interlocutors resorted to rhetorical flights of fancy, but it was perhaps the most flowery and emphatic of all. The romanticization of numbers and statistics had been legitimated months earlier by none other a figure than poet and author Ba Jin (1904–2005). In an essay published in *RMRB*, he acknowledged with approval the description of the ongoing First Five-year Plan as a "song of numbers."[96] Ba Jin accepted that in the past people liked to say that numbers were absolutely boring, but he felt such a formulation was no longer tenable.[97] Numbers, he explained, contained in themselves abundant meaning. Far from being inert, they could speak; they could relay countless touching stories and paint beautiful paintings:[98]

> I have always believed: "Poetry should give people the joy of creation, poetry should spread life." These numbers not only spread life and happiness, they not only depict a happy future, but they also guarantee that happiness. Not only do these numbers gather the sentiments of 600 million people, they also embody their common aspirations and are their signpost [pointing to the future]. With them, correctly, step by step, we shall arrive on the road to [building a] socialist society. They are like a bright lamp, illuminating the hearts of 600 million.[99]

The implication was clear. If Li Ming grasped the power of numbers and statistics in New China, he would not find them boring at all. And the evocation of a collective spirit, of course, was not incidental either; nor was the focus on the people's common aspirations. These themes provided grounds for what was the most concerted and popular attack on Li Ming.

Attacking Individualism, Honoring Model Workers

Eleven of the eighteen responses published in *TJGZTX* and *RWTG* highlighted Li Ming's problem as one essentially stemming from individualism. The broad basis of the critique was provided by the Newsletter Group of the Qingdao Statistics Bureau. In its letter, the group observed:

96 Ba Jin, "'Shuzi de shi,' xinfu de baozheng"; reproduced in Ba Jin, *Ba Jin quanji*, 90–94.

97 As the interlocutors in *RWTG* would remind us, the value of statistics had indeed changed in New China. See *RWTG*, 34.

98 Feng Jixi also acknowledged the revolutionary strength of numbers; see *RWTG*, 10.

99 Ba Jin, *Ba Jin quanji*, 92. A variation of this passage was also quoted in *TJGZTX* and *RWTG*, 60.

It is our opinion that the basic problem can be traced to the fact that Comrade Li Ming has not considered his statistical work in light of the needs of the nation and the people. He has not recognized that every kind of work carries with it great revolutionary significance. . . . Instead, Li Ming persists in separating himself and thinking only of his personal prospects and advancement. . . .[100]

In similar fashion, another interlocutor explained that Li Ming was deceived, even blinded, by personal profit. As a result, he was unable to see the uses and effects of statistics.[101] Indeed, for a revolutionary worker individualism was very dangerous. It was like a noose tying down the worker, making him lose direction, much like a man lost in the darkness of the night. Such a worker would go around and around in his small circle, paying attention only to himself and ignoring the community. He would focus on the small, not the big; on the near, not the distant; and even after the passage of one, two, or three years, he would remain unchanged.[102] Numerous other instances belittling Li Ming's individualism were published.[103] On occasion, being capitalist was added to the charge, making it an error not only of individualism but of capitalist individualism.[104]

Tied to individualism was the implicit tendency toward careerism (野心; yexin), even though it was never articulated as such. The term used instead to refer to the matter was "future prospects." Li Ming had raised the issue in his letter as well. One key problem stemmed from the fact that, unlike various other technical professions, one could not earn an advanced degree or certifications in statistics. *Tantan tongji ganbu de zixue* acknowledged this problem on its first page, observing that the SSB had received numerous letters decrying the absence of advanced degrees for statisticians. Neither advanced doctorate or associate doctorate academic degrees nor any professional qualifications, such as those granted to engineers, were available to statistical workers.[105] Such circumstances naturally raised concerns about career advancement, professional development, and pay scales among a statistical workforce that was extremely young.[106]

100 *RWTG*, 17.

101 *RWTG*, 11.

102 *RWTG*, 12.

103 See, for instance, *RWTG*, 16, 26, 36, 38, 43, 48, 61, 63, 67.

104 *RWTG*, 35, 65; the latter is in Jia Qiyun's summary essay.

105 SSB, *Tantan tongji ganbu de zixue*, 1.

106 For instance, one interlocutor noted that even accountants and other service-based

The responses to problems of careerism began with the assertion that the pay scales were appropriate, but the responses also sought to reframe the goals of study and work. Cadres were told that rather than the attainment of a doctorate, certification as an engineer, or some similar qualification, the goal was to develop and make prosperous the nation's culture, science, and technology, and to bring about rapid industrialization. Statistical cadres were no exception to this call to arms. In other words, once people were clear about this, they would no longer obsess about degrees and status, and their achievements would be recognized appropriately. In the final analysis, accomplishments at work, and not degrees or pay scales, were the only things of importance.[107]

It is at this point that Li Ming, the negative model, had served his purpose. In order to stress achievement over qualification, positive models worthy of emulation were needed.[108] As Jia Qiyun wrote in his December essay, over the past several years there had been countless heroic model workers who had made stellar contributions in the various fields of national development. And, as Jia clarified, statistical workers were very much among such heroes:

> At this year's "National Meeting of Youth Activists for the Construction of Socialism" were there not only model engineers and [industrial] workers but also model shop assistants, model cooks, model pig-farmers? Among our statistical workers was there not also an advanced youth group led by Comrade Chen Jinbao? They all have the same [glorious] future prospects.[109]

Indeed, much like other cadres, statistical workers were the recipients of various model worker awards.[110] Such awards were usually made after year-end

---

professions were doing better on these fronts; see *RWTG*, 24. On the pay and hiring structures among statisticians in the PRC in the 1950s, see *RWTG*, 74–76.

107 SSB, *Tantan tongji ganbu de zixue*, 2.

108 The general promotion of model workers had been occurring at the national level through conferences where activist workers exchanged experiences, their reports subsequently being compiled and disseminated for wider didactic purposes. A discussion on the Mass Line noted: "The leaders must therefore be skilled in uniting the small number of active elements to form a leading group and must rely on this leading group to raise the level of the intermediate elements and to win over the backward elements." See de Bary and Lufrano, comps., *Sources of Chinese Tradition*, Vol. 2, 417.

109 *RWTG*, 69.

110 As early as 1950 Wang Sihua had spoken of the need to master and popularize the experiences of advanced workers. See Wang Sihua, *Wang Sihua tongji lunwenji*, 7.

appraisals were carried out.[111] One of Li Ming's interlocutors reported that because of his stellar work he had been recognized as a Third-Degree Model Worker in 1953. Continued excellence at work resulted in his elevation to a Second-Degree Model Worker in the following year. This honor was not simply limited to within the company or within the city of Taiyuan; he was awarded the label of a Shanxi province advanced worker.[112] Similarly, another statistical worker recalled how she was made a First- and Second-Degree Advanced Worker.[113] The point was thus reiterated: Statistical workers had prospects, and they could become statistical experts and model workers.[114]

By 1956, the time was ripe for a model worker conference dedicated to statistical work. The Beijing Statistics Bureau decided to take the lead. On 18 April 1956, it sent out a notification to all bureaus/departments, communes, associations, and districts within the city to begin making preparations for the city's (and nation's?) first Activist Statistical Workers Meeting.[115] Held the following month, the two-day meeting, bringing together nearly 300 people from 53 different units, recognized and honored advanced workers who were on the battle lines of statistical work.[116] About twenty detailed testimonies were compiled for dissemination in the hope that they would inspire other statistical cadres to give free rein to their enthusiasm and creativity.[117]

Conclusion

Given the PRC's limited resources, the story of capacity building in statistics very quickly turned into a story of how best to handle incapacity. Such incapacity manifested itself in three successive areas. The first incapacity was re-

111 *RWTG*, 37.

112 Ibid., 52–53.

113 Ibid., 59. This is a rare instance where the writer, Ma Xiuying (马秀英) self-consciously identifies as a woman (an illiterate countryside girl), who began participating in revolutionary work in 1949 and was transferred to statistical work in 1954. Although many women were certainly involved in statistical work, a person's gender is not explicitly identified in most of the archival materials I consulted. For more on how rural women experienced the Communist revolution in the 1950s, see Hershatter, *The Gender of Memory*.

114 *RWTG*, 51.

115 BMA 035-002-00203: 2. A similar meeting was organized in Sichuan in March 1957 and graced with an appearance by Zhu De. See *DSJ*, 52.

116 BMA 035-002-00203: 7–9.

117 BMA 035-002-00203: 16–124.

flected in the need to build a training infrastructure, both in terms of schools and training centers and in terms of a statistical curriculum that could accommodate a very wide range of students. Indeed, with the exception of the small minority of cadres who had college- or high-school–level educations, most statistical cadres were in their late teens and had extremely limited educations. A second incapacity was characterized by an inability to train cadres in sufficient numbers and in a sufficiently rapid fashion. Finally, morale was an area in which the SSB encountered tremendous challenges. By the mid-1950s, as work was stabilizing and the PRC had entered the high point of its bureaucratic phase, the SSB could take stock of its achievements and also embark on activities such as self-study and model-worker meetings.

The consequences of some of the rhetorical flourishes one sees with regard to statistics and numbers have already been traced in the previous chapter. In particular, the fetishization of periodic reports—being touted as the "product" that statistical workers were laboring to produce—exacerbated what was already a fairly common problem: the excess issuance of reports. Over time, the problem became increasingly insurmountable as SSB leaders recognized that the very system of periodic reports, just as much as their fetishization, was responsible for reams of paper but no way to actually deal with them.

PART III

Alternatives

7

Seeking Common Ground
Amidst Differences

THE TURN TO INDIA

"WHAT ARE THE COUNTRIES most advanced in statistics? Are you in touch with them?" Zhou Enlai's pointed questions had come at the end of a longer-than-planned visit to the Indian Statistical Institute (ISI) in Calcutta on 9 December 1956.[1] Enamored and intrigued by what he saw, Zhou had become deeply involved in several displays, asking questions and seeking clarifications. In the National Sample Survey Unit, he sat on a table and refused to move until his queries were satisfactorily answered. In responding to Zhou's questions, ISI Director Prasanta Chandra Mahalanobis (1893–1972) listed the United Kingdom, the United States, and the Soviet Union, but went on to note that although the ISI had been in touch with all three, it had not desired to copy from any country: "In India we are trying to adopt and develop the methods to suit our own needs." Zhou approved and informed Mahalanobis that a group of Chinese statisticians would shortly thereafter visit the ISI. "I want them to see everything in detail. We want to learn from you. How long should they stay? . . . Also, we want to send research scholars to work here. What is

---

1 Zhou's visit to India was part of an eight-country tour made from the end of 1956 to early 1957. Accompanying Zhou was He Long (1896–1969), vice premier of the State Council and marshal of the People's Republic of China. Zhou and He Long arrived in New Delhi on 28 November and departed from Calcutta on 10 December 1956. Zhou returned again in late January 1957 and visited Viswa Bharati University in Shantiniketan, where he was awarded the degree of *Desikottama* (Doctor of Letters). He also delivered a nationwide radio address on 6 February. For more, see Waijiao bu (cited hereafter as WJB) 203-00084-01.

the kind of men who should come?" Mahalanobis told Zhou that he wanted a mix of types, some advanced and some with good experience in practical work. He also wanted an economist or two who could share China's planning experiences. Zhou again approved. "Yes, we must have exchange of men. . . . We shall arrange this." Zhou then took Mahalanobis's two hands in his and stated very firmly: "And next year in Peking."[2]

Zhou and Mahalanobis's conversation represented a remarkable *volte-face* in Chinese views, heralding a new willingness to explore solutions to statistical problems in India. Only a few years earlier, the Chinese Foreign Ministry had categorically rebuffed Mahalanobis's invitation to join the International Statistical Association for Asia and the Far East (ISAAFE).[3] In the considered judgment of the ministry, it was felt that the Chinese should "in principle not plan to maintain links with the [said] organization":

> The [stated] Association has extremely close ties with the international statistics society, operates under the influence and control of imperialism, and there is a great probability that in the name of learning and science it will be used to conduct intelligence [related] activities.[4]

The People's Republic was much better off without involving itself in imperialist misadventures, especially when it had radically different views on the definition and purpose of statistics. After all, a close and profitable relationship had already been established with the Soviet Union, and little was to be gained by dealing with what appeared to be suspect organizations like the ISAAFE or the two ISIs.[5]

Zhou's urgency that December morning was real; the results tangible. On 11 December 1956, just two days after his visit, a team of Chinese statisticians arrived at the ISI. Ostensibly there to participate in the silver jubilee celebrations of the institute, they stayed for a month. Led by Wang Sihua, now a

2 Mahalanobis to Pitambar Pant, 16 December 1956, File No. 915-Correspondence with Shri Pitambar Pant, 1953-1956, P.C. Mahalanobis Memorial Museum and Archives (cited hereafter as PCMMMA), 11–12.

3 WJB 105-00229-01: 7–8. For more on the ISAAFE, see also "File 393-International Association for Asia and the Far East," PCMMMA, Kolkata.

4 See WJB 105-00229-01: 44, 45, and 47.

5 In 1953 Mahalanobis had also tried to invite Peking University mathematical statistician P. L. Hsu (Xu Baolu, mentioned briefly at the end of Chapter 4) to spend several months at the ISI. The archival trail runs cold, but from what I can determine, Hsu never visited the ISI. For more, see "File P.L. Hsu," PCMMMA, Kolkata.

deputy director of the SSB, the four-person delegation also included the noted demographer, statistician, and Renmin University professor, Dai Shiguang (1908–1999). Within months after returning to Beijing, deliberations over inviting Mahalanobis to China began. "And next year in Peking" did indeed become a reality. Accompanied by his wife and by the statistician D. B. Lahiri, Mahalanobis spent three weeks in China during June and July 1957. The two statisticians lectured to large audiences at the SPC, the SSB, and various universities, held numerous discussions with statisticians and planners, and planned future cooperation and exchanges. Later that year, two Chinese statisticians, Wu Hui (1927–2018), from the Industrial Statistics Division, and Gong Jianyao (1927–), from the Agricultural Statistics Division, arrived at the ISI and spent over a year in India studying statistical theory and its applications.

At the heart of these exchanges was a desire to learn more about mathematical statistics and its applications, particularly the cutting-edge statistical method of large-scale random sampling. This chapter traces these exchanges, explaining their timing and the motivation behind them. Each set of actors in these exchanges had its own agenda. The Indians were particularly keen to learn more about China's planning methods. The Chinese were mainly focused on learning about an approach to statistics, which, for political and ideological reasons, had been rendered anathema in China. The exchanges, short-lived though they were, lay bare Chinese exasperation with the limits of socialist statistics and their willingness to search for solutions beyond the Soviet Union. They also compel a reassessment of early Cold War scientific networks, which should no longer be dominated only by center-periphery models that place either the USSR or the United States at the center.

Statistics as a Case Study in Disaffection

By 1956, statistical work in China stood at a crossroads. In the seven years since the founding of the PRC, an entirely new system had been established premised on defining statistics as a social science. In the first several years after 1949, statistical treatises written by Chinese statisticians freely reflected both Soviet as well as what were subsequently identified as bourgeois Anglo-American approaches to statistics. Beginning in 1950 and gathering momentum in the following year, Anglo-American statistics was progressively critiqued and repudiated in favor of the Soviet understanding of statistics. By the time the SSB was established in 1952, the Soviet approach held sway, both in the academy as well as in the practical field of statistical work. This Soviet approach attained even greater influence in the PRC following the 1954 Moscow conference on statistics. At this conference, statistics was reaffirmed as a social science, in service to building a socialist society and isolated from the tainted, bourgeois, and socially unproductive pursuit of abstract mathematical statistics.

The practical implications of this definition were twofold. Within the academy, the study of statistics stood divided. The discipline of statistics, now defined as a social science, was taught in Statistics Departments where the focus was primarily on procedure, methodology, and work. Mathematical statistics, that "unproductive" and bourgeois pursuit, was banished to Mathematics Departments. As a result, much as in the Soviet Union, the theoretical and practical arms of statistics were separated, denying the possibility of mutual cross-pollination or the discovery of new theories and methods through application, experimentation, and testing.[6] Outside the academy, in the world of statistical work, the impact was just as significant, and resulted in essentially two methods for gathering statistical information: (1) the establishment of a periodic reporting system based on complete enumeration; and (2) where complete enumeration was deemed impractical, sample surveys employing typical sampling methods (典型调查; *dianxing diaocha*). Accordingly, by 1957 the SSB was able to summarize:

> For the purpose of providing statistical data for drafting and examining national plan [*sic*], the complete enumeration periodical report system has been established in all the economic departments in the country. To supplement these, a sampling body of sixteen thousand farm households and another of six thousand industrial workers families have been selected for survey purpose.[7]

6 In his speech on intellectuals, Zhou [Chou], "On the Question of Intellectuals," 141, would criticize this tendency to emphasize practical concerns over theoretical ones: "It is certainly unquestionable that theory must not be divorced from practice and we must fight against any theoretical study which is dissociated from practice. But the main tendency at present is the neglect of theoretical study." Schmalzer, "Self-Reliant Science," 86, has observed that science's "overdetermined" dichotomy between applied and basic science during the Mao era should be understood by the emphasis on self-reliance, which had at least two implications: the use of local methods and local independence from the center.

7 See WJB 105-00530-05: 11–12. "Complete enumeration periodical report system" is how the Chinese themselves translated 全面统计报表制度 (*quanmian tongji baobiao zhidu*).

As already noted, this was a method of sampling in which the samples are not selected using any random or stochastic method. A precursor to random sampling, it had a long history of being regarded as a valid and worthwhile method. Even into the early twentieth century it was officially accepted in many places and, as such, its use into the 1950s was not altogether unusual.[8]

Within several years, the constraints imposed by the methods of socialist statistics became impossible to ignore. For a country with limited technical resources and a limited supply of trained personnel, yet saddled with an enormous agricultural sector and the largest population in the world, the challenges posed by complete enumeration were legion. And, they were made further intractable by the use of the periodic reporting system. Paper reports, full of tables and summaries, were generated at each level of the administration. Very soon, not only Beijing, but every administrative level in the country complained about the flood of papers and numbers, and their lack of capacity to make any sense of them. The problem of excess issuing (滥发; *lanfa*) of reports drew the attention of all levels of the statistical system. Even though repeated attempts were made to tackle the problem, it remained a constant feature of statistical work into the second half of the decade.

The challenges of excess issuing, incapacity, and morale—described in the previous two chapters—became particularly acute by 1956. The deliberate attempts to address them provide a specific and highly illuminating case of the growing Chinese disaffection with a strategy of overt reliance on the Soviet Union for technical and scientific advice. Even though recent scholarship has highlighted the tensions in the Sino-Soviet story, including specific cases where initial idolization of Soviet models was confronted by the reality of their execution or their sheer inapplicability to Chinese conditions, the causes of the disenchantment that began in 1956 are still largely attributed to two more overarching reasons. As Hua-Yu Li and Thomas P. Bernstein have observed, the general consensus in the scholarship is that the evolution of Sino-Soviet relations strongly influenced Chinese attitudes toward the Soviet model.[9] Accordingly, the causes for the disaffection are still largely framed either as an ideological dissonance, or an egotistical one, or both. Indeed, most studies of the Sino-Soviet split tend to focus on top-down factors, such as the role of ideology, the impact of the Hungarian revolt and its repression, Khrushchev's

8 Bethlehem, "The Rise of Survey Sampling," 8.

9 Bernstein and Li, *China Learns from the Soviet Union*, 3; see also Shen Zhihua, *Sulian zhuanjia zai Zhongguo*, 410.

speech denouncing the cult of Stalin (de-Stalinization), or internal struggles between Mao and other top Chinese leaders.[10]

In his January 1956 speech, "On the Question of Intellectuals," Zhou Enlai recognized other reasons as well. The report acknowledged the continuing central role of Soviet aid for the advancement of science and technology in China, but buried in the middle were also clear words of caution:

> On the question of the study of the Soviet Union, however, in the past there had also been such defects as undue haste, arbitrary learning, and mechanical application. Some comrades even arbitrarily rejected the achievements of the capitalist countries in science and technique. These defects should henceforth be avoided.[11]

Zhou's warning is redolent of the possibility that the disaffection was widespread and substantive. It is also suggestive of a new openness to actively learn from other countries, including putatively capitalist ones.

P. C. Mahalanobis and the Promise of Large-Scale Random Sampling

In Prasanta Chandra Mahalanobis, Zhou would encounter perhaps the most articulate spokesman for the promotion and use of statistics. "A physicist by training, a statistician by instinct and a planner by conviction,"[12] Mahalanobis melded scientific and intellectual heft with a tremendous drive to employ statistics to understand and address the problems of everyday life. As a result, he became intrinsically involved not only with independent India's first forays into economic planning and development, but, perhaps even more significantly, with the development and establishment of statistics, most notably sample survey techniques, as a principal tool for knowing the world.[13] Maha-

---

10 The Sino-Soviet alliance has been the subject of renewed interest in recent years, resulting in studies that approach the alliance and its collapse from a multiplicity of perspectives. For some representative works, see Bernstein and Li, *China Learns from the Soviet Union*; Jersild, *Sino-Soviet Alliance*; Lüthi, *The Sino-Soviet Split*; Westad, ed., *Brothers in Arms*; and Shen Zhihua, *Sulian zhuanjia zai Zhongguo*.

11 Zhou [Chou], "On the Question of Intellectuals," 137. To be fair, this appears to be the only cautionary instance in the long report, which otherwise goes to great lengths to stress the value and appreciation of Soviet technical assistance.

12 Rao, "Prasantha Chandra Mahalanobis," 455.

13 A useful overview of Mahalanobis's polymath activities and achievements is Rao, "Prasantha Chandra Mahalanobis." A shorter assessment, focusing on activities through the early

lanobis's original training had been in physics, and it was a subject he continued to teach into the 1950s. But a chance encounter with statistical journals in a Cambridge University library led to his first exposure to statistics. By the mid-1920s he had abandoned plans to return to Cambridge to continue a PhD in physics, choosing instead to remain in Calcutta and pursue the application of statistical methods to practical problems. In 1932, along with several like-minded individuals, he established the ISI in a tiny room in Calcutta's Presidency College. By the 1950s the ISI had moved to a sprawling campus north of the city and had evolved into one of the premier centers for statistical research and application across the globe.[14]

In his Presidential Address to the thirty-seventh session of the Indian Science Congress in January 1950 Mahalanobis observed that, "at each upsurge of social and political development, or during war, there is a rapid growth and expansion of statistical practice."[15] Later in the same speech, he added:

> Statistics is not only an applied science but is also a public science. It is because of the close connexion with public activities that big developments in statistics have always occurred only when there has been need of unified policy and co-ordinated action in times of war or peace.[16]

To his mind, the 1940s and 1950s were precisely such a time. Wartime exigencies had spurred the application of statistical theory to practical problems and

1950s, can be found in Ghosh, "Mahalanobis and the Art and Science of Statistics." For a longer treatment covering his entire life, see Rudra, *Prasanta Chandra Mahalanobis*.

14 In an interview, long-time ISI professor C. R. Rao recalled receiving an offer in 1946 "from Cambridge University, UK, to do statistical analysis on some skeletal measurements, based on methods developed in India." He added he "was glad to accept the offer considering it as transfer of technology from an underdeveloped country to an advanced country." Rao obtained his PhD in statistics under Ronald A. Fisher at Cambridge before returning to the ISI and carving out a distinguished career. Berze and Mehta, eds., "ISI Honorary Member Interviews: Professor C. R. Rao."

15 Mahalanobis, "Why Statistics?" 196.

16 Ibid., 211. "In statistical research the greatest stimulus has always come from the need of solving practical problems. R.A. Fisher's work on the design of experiments was due to the urgent need of solving the deadlock in agricultural field trials. W. A. Shewhart's work on quality control arose from the need of improving the efficiency of inspection in large scale production." Mahalanobis went on to provide three additional examples to prove his point: Post–New Deal United States, where definite statutory responsibilities were assigned to the Division of Statistical Standards; the exigencies of wartime planning in the UK, which led to the formation of the Central Statistical Organization; and the USSR, where the Central Statistical Administration was an integral part of GOSPLAN, the national planning organ.

now, in the postwar world, statistics had a foundational role to play. In 1947 he brought his expertise and convictions to the newly formed UN Statistical Commission and served as chairman of its Sub-Commission on Statistical Sampling.[17] Mahalanobis was the driving force behind the sub-commission. "With the vision of a pioneer he perceived how sampling methods could overcome the otherwise insoluble problems of obtaining accurate information about a vast and still largely illiterate country like India"[18]

Sampling, of course, has a long history. In his 1950 Presidential Address Mahalanobis had explained that its twentieth-century formulation was a combination of the long history of purposive sampling with that of recent developments in probability and mathematics. Purposive sampling is a form of sampling in which the samples chosen for analysis are selected because of some specific characteristic that they possess. Today, we would identify it more closely with qualitative, not quantitative, research.[19] For a long time, it was regarded as the only proper method of sampling and as late as 1926 it was given considerable prominence by a committee of experts at the International Statistical Institute.[20] But since then evidence had been mounting that it produced results that were neither representative nor accurate. The method of complete enumeration was also brought under increased scrutiny. In a 1950 paper in the *Bulletin of the International Statistical Institute* Mahalanobis had demonstrated that "with an acceptable level of precision the costs of sample surveys are only about ten percent of that of a complete enumeration."[21] In the same issue, Mahalanobis's friend, collaborator, and colleague on the sub-commission, the statistician Ronald A. Fisher (1890–1962), went even further and stated that not only were sampling procedures ideal because of their adapt-

17 Part of UNESCO, the commission is the "apex entity of the global statistical system. It brings together the Chief Statisticians from member states from around the world. It is the highest decision-making body for international statistical activities especially the setting of statistical standards, the development of concepts and methods and their implementation at the national and international level." http://unstats.un.org/unsd/statcom/commission.htm, accessed 29 February 2012.

18 Bethlehem, "The Rise of Survey Sampling," 16.

19 To put it crudely, imagine, for instance, that we were to select the upper Manhattan neighborhood of Morningside Heights for intensive statistical analysis, and then proceed to claim that the results reflected the larger reality of Manhattan. For a discussion of qualitative sampling, see Desrosières, *The Politics of Large Numbers*, ch. 7 ("The Part for the Whole: Monographs or Representative Samplings").

20 Rao, "Prasantha Chandra Mahalanobis," 472.

21 Bethlehem, "The Rise of Survey Sampling," 16.

ability, speed, and economy but they were also more scientific than complete enumeration.[22] Mathematical theory undergirded sampling, and this permitted the "careful design of sample surveys with a preset level of precision."[23]

"Large-scale sample survey techniques as practiced today," declared the statistician C. R. Rao, in his 1973 memoir of Mahalanobis for the Fellows of the Royal Society, "owe much to the pioneering work of Mahalanobis in the forties and fifties."[24] According to Rao, Mahalanobis made three notable contributions to sample survey techniques: pilot surveys, the concept of optimum survey design, and interpenetrating network of samples.[25] He developed these methods during the 1930s and 1940s, starting initially in 1937 by using sampling procedures to estimate jute yield in Bengal. During the next two decades, along with his colleagues at the ISI, he carried out numerous sample surveys, developing new procedures and fine-tuning old ones. Mahalanobis's contributions were not limited to the development of theory and method; he also assisted in establishing the framework and standards through which they could be used throughout the world. As chairman of the UN Sub-Commission on Statistical Sampling, he oversaw the publication of "The Preparation of Sampling Survey Reports," which continues to provide common guidelines for modern sampling procedures in use by statistical bureaus throughout the world today.[26]

22 Ronald Aylmer Fisher (1890–1962) was a statistician, geneticist, and evolutionary biologist whose contributions to statistics include the analysis of variance (ANOVA), the method of maximum likelihood, the derivation of various sampling distributions, and experimental design. As a result of these contributions, many regard him as one of the principal creators of modern statistics. For more, see Yates and Mather, "Ronald Aylmer Fisher."

23 Bethlehem, "The Rise of Survey Sampling," 16.

24 The mathematical statistician Harold Hotelling (1895–1973) observed: "No technique of random samples has, so far as I can find, been developed in the United States or elsewhere, which can compare in accuracy with that described by Professor Mahalanobis." Fellow statistical luminary R.A. Fisher noted: "The I.S.I. has taken the lead in the original development of the technique of sample surveys, the most potent fact-finding process available to the administration." See Rao, "Prasantha Chandra Mahalanobis," 472.

25 Rao, "Prasantha Chandra Mahalanobis," 473. For a discussion of Mahalanobis's critics and alternative viewpoints about best sampling procedures during this period in India, see Gore, "P. V. Sukhatme: A 'Social' Statistician."

26 United Nations Statistical Office, "The Preparation of Sampling Survey Reports." Members of the first subcommission included statistical luminaries such as G. Darmois, W.E. Deming, F. Yates, and R.A. Fisher. A follow-up report was published in 1964, with Mahalanobis serving as the subcommission's vice chairman. Members included F. Yates (chairman), T. Dalenius, N. Keyfitz, V. Monakhov, and P. Thionet. S. Zarkovic participated as a representative of

Much like Lenin before him, Mahalanobis's interest in statistics was almost entirely instrumental, a point he made most forcefully in his essay "Statistics must have a purpose."[27] This desire to use statistics to solve real-world problems also got him involved in economic planning by the late 1940s. His closeness to India's prime minister, Jawaharlal Nehru, led to an appointment on the Planning Commission and before long he was charged with drawing up India's Second Five-year Plan (1955–1959). In preparation for this work, Mahalanobis spent the summer of 1954 traveling in the United States, the United Kingdom, Eastern Europe, and the Soviet Union, conferring with economists, statisticians, and planners.[28] A two-sector growth model that he created in 1953, similar to G. A. Feldman's 1928 model, became the basis for India's Second Five-year Plan.[29] Earlier in the decade, he had succeeded in bringing the biennial conference of the International Statistical Institute to India. The institute's twenty-seventh session was held in Delhi in December 1951.[30] With the exception of Japan, which had played host in 1930, this marked the first time that an independent Asian or African nation had hosted the prestigious conference.[31] By the 1950s Mahalanobis had also built up the ISI to be one of the leading centers for statistical innovation in the world. It was a magnet for all kinds of

---

the Food and Agriculture Organisation, and D. B. Lahiri of the Indian Statistical Institute was present as an observer. P. J. Loftus, director of the Statistical Office, represented the secretary-general and M. D. Palekar served as secretary of the group (United Nations Statistical Office, "Recommendations for the Preparation of Sampling Survey Reports," v).

27 "Statistical data may be required to solve problems of a practical nature or for purely scientific or theoretical investigations. Some of the data may be required for current use; some may have to be collected for future utilization. The point to be stressed, however, is that statistics must have some purpose, practical or theoretical, and for immediate or future use." Mahalanobis, "Statistics Must Have a Purpose," 4.

28 Mahalanobis was also seeking technical and material aid for statistical and planning matters, in particular computers and related technology. See Papers of Pitambar Pant (190 LII), Nehru Memorial Museum and Library, New Delhi.

29 A key feature of the model is changing the pattern of industrial investment in order to generate a domestic consumption goods sector. On alternatives to the Mahalanobis model and on contemporary debates surrounding planning strategies, see Desai, "Development Perspectives."

30 Mahalanobis had tried to bring the conference to India in 1949, but an invitation from Switzerland was preferred. India was assured that its invitation would be given serious consideration for 1951. For more, see File "EA/UN-I/9(58)-UNI, 1949," at the National Archives of India (cited hereafter as NAI), New Delhi.

31 Cairo hosted the ISI's 17th conference at a time when Egypt was a British protectorate.

scientists, from either side of the Iron Curtain and beyond.[32] With the support of UNESCO, in 1950 he had also set up at the ISI an International Statistical Education Centre (ISEC), the principal purpose of which was to train statistical workers and students from Asia and Africa. By 1956, students from fourteen to fifteen countries were receiving training at the center.[33]

The Turn to India

Frustration with socialist statistics and its constrictions was certainly the dominant factor in prompting Chinese statisticians to look past the Soviet Union. Crucially, however, domestic compunctions were coupled with greater global engagement as Chinese foreign policy took a much more open turn following the Afro-Asian Conference at Bandung in the spring of 1955.[34] Relations with India, convivial ever since it became the sixteenth nation to recognize the PRC on the international stage in early 1950, also experienced qualitative changes.[35] As the Indian leader of an India-China Friendship Association delegation to China noted in 1956:

> We are proud to record here another important fact that premier Chou-en-lai was pleased to grant us an interview which lasted for over an hour. We

32 The ISI annual reports include long lists of visiting scientists of varying disciplinary and ideological persuasions. Mahalanobis himself was a great traveler and the ISI annual reports usually devoted several pages to his global travels. By the end of the decade, some exasperated friends and colleagues began to worry that the ISI administration was suffering as a result. In the early 1960s, this issue even drove J.B.S. Haldane, then a full-time professor at the institute, to quit in protest. See Guha, *An Anthropologist Among Marxists*, 134–135.

33 Anon., "Professor Mahalanobis Welcomes Chou En-Lai," 30–31. On the ISEC, see Nixon, *A History of the International Statistical Institute* (Appendix V).

34 An important precursor to the establishment of the Non-Aligned Movement in 1961, the Afro-Asian Conference was held in the Indonesian city of Bandung from 18 to 24 April 1955. Leaders of twenty-nine newly independent countries, including Jawaharlal Nehru, Zhou Enlai, and Sukarno, gathered to promote Afro-Asian economic and cultural cooperation in the face of colonial and imperial ambitions. On Bandung and its significance, see Lee, ed., *Making a World after Empire.*

35 "On 1 April 1950, India became the first non-socialist bloc country to establish diplomatic relations with the People's Republic of China." See: "India-China Relations," Government of India, Ministry of External Affairs, at http://www.mea.gov.in/Portal/ForeignRelation/China_January_2014.pdf, accessed 17 May 2014. Also see Heinzig, *The Soviet Union and Communist China*, 292. K.M. Panikkar (*In Two Chinas*), India's ambassador to the Republic of China, became India's first ambassador to the PRC.

discussed very frankly and unreservedly several problems common both to India and China and the general feeling on both sides was that it would be in the interests of both the countries if an exchange on a more lasting basis between scholars and Scientists or technicians of both countries is made possible. This of course, it was recognized, requires [*sic*] to be done at Government level.[36]

This new emphasis in areas of science and technology appears not to have been mere lip service.[37] As many as three Indian Planning Commission delegations visited China in the summer of 1956 with the explicit aim of studying Chinese planning techniques and experiences. The first was actually to have been led by Mahalanobis himself, but owing to his ill health, it was led by his protégé Pitambar Pant (d. 1973), who served as secretary to the Planning Commission.[38] The goal of this delegation, which arrived in China in July, was to study the Chinese economy and planning methods.[39] That same month, two other delegations arrived in Beijing; the first to study agrarian cooperatives and the second to study Chinese agricultural planning and techniques.[40] The Chinese, in turn, sent a team to India in August to study "various technique[s] of using Bamboo in House Construction."[41] Zhou Enlai summed up this new, more substantive enthusiasm for engagement in a letter he sent to Jawaharlal Nehru on the same day that he visited the ISI: "Through the present visit, we have been able to know more [of] India and have realized that in many respects the Chinese people can learn from your people. The Chinese Govern-

36 NAI EA FEA_4(4)-FEA/56: "Report of the Leader, Cultural Delegation sent by the India-China Friendship Association to China in April–May 1956 under the leadership of Prof. P. V. Bapat," 41. On the still largely under-explored history of China-India scientific contacts in the 1950s, see Ghosh, "India and Science and Technology in the Early PRC."

37 As I have argued elsewhere (Ghosh, "Before 1962"), our understanding of Sino-Indian relations during the 1950s is overwhelmingly dominated by diplomatic history. But if we shift out lens to include scientific, economic, and cultural life, we stand to discover different sources and different historical arguments that have much to contribute to an understanding of global, Chinese, and Indian history.

38 For more on Pant, see Vaidyanathan et al., "Pitambar Pant."

39 NAI EA_FEA_4(18)-FEA/56; see below for more on this delegation and its possible importance.

40 Government of India, Ministry of Food and Agriculture, *Report of the Indian Delegation to China on Agricultural Planning & Techniques*; Government of India, Planning Commission, *Report of the Indian Delegation to China on Agrarian Cooperatives*.

41 NAI 7(10)-FEA, 1955 (Secret).

ment will consult the Indian Government through diplomatic channels about sending people over to study."[42]

While the reengagement in statistics occurred within this improved bilateral milieu, it was the delegation led by Pitambar Pant that provided the initial spark. By the time it arrived in Beijing on 12 July 1956, the Chinese may already have had some sense of the statistical methods being employed in India. During his 1955 China tour, Nehru had informed Zhou Enlai of the work that Mahalanobis had been doing in the field of statistics and planning.[43] In all, the delegation spent a total of forty days in the People's Republic.[44] During that time, they held over thirty-six meetings with officials, planners, and statisticians from several bureaus and they also delivered a handful of lectures.[45] In addition to Beijing, they also toured the famous Anshan Steel Works and visited the cities of Tianjin, Changchun, Shenyang, Shanghai, and Guangzhou. Such travel "gave an opportunity to the team to see the work of planning and statistical organizations at the provincial and municipal level."[46]

If they had not been so before, the benefits of mutual exchange and learning seem to have become evident by the end of the trip. Just before departure, the delegation traveled to the CCP coastal resort town of Beidaihe on northeast China's Bohai Sea and spent two hours in conversation with Zhou Enlai. During the conversation, Zhou stated that he was particularly impressed with Indian statistical methods in the field of agriculture. He lamented that the Chinese had been using purposive sampling, which had led to nonscientific

42 WJB 203-00084-01: 129.

43 A *People's Daily* article published on 26 January 1956 (India's Republic Day) praised India's achievements in economic construction and mentioned Mahalanobis's role in drawing up the Second Five-year Plan. See Feng Zhidan, "Yindu renmin de jingji jianshe chengjiu."

44 They entered the PRC through Shenzhen on 8 July 1956 and departed on 17 August. In addition to Pant, the delegation included the statisticians Moni Mukherjee and R. Natarajan. See WJB 105-00487-01: 1–8.

45 "Pitambar Pant spoke on Planning in India. R. Natarjan talked about Industries in India and M. Mukherjee spoke on the Statistical set up of India. M. Mukherjee gave another talk to a small group of technical workers on the method of estimation of national income in India and the logical basis of income estimation work in the West and in the East." Indian Statistical Institute, *Annual Report, April 1956–March 1957*, 84.

46 Ibid. The Indians provided the Chinese with a list of eighteen questions/subjects of interest, responses to which had begun to be drafted. This list was later supplemented by an additional ten questions delivered on 27 July. In particular, the Indians were interested in ascertaining the nature and extent of Soviet aid to the PRC. WJB 105-00487-01: 13–14, 21, and 32–34.

results. In this matter, he said, India was more advanced than China, and China ought to study India.[47] Zhou went on to add that the history of planning in both countries was very short. Their experience was extremely limited, and they had committed errors along the way. But he "hoped that henceforth [the two nations] could mutually learn from each other."[48] Pant's suggestion that an Indian team visit the following year was warmly received, as was his suggestion that one to three Chinese planners be sent to the Indian Planning Commission for a period of six to twelve months, though the internal memos went on to note that Pant did not make clear exactly what he expected the Chinese planners to do.[49] Finally, Pant invited a one-to-three-member team to Calcutta in December of that year to participate in the silver jubilee celebration of the ISI. Zhou not only confirmed Chinese participation but added that a study team would also be sent to the ISI. Then, for good measure, he added: "I myself hope that I [too] will be able to meet with Prof. Mahalanobis when I am in Calcutta in mid-December."[50]

Silver Jubilee Assessments

Not only did Zhou find time to personally visit the ISI but within two days he was followed by a four-member delegation led by Wang Sihua, now a deputy director of the SSB. Other members of the team were Huang Jiantuo (??–??), director of the SSB's Agricultural Statistics Department; Dai Shiguang; and Sun Shizheng (1919–??), identified as a research officer at the SSB.[51] Their first task was to attend the ISI's twenty-fifth anniversary celebrations. Wang later reported that the People's Republic and the Soviet Union were the only two countries that sent official delegations. Representatives of twelve other countries—all "capitalist"—who were present in a personal capacity, included such prominent statisticians as Ronald A. Fisher of the United Kingdom and the Americans Jerzy Neyman (1894–1981) and Morris H. Hansen (1910–90).[52]

---

47 WJB 105-00487-02: 5.

48 WJB 105-00487-02: 9.

49 WJB 105-00487-01: 30.

50 WJB 105-00487-02: 3.

51 *Samvadadhvam* 1, no. 3 (1957), 20. Altogether, the delegation was in India from 11 December 1956 to 16 January 1957.

52 Wang Sihua, "Kaocha Yindu tongji gongzuo de baogao," 90. The other countries were the Netherlands, the United States, Lebanon, Thailand, Pakistan, Japan, Switzerland, and the United Kingdom. See *Samvadadhvam* 1, no. 3 (1957).

The festivities included boat rides on the Hooghly River, paper presentations, and much feting of the institute's success. In his congratulatory message, Wang Sihua lauded the achievements of the institute, reserving special praise for advances in sample surveys and agricultural statistics:

> In these fields, the Indian Statistical Institute has accumulated not only data for planning and policy making, but also a great amount of valuable experience and methods, which will not only be useful to India but also to other countries including China.[53]

With the niceties taken care of, the delegation then proceeded to spend over a month studying statistics and statistical methods.

Within a month after their return to Beijing, Wang Sihua published a report in *TJGZ*, summarizing the delegation's findings and putting forward its recommendations.[54] Wang described India as an agricultural nation with over fifty million households, immediately drawing strong links to the reality as it obtained in the People's Republic as well. However, in line with the federal structure of the country, eleven states carried out their work independently, while the work of three states was jointly coordinated. The center only collected, coordinated, and then summarized material it received from the various state statistical offices.[55] Another worrying feature was the inefficiency of the system—the rate of not filing reports was as high as 10 to 15 percent, and the incompleteness of the data was also a concern. The national income estimate had a large error that ranged from 10 to 30 percent for different sectors of the economy. Under such circumstances, complete enumeration not only required a huge number of investigators but also was uneconomical. The entire process, from designing a complete enumeration to collection, analysis, and publication, could take anywhere from one to two years. Compared to this, sample surveys were much quicker and only required a small number of investigators

53 *Samvadadhvam* 1, no. 3 (1957), 16.

54 Wang Sihua, "Jieshao Yindu tongji gongzuo." The article is reproduced in Wang Sihua, *Wang Sihua tongji lunwenji*, 95–102. Dai Shiguang, Wang's fellow delegation member, also wrote an article on India's sample survey system but it was entirely factual in nature and did not seek to make recommendations for the PRC (Dai Shiguang, "Yindu de quanguo chouyang diaocha").

55 According to Wang's reckoning, this Central Statistical Office (CSO), also set up in 1950, had 250 workers, divided into six divisions: publications, training, coordination, standardization, planning, and national income. The total workers in the national system (excluding those under different ministries) numbered 2,400. Wang Sihua, *Wang Sihua tongji lunwenji*, 95–98.

who would be sent to several sample households.[56] Wang told his readers that
the Indians believed sample surveys were cheaper, quicker, and more accurate
(较经济, 较快, 又较正确; *jiao jingji, jiao kuai, you jiao zhengque*) and that
random sampling was the best way to avoid bias. "They believe that if the scope
of the survey is very large, the survey area is very extensive, and if one wants
to obtain data within a short period of time, then the only practical and eco-
nomical means of achieving this is through sampling."[57]

Accordingly, except for a decennial census, a quinquennial farm and livestock census, and annual censuses of twenty-nine heavy industries, all other
kinds of data relating to income, production, economic conditions in rural and
urban areas, urban employment, and so forth, were collected using large-scale
random sample surveys. The delegation was impressed with the numbers involved in statistical work and the relative abundance of published material on
sampling work. For instance, 850 out of the ISI's staff of 1400 were directly
involved in sampling-related work.[58] Statistical training also seemed to be
more widespread than in China. Fifteen of the 33 national universities possessed Departments of Statistics or offered training programs, and a total of
2,000 students were engaged in the study of statistics. The ISI trained advanced
statisticians, focusing in equal measure on theory and practical application.
Error checking was carried out meticulously, with machines central to the
process. The ISI already had 250 machines with a dedicated staff of 400.[59]

Wang identified three major characteristics of sample surveys in India. The
first was that they were in widespread use, often national in scope. This was
significant because India stood far ahead of other capitalist countries in its
employment of sampling technology. National sample surveys had been carried out in industry, agriculture, animal husbandry, household industry, transportation, commerce, and services, as well as in urban workers' livelihood and
rural costs of living and income. A second characteristic was that sample surveys were carried out continuously, using the method of interpenetrating
samples. Accordingly, it was possible to check and cross-check results and
easily discover errors. In addition, all methods of examination and verification,

---

56 Wang also spoke of errors of two types: i) an error that is peculiar to sample surveys, and
ii) an error that is common to all kinds of surveys, including complete enumeration. In India,
the latter were relatively more common, since personnel who did a complete enumeration were
generally less able and less well-trained. See Wang Sihua, *Wang Sihua tongji lunwenji*, 99.

57 Ibid., 99–100.

58 Ibid., 96.

59 Ibid., 98.

including purposive ones (典型研究; *dianxing yanjiu*), were used to raise the quality of the data. Finally, sampling work was closely tied to policy making in India, and thus it received attention and support from the government. To illustrate his point, Wang explained that when Bengal was struck by a famine in 1943, the ISI had been able to investigate the effects of the famine in a detailed way.[60]

Having established the usefulness of sampling, Wang offered his recommendation: "[to] energetically launch national-scale sample surveys in our country, and extensively collect [data] on every aspect of social and economic activity."[61] Wang bemoaned the recent status of statistical work in the PRC. On the one hand, there were too many statistical forms to the point of overflowing, while on the other, the statistical materials so produced were unable to completely satisfy the needs of the party leadership. According to Wang, such an irrational situation had to be swiftly remedied. The only way to stop or limit the excess issuance of unscientific or fiscally wasteful reports by various departments was to speedily adopt the use of scientific sample surveys. Wang conceded that the complete enumeration periodic reporting system remained useful for tracking production and circulation. But when it came to economic activities on a national scale, especially those activities dealing with the complex issues involving allocations and expenditures, pursuit of complete enumeration at the ground level only resulted in the data and indicators being flooded in needless trivial details (十分繁琐; *shifen fansuo*). The resultant workload was too onerous, and it was inconsistent with the principle of spending less material and human resources for greater results, that is, it was inefficient. In conclusion, Wang declared that only by organizing specialized scientific sample surveys could the Chinese achieve twice the results with half the effort (事半功倍; *shiban gongbei*).[62]

"And Next Year in Peking"

At about the time that statisticians, planners, and leaders were reading Wang Sihua's exhortation to adopt sample surveys, deliberations were initiated at the SSB to invite Mahalanobis to the PRC for a lecture and discussion tour.[63] The

60 Ibid., 100.

61 Ibid., 100–101.

62 Ibid., 101–103.

63 Unless otherwise specified, the primary-source material for this section is WJB 105-00530-06. This archival file consists of a series of fourteen bulletins issued by the office in

following month, the Chinese Foreign Ministry informed its embassy in Delhi that the SSB, Renmin University, and the Economics Research Institute of the Chinese Academy of Sciences would jointly invite Mahalanobis to the People's Republic.[64] Sometime toward the end of May the three designated host institutions sent a letter to the State Council identifying two principal topics for the visiting experts to lecture on: problems relating to statistical sample surveys, and the achievements of mathematical statistics and its applications in practical work.[65] A memorandum from earlier in May, meant to be forwarded to Mahalanobis, provides further details about the questions that the SSB wanted to raise:

1. The concrete procedures of the sample survey concerning crop yield, such as the criteria of selection of survey units and the adequate number of such units.

2. How can one guarantee that the farm budget statistics as a result of such sampling method will, on the one hand, represent the [sic] national or provincial characteristics, and on the other hand, satisfy the needs of the administrations on [sic] the county level? What are the ways and means of controlling and investigating the representative character of the sampling of such a coordinated and complex nature?

3. In order to study the economic activities such as the distribution of income of the agricultural producers' cooperatives, sample survey is necessary. But how are the proper measures of sampling and the adequate number of sample units determined? How can such sampling be representative of the whole country and the various provinces and at the same time satisfy the requirements of the counties?[66] [Original in English.]

The memorandum further observed that sampling bodies of 16,000 farm households and families of 6,000 industrial workers had been selected, but it wanted to know "to what other fields should we extend such sample surveys

---

charge of handling Mahalanobis's affairs (接待办公室; *jiedai bangongshi*) and it covers his activities during the entire duration of his stay in China.

64 WJB 105-00530-05: 3.

65 WJB 105-00530-05: 16–18. The letter was received 31 May 1957.

66 WJB 105-00530-05: 11–12.

in our country?" Another letter, dispatched on 28 May 1957 to the Chinese Foreign Ministry by Wang Sihua, articulated additional topics: Mathematical statistics, its concrete uses and its development trends; India's experience with planning nationwide sample surveys; and [Statistical] Quality control of industrial products in India.[67]

In return, the SSB acknowledged that Mahalanobis hoped to resolve three requests of his own while in China. The first was to organize visiting students who would travel to India for study; the second was to have specialists sent to the ISI to help with translating materials from the Chinese; and the third was that he was interested in understanding the training of health-services personnel in China. The letter went on to acknowledge that Mahalanobis was not only India's premier statistician and the moving intellectual force behind its Second Five-year Plan but also that he enjoyed much international fame. In receiving him, an attitude of sincere and genuine friendship was to be adopted, thereby advancing Sino-Indian friendship generally and closer ties in statistical science in particular. Ideological differences were to be set aside. The guiding principle was to be to "seek common ground while accepting differences" (求同存异; *qiutong cunyi*). The Chinese would learn with an open mind (虚心; *xuxin*), and both achievements and shortcomings would be presented honestly. A list of preparatory materials, including published statistics, was drawn up so that they could be presented to the visitors for context and background.[68] Wang Sihua was designated the leader of the team responsible to receive the visitors.[69]

Mahalanobis entered the People's Republic via Shenzhen on the morning of 19 June 1957. By three in the afternoon of the same day his party, including his wife and the specialist in sampling methodology, D. B. Lahiri, arrived in the southern metropolis of Guangzhou (formerly Canton). The following morning they visited Sanyuanli Agricultural Production Cooperative within the city and "saw the system of maintaining the village statistical records, and got acquainted with the method of 'purposive sampling' (on the Soviet model)

---

67 WJB 105-00530-05: 13–14.

68 These included materials on national income, statistical data on the achievements of socialist construction, farm family income, nationwide family income of workers, and other materials that Mahalanobis might request, assuming they were deemed safe to share.

69 Other members included: Huang Jiantuo (head of the Agriculture Statistics Department), Di Chaobai (head of the Economics Research Institute of the Chinese Academy of Sciences), Qi Luming (director of Renmin University's Statistical Planning Department), Dai Shiguang (Renmin University professor), and six others.

which was being extensively used for estimating agricultural production and for family budget enquiries."[70] Later that day, Chen Yingzhong (1904–??), director of the Guangdong Statistics Bureau, and Guo Ling (??–??), director of the Guangzhou Statistics Bureau, gave Mahalanobis and Lahiri a short presentation on the nature of provincial and municipal sampling work. They described the methods used to estimate agricultural production, rural and urban incomes, and livelihood measures. Mahalanobis was immediately suspicious when he was told that the error in the sample surveys was barely 2 percent. He remarked that for such [purposive] sample surveys the error ought to be much higher, adding that in the Soviet Union it was on the order of 15 percent, in the United States 12 percent, and in countries such as Germany and Japan it was above 10 percent. The only way to carry out reliable surveys was to use random sampling, he exclaimed.

Over dinner later that evening Mahalanobis urged his hosts not to be too dogmatic in their approach to statistics. He told them that "even though the Soviet Union was the first nation to establish socialism, many of its statistical methods are too old and not easy to change anymore."[71] Based on his visits to the Soviet Union and his numerous interactions with its leading statisticians and economists, Mahalanobis was well aware of the status of Soviet statistical research and work. Three years earlier, he had written to Pitambar Pant noting that "one orthodox viewpoint in USSR has been to exclude mathematics from economics; to insist on statistics to be almost exclusively the handmaiden of economics; and to separate mathematical statistics as a separate (and somewhat abstract) subject under probability and mathematics."[72] In his unpublished report on the China tour he would add that in spite of much discussion, "statistical practice in USSR, however, still continues to be based on 'purposive sampling.' "[73] The situation he encountered in China was similar, but his presence there suggests that the Chinese were aware of the limitations of their methodology. Discussions over the following three weeks would indeed indicate to him that the Chi-

70 P.C. Mahalanobis, "Some Impressions," 1.

71 FMA 105-00530-06: 3.

72 Mahalanobis to Pitambar Pant, 3 July 1954, Papers of Pitambar Pant (190 LII), Nehru Memorial Museum and Library, New Delhi, 49.

73 Mahalanobis, "Some Impressions," 5. Later in the trip he would inform his hosts that since 1947 the Soviet Union had not attended any sample surveys activities organized by the International Statistical Institute. FMA 105-00530-06: 15.

nese were "keeping an open mind" to the possibilities offered by mathematical statistics.[74]

Xue Muqiao's Briefing

On the following day, 21 June, Mahalanobis and company flew to Beijing, arriving in the capital at three in the afternoon. The welcoming committee at the airport included statistical luminaries of the PRC such as Xue Muqiao, director of the SSB, as well as the familiar faces of Wang Sihua and Dai Shiguang. Wang and Dai personally accompanied the visitors to the Beijing Hotel, where they were to stay. Over the course of the next day, Mahalanobis and Lahiri met with representatives of the three host institutions to decide upon the goals and tasks for the coming weeks. In addition to offering his knowledge and experience, Mahalanobis was keen to learn about all aspects of socialist construction in the PRC. Naturally, he had an interest in statistics and the statistical system, but his long list of questions and topics touched upon all aspects of planning, training of personnel, and national income. He was also keen to see a copy of the PRC's "Twelve-year Plan for Scientific Research," which had been announced the previous year.[75] When it came to planning his lectures, Mahalanobis wanted to know the general level of the audience so that he could better calibrate the technical parts of his explanations. A formal welcoming party was held for the visitors on the evening of 23 June, with an even bigger and more impressive list of Chinese statistical luminaries in attendance. In his welcoming speech at the party, Xue Muqiao stressed the advances India had made in developing and applying sample surveys. It was precisely this experience that the People's Republic wanted to mine, and therefore it had invited Mahalanobis, or Prof. Ma as he came to be called, to visit. Usual pleasantries from both sides followed. At the end of the evening, D. B. Lahiri turned to his interpreter and expressed the hope that future dinners would be smaller in scale so there could be more meaningful

74 Mahalanobis, "Some Impressions," 6.

75 This last request seems to have been denied (WJB 105-00530-06: 77). In general, a decision had been taken to freely share those materials and data that were already in the public domain. Requests for data that were not would be handled on a case-by-case basis. Statistical data were still considered secret in the People's Republic and, as a result, many of Mahalanobis's requests were eventually denied. On the twelve-year plan, see Wang, "The Chinese Developmental State During the Cold War."

conversations. The strongly political flavor (政治味道; *zhengzhi weidao*) of the party had made him uncomfortable.[76]

The delegation's work began in earnest on 24 June, with a briefing by Xue Muqiao on statistical work in the People's Republic.[77] Xue explained that over the course of the preceding five years an integrated statistical system modeled after that in the Soviet Union had been established, reaching down to the level of the village cooperatives throughout the country. With a staff of about 650 personnel, SSB headquarters was at the apex of the system and was responsible for the collection and auditing of all statistical data and their subsequent delivery, as required, to various government agencies. Below, at the provincial level, statistical offices coordinated and guided work across the roughly 2,200 counties in the country. Under the counties, village administrations coordinated the statistical work of about 750,000 village cooperatives. At this level, much of the work was carried out by part-time workers. In line with the Soviet model, the scope of statistics was "comprehensive," covering population, agriculture, industry, transport, trade, consumption, health, education, finance, national income, employment, and so on. Unlike in the Soviet Union, however, the provincial statistical bureaus were under the administrative control of the provincial people's committees and their expenses were paid out of the provincial budgets. Xue explained that this made the Chinese system more administratively decentralized but he still felt that it was "probably as integrated as the Soviet system."[78]

Xue then proceeded to explain the methodology employed by the SSB. The preferred choice was complete enumeration. This worked rather well in several sectors. In the industrial sector, 3,000 units sent monthly returns, and another 60,000 units submitted annual reports. In construction, about a 1,000 of the most important projects reported monthly, with the remainder reporting on an annual basis. Reports from railways, shipping, and trucking were submitted monthly, as were data on government purchases and the consumption of important commodities. The agricultural sector, however, posed a much larger problem. Data of interest included area sown, yield per unit area, "outturn" (that is, output) of important agricultural crops, livestock, and so on. The provincial bureaus had their hands full trying to completely enumerate the 750,000 village cooperatives. Delays were the norm, and the accuracy of the

76 WJB 105-00530-06: 10.

77 Much of the day on Sunday, 23 June, was lost as Mahalanobis came down with a stomach ailment. Lahiri took the opportunity to visit the Great Wall at Badaling and the Ming Tombs.

78 Mahalanobis, "Some Impressions," 1–3.

results was not always high. As a result, in 1955 an initial attempt was made to calculate the yield of crops per unit area over a large part of the country using sample survey techniques. In addition, sample surveys had been conducted to better assess the standard of living of rural, working class, and urban households. For instance, a rural survey in 1955 covered 15,000 farm families in 23 provinces and an urban survey in 1956 included 6,000 families. Xue clarified that the technology used was purposive sampling, and he concluded by observing that through increased contacts with India, and in particular following the visit to India of Wang Sihua's delegation, the Chinese had become particularly interested in learning from India's expertise and experience in carrying out sample surveys.[79] At the end of the briefing, Lahiri, apparently frustrated by the PRC's refusal hitherto to adopt more advanced sample survey techniques, pointedly told his interpreter that he could not fathom why the Chinese lacked confidence in sample surveys.[80] Mahalanobis would later claim that after this "brilliant survey," both he and Lahiri had a much clearer sense of why the Chinese "were so deeply interested in sample surveys and mathematical statistics."[81]

"Quite a Heavy Programme"

With their brief firmly established, Mahalanobis and Lahiri set about preparing their lectures. Over the course of the following two weeks, Lahiri and Mahalanobis led eight group discussions on sample surveys and delivered a total of nine lectures. The group discussions usually lasted from three to four hours and were frequently quite technical in nature. Of the nine lectures, Mahalanobis delivered five and Lahiri four. In the lead-up to each lecture, each of them had to prepare a synopsis, which was then translated and precirculated, leading

79 Much of the description here is from Mahalanobis, "Some Impressions," 3–4.

80 WJB 105-00530-06: 16. Mahalanobis also described and dismissed such sampling, observing: "An attempt is made to select sample-units (agricultural fields, households, etc.) which are typical or have characteristics close to the average. . . . It is now recognized that the method of 'purposive sampling' is not scientifically correct. The only sound method is 'random or probability sampling' which can be relied upon to supply truly representative results; and which alone makes it possible to calculate valid estimates of the sampling error." Mahalanobis, "Some Impressions," 5.

81 "We were deeply impressed by the seriousness of the purpose with which the Chinese discussed these questions and by the amount of details into which they entered to clear up obscure points and to formulate a concrete program of work." Mahalanobis, "Some Impressions," 1, 11.

Mahalanobis to remark: "This was quite a heavy programme." The lectures were usually day-long affairs, divided into two three-and-a-half hour sessions. Mahalanobis delivered two lectures on "Sample Surveys" at the SSB, one on "Planning in India" at the Planning Commission, and two on "Recent Developments in the Theory and Applications of Mathematical Statistics" at Peking and Renmin universities. During most of these lectures, the audience—comprised as many as 300 people—included statisticians, economists, senior students of mathematics, and members of the mathematics staff at the two universities.[82] Lahiri's four lectures were similar in format, though he spoke exclusively on "Applications of the Theory of Mathematical Probability in Sample Surveys." Two of his lectures were delivered at the SSB, one at Renmin University, and one at the Chinese Academy of Sciences. His lectures were especially technical, and the audience was restricted to thirty to thirty-five persons who were familiar with statistical work and knowledgeable about mathematics.[83]

On 24 June, after Xue completed his briefing, Mahalanobis offered a three-stage strategy for the discussions. Each successive stage would begin only after he received the approval of his audience. If it became necessary, he even offered to extend his stay in Beijing and cancel the sightseeing tours to allow for additional time for discussions and lectures. During the first stage, he wanted to introduce the general principles of sampling. If everyone was satisfied with this discussion, he would then proceed to the next stage and more thoroughly discuss these principles. The final stage would be more practical: Starting with a particular province, they would attempt to draft a detailed and concrete sampling program.[84] Having clarified this plan, Mahalanobis began the first stage. He spoke about what a sample was, about errors, the survey object, the differences between random sampling and purposive sampling, and outlined their histories. In discussing the overall structure of statistical work, Mahalanobis stressed the importance of scientific and intellectual freedom, and the need to engage with other disciplines since statistics was essentially an applied science. He also spoke of the importance of training. By this he meant training at all

82 Ni Jiaxun, then an undergraduate at Renmin University, recalled attending one of these lectures. According to him, most of the audience could not follow much of what was being said. However, he remembered it as an important event at the university.

83 Mahalanobis, "Some Impressions," 6–7. On Lahiri's lectures, the bulletin from 5 July stated that there were over twenty people present and that the "discussion was a little too specialized, a little too technical in nature." See WJB 105-00530-06: 62–63.

84 WJB 105-00530-06: 18.

levels—not only that of middle-school students who could become statistical workers. Just as important was the training of statistical researchers and instructors. To make his point clearer, he drew comparisons with the operations in a hospital. "The medical school must maintain an intimate relationship with the hospital. Similarly, statistical research organs must remain in touch with the organs carrying out actual statistical work."[85] Based on experience in India, he estimated that if the Chinese were to conduct a national sample survey with the sole purpose of satisfying the center's requirements, all they would need was 1,200 surveyors. If, at the same time, they also wanted to satisfy the needs of the various provinces, 2,500 surveyors would be required.[86]

In the morning of 27 June, Mahalanobis delivered his first lecture at the SSB on sample surveys and followed that up with a second lecture in the following afternoon. On each day, the audience numbered about 300. At the end of the second day, Xue reported that even though the lecture was at times too specialized, it was still very helpful for the audience.[87] Mahalanobis also handed over a much more detailed list of questions regarding planning and statistical work in the PRC. This list contained a total of seventy-three items, covering topics such as the organization of long-term planning, the structure of planning, the technology used and its improvement, population and labor, gross domestic product and national income, production of goods, natural resources, trade, capital-goods industries, the training of scientific personnel, and so on. The final item was potential Sino-Indian cooperation on planning and statistical matters.[88] The next day, which happened to be Mahalanobis's birthday, both he and Lahiri continued the group discussion on sample surveys at the SSB.[89] After a day of rest, preparation, and sightseeing, the discussions continued on the afternoon of 1 July at the SPC. On 2 July Mahalanobis delivered a lecture on the Indian planning experience. This lecture was attended by over 250 people from the SPC, the State Economic Commission, Renmin University, the Chinese Academy of Sciences, and the SSB. Mahalanobis also delivered lectures to similarly large audiences at Peking and Renmin universities, on 5 and 6 July, on "the development of mathematical statistics and its applications in the social sciences."[90] Over the next several days, Mahalanobis and Lahiri

85 WJB 105-00530-06: 22.
86 WJB 105-00530-06: 20.
87 WJB 105-00530-06: 31–35.
88 WJB 105-00530-06: 38–48.
89 WJB 105-00530-06: 36–37.
90 WJB 105-00530-06: 64–67.

then held independent group discussions and visited the three host institutions as well as the Bureau of Public Health.

The principal content of all the lectures and group discussions—the lecture synopses, lecture transcripts, subsequent questions and answers, as well as the minutes from the group discussions—was compiled, translated, edited, and published one year later in May 1958 by Tongji Press.[91] The foreword to this publication granted that there were specific points on which the Indians and the Chinese did not see eye to eye, but in spite of those differences the Chinese needed to pay greater attention to the science and technique of sample surveys. Therefore, the book was produced as a reference for future research and discussions on mathematical statistics and sampling. Its distribution, however, was most likely limited to statistical bureaus and statistical offices throughout the country.[92] Whether it was used as teaching material or was used in any other specific way remains unclear.

Meeting Vice Premiers and the Premier

Mahalanobis had the opportunity to discuss statistics with the PRC's political leadership on two separate occasions. On 3 July Mahalanobis and Lahiri met with vice premiers Li Fuchun and Bo Yibo (1908–2007).[93] Li and Bo were also chairmen respectively of the SPC and the State Economic Commission, and Li was also the minister of Statistics. At the meeting, Mahalanobis reiterated the advantages of random sampling—it was faster, cheaper, and more accurate, and its errors were easier to ascertain. With a national sample survey in place, he claimed that one would need only four to eight weeks to get a sense of socioeconomic conditions in the country. On specific techniques, he recommended the use of interpenetrating samples and probability sampling. Given China's 2,200 counties, Mahalanobis estimated that the sampling bureau should begin with a staff of about 2,000. Work could initially begin in five or six provinces, and as it intensified each county would be assigned an average of two surveyors, resulting in a total staff of between four and five thousand. Transcribers, human computers (that is, workers with adding machines), and statisticians would, of course, also be required. Mahalanobis urged that a training institute along the lines of the ISI be set up so that both statistical workers

91 SSB, *Yindu Mahalanuobisi jiaoshou fangHua tongji baogaoji.*

92 Underneath the title were the characters "内部发行" (*neibu faxing*, restricted publication).

93 WJB 105-00530-06: 53–60.

and a highly specialized cadre of statisticians would be trained. This institute would also foster closer ties with the Chinese Academy of Sciences and, more generally, between statistics and the disciplines of mathematics and economics—"this was the way of the future."[94]

Li Fuchun and Bo Yibo were very receptive to Mahalanobis's suggestions. They requested that he bring them up during his conversation with Premier Zhou later in the week. Both Li and Bo were particularly keen to see sample survey work begin as soon as possible. Yet lingering doubts about sampling technology remained. Li enquired specifically whether "on the matter of the accuracy of sample surveys, Prof. Ma could once more carry out discussions with the SSB."[95] Bo wondered if India could send a specialist or two to help with a sample survey of cotton production in Hebei in September or October. Mahalanobis replied that a September target date was too soon and would not allow enough time for planning the survey. To be better prepared, he suggested carrying out pilot surveys on the paddy crop. He also recommended that the Chinese immediately send one or two people to the ISI so that planning work could begin at once.

Mahalanobis also took the opportunity to reiterate his specific proposals for enhanced Sino-Indian cooperation in statistics and planning. He requested that the SPC and the SSB arrange for the regular forwarding of technical data to the Indians. He suggested that a small "China Unit" be set up at the ISI. This group would help translate into English important Chinese publications, articles, and technical data, teach Chinese to a few Indian statisticians and professors, and maintain direct links with Beijing. He requested that China send one or two people to oversee this unit, and he gave assurances that all their costs would be borne by the Indian side and that they would be treated as guests. In return, he offered to share materials for the teaching of statistics, and also to impart training to Chinese who visited the ISI. The hope was that three to four people would be sent to India for further study and that this would later become an annual practice. Indian specialists working on sample surveys, mathematical statistics, and other areas, could also be sent to China to help further research, training, or drafting of technical schemes.[96] Li Fuchun and Bo Yibo agreed with this advice and indicated that the details could be worked out with Xue Muqiao and Wang Sihua, both of whom were also present at the meeting.

94 WJB 105-00530-06: 56.
95 WJB 105-00530-06: 57.
96 WJB 105-00530-06: 47–48.

About one week later, on 9 July, Mahalanobis, his wife, and Lahiri were invited to dinner with Zhou Enlai.[97] Xue Muqiao and Wang Sihua were also present, as were the Indian ambassador, R. K. Nehru, and Mrs. R. K. Nehru. Mahalanobis repeated much of what he had told Li Fuchun and Bo Yibo. He was pleasantly surprised to discover that Zhou Enlai had been kept informed of his activities and had "studied some of the rather technical points in my recommendations as he [Zhou] asked some searching and critical questions on the methods of assessing the accuracy of the results of sample surveys."[98] Zhou agreed in principle with most of Mahalanobis's suggestions. The setting up of an institute might take some time, he observed, since the PRC still lacked enough trained personnel. But he welcomed Indian help in the organization of statistical work and research, and he expressed interest in continuing the exchange of personnel, expertise, and experience. He also told Mahalanobis that although it might be difficult to immediately send experts to lecture on Chinese statistics and planning, the two SSB personnel who would be sent later in the year could help begin the work of the Chinese unit. The dinner was a pleasant affair and was subsequently reported in the *People's Daily*.[99] On 12 July Mahalanobis and company left China the same way they had entered, via Shenzhen, assured that exchanges and future cooperation were on a solid footing.

Domestic Difficulties

A couple of months after Mahalanobis's return to India, Wang Sihua addressed the Conference on National Statistical Work. His speech, entitled "Report on an Inspection of Statistical Work in India," started by referring his audience to his article from earlier in the year.[100] Wang stated that in that piece he had covered the general outlines of statistical work in India and he would not revisit already trodden terrain. Instead, he wanted to focus exclusively on sample surveys. His report was less enthusiastic, more guarded, and more strategic than his February article.

Wang took pains to stress that Mahalanobis was aware not only of Soviet statistics but he also knew that Chinese statistics were largely a product of learning from the Soviet experience. Furthermore, Mahalanobis, he assured

97 WJB 105-00530-06: 68–71.

98 Mahalanobis, "Some Impressions," 10.

99 Anon., "Zhou zongli jiejian Yin tongji xuejia Mahalanuobisi," *RMRB*, 10 July 1957, 1.

100 Wang Sihua, "Kaocha Yindu tongji gongzuo de baogao."

his listeners, actually thought very highly of the Chinese periodic reporting system. It was only in those instances where original records were missing, or where the numbers were too high or too many, such as in matters related to the people's livelihood, that the use of sample surveys was recommended. As a socialist country, Wang acknowledged, the PRC should of course use a system of complete enumeration as its main methodology, but it was also important to study other methods, something that previously had not been done sufficiently. He then addressed the doubts, fears, and criticisms head-on:

> There are those who are suspicious: Does our visit to India mean that we have changed our policy of learning from the Soviet Union? Such an understanding is in error. Our India visit and Prof. Ma's [Mahalanobis's] coming to the People's Republic absolutely do not mean that we have abandoned [the policy of] learning from the Soviet Union. In reality, sample surveys have already been in use in the Soviet Union from an early time. It can be confidently stated: in the past, today, and in the future, we want to learn from the Soviet Union. But, at the same time, we must also study those methods of bourgeois countries that are relatively good, especially the methods of sample surveys and the problems related to statistical technology. As far as methods are concerned, if other capitalist countries, and not only India, have even better methods, we can study those as well. Our goal is to provide better services in the construction of a socialist society.[101]

Wang then went on to outline the nature of sampling work in the Soviet Union, providing evidence of its use for decades. He reminded everyone that Soviet specialists had designed the rural livelihood survey (农家收支调查; *nongjia shouzhi diaocha*). It was only thereafter that he began a detailed discussion of India's sampling work, concluding with a list of four advantages. The first was that the scale and nature of sample surveying in India were larger and more expansive than that in England, the United States, or the other capitalist countries. The second was that Indians had made new contributions to the technology of sample surveys, namely the methods of continuous sampling and interpenetrating samples. Third, the Indians had carried out detailed comparative research and experimentation using various survey methods. Finally, the contents of the surveys were tailored to meet practical requirements.[102]

101 Ibid., 91.
102 Ibid., 94.

What was known to Wang Sihua and many of his comrades in Beijing had to be communicated more effectively, and subtly, to the various regional representatives at the conference, and through them to the vast numbers of statistical workers throughout the country. This was that sample surveys were a general category that included purposive sampling as well as random sampling. What was new and particularly relevant to the PRC were the Indian random sampling methods, not the prevailing Soviet methods. Wang adopted a two-stage strategy to bring home this point without actually stating it. He first differentiated sample surveys from complete enumeration, highlighting the already familiar advantages: Sample surveys were cheaper, required fewer personnel, and since their scale was smaller than that of complete enumeration, they could accommodate many more indices and much greater complexity.[103] Having established the usefulness of sample surveys, Wang proceeded to differentiate what he called subjective (主观; *zhuguan*) sampling from random or stochastic (随机; *suiji*) sampling. For this purpose, he resorted to a laundry list of advantages: (1) If the survey design is good, it can completely reflect every characteristic of the totality; (2) random sampling avoids the subjective influence of the surveyor and therefore is more accurate; (3) subjective sampling only allows for one or two indices, far fewer than random sampling; (4) random sampling is more objective; (5) random sampling is a simple and easy method; and (6) random sampling can provide accurate estimates of sampling errors and allows for an estimation of the overall accuracy of the survey.[104] But Wang did not advocate the abandonment of other forms of sampling. As long as a method is scientific, it should be used. What then was the definition of scientific? For Wang, "[a method] is scientific only if it can truly reflect the objective real situation."[105] Thus, subjective sampling, under the right conditions of the party's mass-work line (群众工作路线; *qunzhong gongzuo luxian*), could also be scientific.[106] He further comforted his audience by invoking the authority of Mahalanobis:

103 Ibid., 95.

104 Ibid., 95–96.

105 Ibid., 96.

106 The theory of the mass line is attributed to Mao Zedong and traditionally dated to the adoption of a resolution on Leadership that he authored on 1 June 1943. Its core tenets are encapsulated in the phrase "from the masses, to the masses." Mao went on elaborate ("Some Questions," 119): "This means: take the ideas of the masses (scattered and unsystematic ideas) and concentrate them (through study turn them into concentrated and systematic ideas), then go to the masses and propagate and explain these ideas until the masses embrace them as their

In the past, some comrades believed that the difference between a purposive survey and a sample survey rested on whether or not it was random.
But, on the basis of Prof. Ma's introduction [to sampling], even within sampling subjectivity may exist. Setting aside random sampling, to varying
degrees both mechanical sampling and sampling by classification have subjective elements. Therefore, I believe the principal difference between
sample surveys and purposive surveys is that sample surveys use a small
number of objects [对象; *duixiang*] to calculate the entire object. That is,
the entire sample must represent the whole and the totality [全及总体;
quan ji zongti]. [In contrast,] the principal goal of purposive surveys is not
to analogize for the totality.[107]

Wang pointed out that the method of purposive (or typical) surveys had not
been effective in the past because the selection of typical cases (选典; *xuan-
dian*) had not reflected the objective reality. Therefore, the task now was to
carry out small-scale experiments to determine which method would be most
suitable.

The transformation between February and September 1957 was stark. From
the vigorous promotion of establishing nationwide random sample surveys,
the mood shifted to one of qualified, even muted, endorsement. After several
years of denigrating "bourgeois" statistics, in February Wang Sihua had un-
characteristically written favorably about them. This speech, seven months
later, suggests he did not strike enough of a conciliatory tone in his earlier
article and, therefore, needed to make amends. Indeed, Wang's speech at the
conference can be read as an exercise in persuasion meant to assuage reactionary cadres who feared not only ideological compromises but, perhaps even
more importantly, being shifted out of their comfort zones of established statistical practices and habits. Larger domestic political currents no doubt also
played a part. Starting with the Hundred Flowers movement in 1956 and continuing through May 1957, during which time free expression by intellectuals
was encouraged, the political climate had been relatively conducive to the
exploration of new possibilities. The beginning of the Anti-Rightist movement
in June 1957 led to a narrowing of such possibilities and, no doubt, to the much

---

own, hold fast to them and translate them into action, and test the correctness of these ideas in
such action. Then once again concentrate ideas from the masses and once again go to the masses
so that the ideas are persevered in and carried through. And so on, over and over again in an
endless spiral, with the ideas becoming more correct, more vital and richer each time."

107 Wang Sihua, "Kaocha Yindu tongji gongzuo de baogao," 96.

more cautious tone that Wang adopted in September. Several months later, the sent-down movement began as cadres were sent to factories and villages to engage in manual labor. The movement was later extended to intellectuals. Under the circumstances, advocating capitalist methods could no longer be only a matter of their efficacy.

The Study Tour of 1958

It was in this somewhat confused milieu, when intellectuals and city-dwellers faced the prospect of manual labor in the countryside, that Wu Hui and Gong Jianyao found themselves on a mission to India.[108] Born in 1927, Wu and Gong had in 1950 received degrees in statistics from Chongqing and Nanjing universities respectively. They were part of an entire generation that had been introduced to and trained under an inclusive understanding of statistics, which drew upon texts and materials from the United States, Britain, and the Soviet Union. After 1949, they had to recalibrate their training to serve the PRC's newer and narrower definition of statistics. By the mid-1950s both were working at SSB headquarters in Beijing. Gong was assigned to the Agricultural Statistics Division and Wu to the Industrial Statistics Division. In 1957 they were among ten young cadres shortlisted for additional testing to determine who would be sent to India. Each shortlisted candidate was asked to write an essay demonstrating their English-language skills. In the end, their facility in English coupled with their specialization in key areas—industrial and agricultural statistics—won out.

Wu Hui and Gong Jianyao arrived in Calcutta on 4 January 1958. They were immediately invited by Mahalanobis to travel to Madras (present-day Chennai) to attend a conference on statistical work and planning. Upon their return to Calcutta, they began their studies at the ISI. At that time, the ISI had several tens of students from various Asian countries who were receiving training at the ISEC. As visiting scholars, Wu and Gong's official status was different from that of these students, and they were not required to attend classes. Instead, they were given study space and permitted to conduct their own independent reading and research. The primary focus of their work was the study of Indian sampling techniques. Thus, they began by reading all previous documents and

108 Unless otherwise noted, this section draws primarily from oral history interviews conducted with Gong Jianyao and Wu Hui. I met Wu Hui in Beijing on 8 September 2011. I met Gong Jianyao one week later, on 14 September 2011, at his home in Guangzhou.

publications held in the ISI library related to the ISI's statistical sampling activities. Thereafter, they moved to a close study of the theory and methods of sampling. In the process, they studied simple random sampling, stratified random sampling, cluster sampling, double sampling, and multistage sampling, among other sampling methods. Periodically, they would meet with Mahalanobis, Lahiri, or C. R. Rao to discuss their progress and problems.[109] Lahiri, in particular, was their principal guide and interlocutor throughout their studies. Exposure to the latest Western texts on statistics and sampling rounded off their readings.

Although the bulk of their work at the ISI was theoretical or focused on book-learning, Wu Hui and Gong Jianyao also had an opportunity to gain firsthand experience in the practice of sampling by participating in two sample surveys. The first was a land-use sample survey where they witnessed the Mahalanobis-devised method of cutting circular crop samples in fields. The second was a family budget survey carried out in Calcutta. Other activities included sitting in on meetings of the National Sample Survey (NSS), where sampling teams would report their recent experiences, problems, and discuss solutions for the next round of the NSS. Wu and Gong also had the opportunity to interact with the various Indian and foreign statisticians present at the ISI during 1958. Both remembered, in particular, the visit of the Soviet statistician A. Ezhov, who, from 1950 to 1952, was the first Soviet statistical expert in the PRC. Gong also became close friends with C. R. Rao, who at that time ran the ISI's training school.[110] During the year, Gong and Wu visited Bombay, Pune, and Delhi, as well as several paddy and jute sample survey sites in the country.[111] While in Delhi, they also met with Pitambar Pant, who was still serving on the Indian Planning Commission.

During the one year and five days that Gong and Wu spent in India, events within the PRC would push statistics, as just about everything else, in new

109 Both Wu and Gong recalled Mahalanobis fondly, observing that he left a rather deep impression on them. In a 2001 article, "Shidao qinghuai," published in *Zhongguo tongji* (China Statistics), Gong went so far as to claim that two professors with the surname Ma had played a hugely inspirational and instrumental role in his life. The first, he told his readers, was Professor Ma Yinchu (1882–1982), the well-known economist and one-time president of Peking University, who had been persecuted during the late 1950s for his championing of birth control. The second Professor Ma was none other than Mahalanobis.

110 Based on Gong Jianyao's memory; I have not been able to independently confirm whether Rao was indeed running the ISI training school at this time.

111 Gong Jianyao, "Woguo chouyangfa yanjiu yu shijian de fazhan," 10.

directions. Gong recalled reading about the Great Leap Forward (GLF) in the papers and receiving letters from friends and family describing the push toward collectivization. Wu also remembered discussing the agricultural growth results during the GLF, claiming that Gong, the agricultural statistician, was openly suspicious of them. Statistics, they were informed, was changing. As Gong summarized in an essay written decades later:

> [With] the collectivization of peasant production and consumption, all [statistical] data could basically be obtained [directly] from the communes. The peasant family revenue and expenditure survey, which had been set up on the basis of surveys of sample households, had fulfilled its historical obligation/role [历史使命; lishi shiming].[112]

In one fell swoop, established statistical systems as well as future potential surveys based on random sampling were sidelined. Toward the end of their stint in India, Wu and Gong worked hard to draft a lengthy report containing a summary of their activities and their recommendations for future statistical work in China. Upon their return in early 1959, it "too had to be rewritten and was converted into a simple report and submitted to the SSB leadership to read and dismiss (参阅了事; canyue liaoshi)."[113] We shall turn to these dramatic changes in the following chapter.

Things Fall Apart

The Sino-Indian exchanges and the much-touted statistical cooperation were to become a victim of the GLF. Worsening diplomatic relations between the two countries also played a part. Disagreements over the Sino-Indian border, which had existed since the start of the decade, began to dominate diplomatic exchanges as each side hardened its position. The ensuing frosty relations no doubt served to reduce some of the post-Bandung enthusiasm and to endanger the spirit of peaceful coexistence espoused in the 1954 *Panchsheel* Treaty.[114]

---

112 Ibid., 10–11. During my interview with him, Gong recalled the negative effect the 1958 Baoding meeting had on statistical work during the GLF.

113 Gong Jianyao, "Woguo chouyangfa yanjiu yu shijian de fazhan," 11. Both Gong Jianyao and Wu Hui would play major roles in statistical work in subsequent years and were principal actors in the post-1978 reform of statistics as a discipline and as a profession.

114 The "Agreement on Trade and Intercourse between the Tibet region of China and India" was signed on 29 April 1954 in Beijing by Indian Ambassador N. Raghavan and Chinese Deputy Foreign Minister Zhang Hanfu. Its five principles of peaceful co-existence included mutual

It is unclear to what extent Mahalanobis and the ISI had been aware of these differences or of the major upheavals within the PRC during 1958. In October 1958, he lobbied the Chinese to send several planners with expertise in heavy industry to help with the drafting of India's Third Five-year Plan.[115] He lamented that the halfhearted development of heavy industry in India during the period of the Second Five-year Plan had been a big mistake, and he wanted to take advantage of China's direct experience in this area. The Chinese Embassy in New Delhi duly forwarded Mahalanobis's request to the Chinese Foreign Ministry in Beijing and sought further instructions. The Chinese Foreign Ministry, in turn, asked the SPC for advice on the matter. But it also added that if Chinese experts were to be sent, then their primary goal would be to "expand our country's influence."[116] After additional deliberation involving the Foreign Ministry and the SPC, it was eventually agreed to send a three-man team. A telegram sent to the Chinese Embassy in New Delhi on 26 February 1959 listed the members of this team: Chen Xian, deputy director of the SPC's Long-term Comprehensive Planning Bureau; Wang Yilin, deputy director of the Comprehensive Department of the Capital Construction Planning Bureau; and the recently returned Wu Hui, who would serve as interpreter for the team. The telegram made clear that these experts would only serve as consultants to introduce recent experience in heavy industry. Under no circumstances would they advise the Indians on their problems. The Chinese Embassy in New Delhi responded several days later, on 6 March, stating that Mahalanobis welcomed the Chinese experts and hoped they would visit from April to June. Pitambar Pant and M. M. Mukherji, both of whom had been part of the 1956 Planning Commission delegation and were now involved in comprehensive planning, would receive them and arrange their activities.[117]

The archival trail turns cold after this. On 10 March an uprising occurred in Lhasa, and by the end of the month the Dalai Lama and several tens of thousands of his followers had escaped to India. They were granted asylum and allowed to settle in the northern hill station of Dharamshala. At a scientific conference held in Moscow in May, a little over one month after the Dalai

respect for each other's territorial integrity and sovereignty, mutual nonaggression, mutual noninterference in each other's internal affairs, equality and co-operation for mutual benefit, and peaceful co-existence.

115 WJB 105-00647-02: 1.
116 WJB 105-00647-02: 4.
117 WJB 105-00647-02: 5–9.

Lama's flight from Lhasa, Mahalanobis bumped into the Chinese physicist
Zhou Peiyuan (1902–1993), and the two engaged in a long conversation.[118]
Zhou later filed a report about their meeting with the Chinese Foreign Ministry.[119] Eight of the report's nine pages were devoted to a discussion of Tibet
and geopolitics. Only on the last page did Zhou dwell on a few scientific points
that had also come up. As for the three-member Chinese team, it never visited
India.

118 Zhou and Mahalanobis were both part of a delegation of the Association of International Scientists (世界科学工作者协会代表团; *shijie kexue gongzuozhe xiehui daibiaotuan*)
that had written to Khrushchev requesting a cessation of nuclear testing. See Anon., "Heluxiaofu
fuhan shijie kexue gongzuozhe xiehui daibiaotuan."

119 WJB 105-00656-02: 1–9.

8

A "Great Leap" in Statistics

THE YEAR 1958 marked a watershed in the history of the early PRC. The First Five-year Plan had been a success, meeting and in many cases exceeding the targets that had been set. And yet, instead of a Second Five-year Plan that built upon the first one, China was launched down the path of the Great Leap Forward (GLF).[1] At its most basic level, the GLF was a radically different approach to economic and social organization. It retained planlike characteristics of setting specific production and agricultural targets, but it did so by implementing a far-reaching reorganization of labor and production technologies. In the agricultural sector, farming went through another round of collectivization, and the people's communes thus created held several thousand households that ate in collective kitchens and labored in nationalized fields, at backyard blast furnaces, or on thousands of infrastructural projects such as dams and canals. Similar labor-intensive practices were introduced in the industrial sector as China sought to overtake England's steel production within fifteen years.[2]

Much of the impetus behind the GLF came from Mao's disavowal of the Soviet Union and her "advanced experience." At the Chengdu Conference in March 1958, "Mao applied the label of 'dogmatism' to previous Chinese policies towards heavy industry, planning, banking, and statistics, explaining that because China lacked native experience and competence in these areas, she

---

1 That a Second Five-year Plan was in the works by 1956 is evident from propaganda publications. See, for instance, Anon., *Cong di yige wunian dao di erge wunian*. The publisher, Hong Kong–based Full Wind Publishing Co., was quite possibly a CCP publishing front; the colophon of the book notes that the text's original publisher was the Central Printing House.

2 For succinct summaries of the GLF, see Riskin, *China's Political Economy*; and Meisner, *Mao's China and After*.

had rigidly copied Soviet practices regardless of their appropriateness to Chinese conditions."[3] Like almost everything else, statistics too did not escape the consequences of a rejection of Soviet practices.

Today, any mention of statistics and the GLF brings to mind what is arguably the worst man-made disaster—at least in absolute terms—in human history.[4] From 1959 to 1962 China endured a famine that brought the country to a standstill and caused the death of anywhere between 15 and 45 million people. The famine became an object of study in the early 1980s, when the PRC released demographic data for the preceding decades.[5] Scholars quickly noticed sharp fluctuations in both fertility and mortality rates during the GLF years, prompting closer analysis and initial attempts to arrive at an accurate number of unnatural deaths. During the ensuing years, a range of Chinese and non-Chinese scholars—economists, demographers, sociologists, and, most recently, historians—have progressively revised, not without controversy, the official Chinese estimate of 15 million unnatural deaths (Table 8.1). This is the immediate context within which statistics is inescapably intertwined with any discussion of the GLF.

But outside of the actual number of deaths, there has also been an enduring focus on understanding how such an event came to pass. Early attempts focused on organizational factors and elite politics. More recently, scholarship has expanded to include regionally specific detailed studies that emphasize the variability in the lived experience of the famine. Specific causes proposed over the past several years include the following: (1) highly exaggerated production figures during fall 1958 encouraged unrealistic plans; (2) procurement rates that, as a result, were too high; (3) a massive transfer of manpower from agriculture to industry created imbalances that were impossible to cope with, and

3 Riskin, *China's Political Economy*, 114. For more on Mao's critique of the Soviet Union and of Soviet experience, see Mao Zedong, *A Critique of Soviet Economics*, in particular James Peck's introduction, 24, that examines the evolution of Mao's views, which eventually came to consider the Soviet experience "utterly inadequate as a positive model for China's drive to build socialism."

4 There is a large body of ever-expanding scholarship on the famine that ensued in the wake of the GLF. Some of it is discussed below. For a trenchant review of some of the most recent works, see Garnaut, "Hard Facts and Half-Truths."

5 As Becker, *Hungry Ghosts*, 301, has explained, in the years "after the Great Leap Forward, China published few statistics and those figures that were made available merely consisted of percentages, none of which could be verified, let alone measured, against independent research." It was only after the 1980s that this became possible.

TABLE 8.1. Estimates of Unnatural Deaths during the GLF

Year	Scholar	Unnatural deaths (millions)
1981	Ansley Coale	16.5
1982	John Aird	23
1984	Ansley Coale	25
1984	Ashton et al.	29.5
1987	Peng Xizhe	23
1987	Judith Banister	30
1994	Chen Yizi	43
1996	Ding Shu	35
2005	Cao Shuji	32.5
2008	Yang Jisheng	36
2010	Frank Dikötter	45

Source: Based on discussions in Riskin, "Seven Questions about the Chinese Famine," 113; Yang Jisheng, Mubei, 394–430; and Dikötter, *Mao's Great Famine*, 324–334. It should be noted that these figures do not include the fall in population due to lower birth rates, which Yang estimates at another 40 million.

were exacerbated by grain exports and wasted food supplies in public dining halls in the fall and winter of 1958; (4) radicalism of provincial leaders and Mao's refusal to change policies in the summer of 1959 heightened the crisis; and (5) bad weather, the official CCP verdict, which contributed to lower production yields.[6]

The famine has also drawn the attention of scholars working in comparative frameworks. Most notable is work by the economist Amartya Sen, who has pointed out that famines are less likely to occur in democratic societies where the flow of information is relatively unrestricted.[7] Others such as Felix Wemheuer, Cormac Ó Gráda, and Andrea Graziosi have sought to study the GLF famine by comparing it to other famines across time and space, most notably the Soviet famine of 1932–1933 and the Bengal famine of 1943–1944.[8] It is in this context that statistics makes its second claim to relevance, since it was the very quality and availability of data that were at the heart of the escalating and

6 Summarized in Wemheuer, *Famine Politics in Maoist China*, 12; most of the works offering these explanations are cited below.

7 Sen, *Poverty and Famines*; see also Sen, *Development as Freedom*.

8 Ó Gráda, "The Ripple That Drowns?"; Ó Gráda, *Famine: A Short History*; Wemheuer, *Famine Politics in Maoist China*; Graziosi, "Political Famines in the USSR and China"; Yang, "China's Agricultural Crisis"; and Bianco, *La récidive*.

mutually reinforcing spiral of ever-higher production reports and ever-higher targets.

Such scholarship typically acknowledges—though often only in passing—that the GLF was accompanied by the dismantling of statistical work, which fundamentally crippled the ability of the leadership to access reliable data. There does not exist, however, any detailed account of what precisely such "dismantling" entailed. Yet, a focus on what precisely happened to statistics is important because it shifts our attention away from elite politics and how much Mao and other leaders could have known or did indeed know (a question that is often asked to assign culpability). That manipulation of reports took place is well established and beyond dispute. But a sole focus on manipulation obscures how changes in statistical practices also significantly reduced the capacity of the statistical apparatus to gather data and independently assess and verify claims. As this chapter shows, these changes included eschewing the use of the periodic reporting system and quickly disavowing the more recent dalliance with random sampling. What was endorsed instead was the one remaining method, which was fundamentally ethnographic and localized. It is this reduction in capacity, technical in nature and distinct from the infrastructural incapacity identified earlier, that made it virtually impossible to gather any reliable aggregate data estimates.

Picking up our story in early 1958, this chapter traces how, over the ensuing months, the tussle between socialist statistics and its probabilistic alternatives was largely overwhelmed by the rise to dominance of this third, self-consciously ethnographic, method. Driven by the momentum generated by Mao's antiexpert desire to promote mass science, typical sampling (that is, typical surveys), the poor cousin of the periodic report system, was rebadged and valorized as the only true way to carry out social research.[9] The chapter begins by first uncovering the place of statistics in existing GLF scholarship. It then looks at what exactly happened to statistics and statistical work during 1958 and 1959, exploring an on-the-spot meeting in the northern city of Baoding in Hebei province, where a great leap forward in statistics was proclaimed. It then tracks writings on the power and efficacy of typical sampling during the 1950s, situating its eventual valorization during the GLF within a longer trajectory of discussion and debate.

9 A case could be made for translating *dianxing diaocha* (典型调查) as "paradigmatic" surveys. Such a translation would better align with Mao's own concerns in his 1927 report, where the goal was not to locate representative sites but rather sites that were exemplary for the illumination of key concerns and for devising policy measures.

Collapse and Quality: Scholarly Assessments
of Statistics during the GLF

The 2009 edition of *DSJ*, which chronicles all statistical activities during the first sixty years of the PRC, does not record any break in statistical activity in 1958.[10] This depiction of a seamless evolution of events, however, runs counter to the way in which most scholarship has understood statistics during the GLF. In perusing this scholarship, which can be dated back to the early 1980s, two related narratives emerge. The first is about the collapse of the statistical system. The second deals with issues of information and how much could be reliably known.

Collapse

In the second of his three-volume history on the origins of the Cultural Revolution, Roderick MacFarquhar noted that "although the FYP [Five-Year Plan] system was not thrust aside, article 9 of the Sixty Articles [January–February 1958] contained the seeds of the undermining of the basis of careful planning, the statistical system."[11] The result was the setting of multiple increasingly euphoric targets that "led directly to the statistical fiasco of 1959."[12] For MacFarquhar, it was the Chengdu Conference in March 1958 that rang the death knell for the SSB.[13] For a bureau structured around professionalization and the observation of standards, the charge of dogmatism proved impossible to disavow.[14]

Writing the following year and focusing on food grain procurement, the political economist Kenneth Walker echoed some of MacFarquhar's concerns about statistical activity, observing that "the country's statistical services disintegrated so that the Government could no longer calculate the correct level of grain production."[15] In 1987, the demographer Judith Banister offered a similar assessment. In her landmark book on China's population,

---

10 *DSJ*, 56–62.

11 MacFarquhar, *The Origins of the Cultural Revolution*, Vol. 2, 31.

12 Ibid., 32.

13 Held from 8 March to 26 March 1958 in Chengdu, the conference was convened by the CCP Central Committee and attended by leaders of related departments of the CC, the central government, and party bosses from the provinces, municipalities, and autonomous regions.

14 MacFarquhar, *The Origins of the Cultural Revolution*, Vol. 2, 41.

15 Walker, *Food Grain Procurement*, 58, 167.

she maintained that "China's statistical system began to disintegrate in early 1958 and did not fully recover until over two decades later." Any reforms that were being contemplated were soon abandoned as the statistical system "became one of the many casualties of the Great Leap Forward. . . . Statistical operations at all levels were still seriously defective as of 1962."[16] By the end of the 1980s a pattern had emerged wherein the key words describing statistical work during the GLF were without fail variations on collapse: fiasco, undermining, weakness, disintegration, and defective.

In 1996, Jasper Becker's *Hungry Ghosts* brought the detailed story of the famine to a much wider audience.[17] Becker's assessment of statistical activity followed the pattern already set: "in fact there was no way of knowing the real size of the harvest since the State Statistical Bureau had been dismantled and its local offices replaced by 'good news reporting stations.' "[18] The decade also witnessed renewed attempts at addressing the causes and scope of the GLF famine. In the same year as Becker, political scientist Yang Dali published a study of rural China in the years following the GLF, observing that "any semblance of serious planning was abandoned [during the Leap]. The virtual shutdown of the state statistical system at the time would in any case have ensured chaos."[19] Yang concluded that "it is now generally recognized that the campaign atmosphere of the Great Leap, the intense political pressures on and competition among local party officials, and the extensive disruption of the statistical system led to extreme exaggerations in reporting agricultural output."[20]

Toward the end of the 1990s, *China Economic Review* dedicated a special issue to the Great Leap famine.[21] In his introductory essay, D. Gale Johnson observed that the problem of data was an enduring one: "Much available data relevant to the understanding of the famine were assembled in the late 1970s and early 1980s. . . . What is not clear is how much of the evidence was derived from actual records and how much was based on recall."[22] The economist Carl Riskin directly addressed the question of statistical data in his article in the

16 Banister, *China's Changing Population*, 13.

17 Becker, *Hungry Ghosts*.

18 Ibid., 79. Elsewhere in his book, 84, Becker wrote that "all pretense of dealing with genuine statistics had been abandoned."

19 Yang, *Calamity and Reform*, 34.

20 Ibid., 65.

21 Johnson, "China's Great Famine."

22 Ibid., 106.

same issue.[23] He cited works by Yang Dali and Judith Banister to reiterate Johnson's fears: "Where did these retrospective statistics come from, how were they put together, and how reliable are they? The Great Leap Forward period was one of social disorganization and the state of government agencies was chaotic. The State Statistical Bureau itself was in disarray and essentially moribund—one scholar (Yang Dali) referred to its 'virtual shutdown.'"[24] Riskin thus provided additional descriptors to go with our 1980s list: shutdown, dismantled, moribund, disarray, complete politicization.[25]

In the twenty-first century, scholarship on the GLF has expanded in new directions, but the consensus on a statistical breakdown has persisted. Relaxation in archival access, which began in the 1990s, has permitted the writing of several granular histories of the GLF, recognizing the diversity of experiences across China. Many of these studies are the result of trawling through vast amounts of local data to produce fresh estimates of grain production and famine deaths. Notable examples include Cao Shuji's *Da jihuang* and Yang Jisheng's two-volume *Mubei*, which was translated into a single-volume English edition in 2012.[26] Yang's reliance on contemporary county- and provincial-level statistics was tempered by an awareness of the rampant problem of inflation and exaggeration, and the resultant stresses placed on the SSB.[27] Published several years after Yang's original Chinese edition, Frank Dikötter's history of the famine endorsed earlier assessments: "As a result of political pressure the statistical work of the Bureau of Grain disintegrated from 1958 to 1962, to such an extent that the state itself could no longer calculate a realistic level of grain production."[28] Even more recently, Felix Wemheuer has observed that "during the famine the system of statistics actually collapsed altogether."[29]

23 Riskin, "Seven Questions about the Chinese Famine."

24 Ibid., 112.

25 Writing about the perceived success of the dining halls, Riskin explained that "what made possible this tragically mistaken perception was the ideologically motivated destruction of the information system and the total abandonment of objective reporting" ("Seven Questions about the Chinese Famine," 115, 116, 119).

26 Cao Shuji, *Da jihuang*; and Yang Jisheng, *Mubei*. See also Cao Shuji and Bin Yang, "Grain, Local Politics, and the Making of Mao's Famine"; and Manning and Wemheuer, eds., *Eating Bitterness*.

27 For his use of local data, see Yang, *Tombstone*, 120, 160, 200, 240, 254, 288. On exaggeration and the SSB, see Yang, *Tombstone*, 241, 258.

28 Dikötter, *Mao's Great Famine*, 130.

29 Wemheuer, *Famine Politics in Maoist China*, 50.

Even as each of the three waves of GLF scholarship identified here has broken new ground in our understanding of the famine, they have all continued to summarily dismiss actual statistical activity during the GLF. That such judgment has been passed is understandable because of the frustrations of limited and uncertain access and the challenges of parsing in meaningful ways the data that can be assembled. For our purposes, it continues to highlight our ignorance about what specifically happened to the statistical apparatus.

Quality of Information

Scholars who dismiss statistical activity during the GLF also highlight the importance of the quality of information that circulated and worked its way up the party and state leadership structure, and the role of manipulation in that process.[30] In many ways, this provides an epiphenomenal gloss on the breakdown thesis by describing one of the key features of statistics during the GLF—the constant upward estimation of production figures. Carl Riskin drew attention to this question in 1998, noting that "much more needs to be known about what and how much information the central government had about conditions in the countryside and about how this information was treated."[31] For others, there also remains the enduring hope that the tragedy might have been averted if only the senior leadership had a clearer picture of the reality on the ground. Ralph Thaxton calls this the new Maoist version of the GLF tragedy, which "nearly always implied that genius Mao would have solved the problem if only he had received the right information in time. . . ."[32]

In an article published in 2006, Thomas Bernstein addressed how much Mao could have reliably known, focusing in particular on the flow and recep-

30 For a comparative discussion on the quality of data between the Soviet Union and China, see Graziosi, "Political Famines in the USSR and China," 98–101.

31 Riskin, "Seven Questions about the Chinese Famine," 119. Even more recently, literary scholar Daniel Vukovich, in China and Orientalism, 72, has made the much more sweeping claim that adequate records simply do not exist. While in many places and instances this may well be true, the two reasons he ascribes—the absence of significant statistical activity and the lack of a preexisting colonial archive—do not withstand historical scrutiny. This book has demonstrated the vigorous engagement of the early PRC state in statistical activity; and to assume that only a colonial state was capable of collecting data does a great disservice not only to the long history of data-gathering in imperial China but also to the diverse range of social science research produced by state and private actors during the Republican era.

32 Thaxton, Catastrophe and Contention in Rural China, 298.

tion of information.[33] Given the impossibility of archival access, Bernstein relied upon the biography of Mao (*Mao zhuan*) and other published materials detailing the day-to-day activities of the great helmsman during that period.[34] A close analysis of these materials led him to conclude that in the earlier phases of the Leap Mao was frequently uncritical of boastful reports and eager to disseminate them across the country.[35] But as the Leap wore on and the situation on the ground worsened, Mao's own certitude deserted him and he began to recognize that he was unable to have any comprehensive picture at any level: "The Chairman observed that neither the center nor the provinces knew the real situation."[36] According to Bernstein, even the biography of Mao (*Mao zhuan*) sought to ascribe blame for this outcome to Mao's age and to the false reports that were being produced at lower levels.[37]

Working primarily with local-level data, Felix Wemheuer has gone one step further than Bernstein to argue that many of the practices of concealing production can be traced to the introduction of the state monopoly for the purchase and sale of grain (统购统销; *tonggou tongxiao*) in 1953. Acknowledging the universal tendencies for peasants to underreport and for the state to overestimate, Wemheuer has explained that the GLF famine "can also be considered a result of the inability and unwillingness of the state to guess the truth within the narratives of difficulties and hunger told by the peasants."[38] Therefore, for Wemheuer the conflict between the villages and the state was a crucial element contributing to the poor quality of statistical information.[39] In their introduction to Yang Jisheng's *Tombstone*, Edward Friedman and Roderick MacFarquhar observed that Yang shows quite clearly that "while junior officials did falsify data to benefit their own careers, Mao had enough reports from senior colleagues to know that his policy of extracting an increasing percentage

33 Bernstein, "Mao Zedong and the Famine of 1959–1960"; see also Yang Kuisong, "Mao Zedong shi ruhe faxian da jihuang de."

34 Bernstein, "Mao Zedong and the Famine of 1959–1960," 424.

35 Ibid., 427, 435.

36 Ibid., 439–440.

37 "The sources from which he derived his understanding of the situation narrowed more and more, and it became very difficult to discover the true situation at the grassroots. Failure to investigate what was actually going on in the villages was a result of his age and various objective circumstances. He relied on reading reports from below, which not only were sloppy but contained many false reports." *Mao zhuan*, 1073, cited in Bernstein, "Mao Zedong and the Famine of 1959–1960," 441.

38 Wemheuer, *Famine Politics in Maoist China*, 49.

39 Ibid., 61.

of grain from the countryside was causing millions of deaths."[40] And yet, even as late as 1961, Mao insisted that it was the very dynamism of GLF activity that rendered statistics incapable of keeping pace:

> We always demanded statistics on how much of what had been planted, how much was produced, how much fertilizer had been spread today, what would be done tomorrow. *With all that reporting and calculating, it was impossible for statistics to keep up.* That's how things go: you issue chaotic directives and I give a nonsense report, and the result is exaggeration that makes no sense at all (Emphasis added).[41]

"On the Spot" in Baoding

Was the impossibility of keeping up that Mao bemoaned merely an extension of the kinds of incapacities we encountered in chapter 6 or was it something altogether new? How might we move beyond these narratives of collapse and poor quality and add to our understanding of statistics during the GLF? Indeed, a closer look at the published and archival record suggests that statistics did not collapse but rather that the composition of statistical activity was completely transformed. This transformation had clear implications for the quality of data that were generated.

In his organizational study of statistical activity during the 1950s, the economist Li Choh-Ming provided some early clues about this transformation.[42] Li pointed out that among the key changes wrought by the GLF was an intense push toward decentralization of statistical activity, essentially upending the SSB's commitment to prioritizing regular services from the center down to the county (commune) level.[43] This decentralization favored local statistical activity, in "opposition to unified computing methods and statistical technology," which were attacked as dogmatism run amok.[44] It is within this milieu that model surveys (i.e., typical sampling) re-entered discussions of statistical methodology.[45] According to Li, these changes had two major components. The first was the promotion of emulation drives in which the generation of

40 Yang, *Tombstone*, x.

41 This is taken from Mao's speech to an enlarged Politburo meeting, held 5 March 1961. See Yang, *Tombstone*, 258.

42 Li, *The Statistical System of Communist China.*

43 Ibid., 69–70.

44 Ibid., 72–76.

45 Ibid., 76–77.

local data dominated national data in a ratio of nine to one and was composed of background, progress, and comparative statistics. This dramatic reversal in relative importance between local and more aggregated data was achieved by encouraging what Li termed "indigenous methods."[46] "All methods of investigation hitherto prescribed by the State Bureau were generally discarded as inconsistent with local needs, which required that progress statistics be compiled within a matter of hours. The method that meets this requirement is the model survey."[47]

A change as drastic as this had to be properly communicated. Between 26 June and 8 July the SSB convened a national "on-the-spot" conference (现场会议; xianchang huiyi) in the northern city of Baoding to promote its vision of a great leap in statistics.[48] The DSJ notes that discussions on organizing this meeting were initiated in the spring of 1958, and the SPC gave the SSB the go-ahead on 4 May.[49] Consisting of regular meetings as well as extended site visits, the Baoding conference had two explicit aims: to criticize statistical practices during the past several years, especially given the new dictum of "politics in command," and to promote statistical work in Hebei province as the exemplar of a new approach to statistics.[50] Following two days of meetings on 26 and 27 June, the delegates visited an exhibition of statistical work, which had been jointly curated by the SSB and several provincial and municipal statistical bureaus, including those in Beijing, Shanghai, Shanxi, and Sichuan.[51] After that, approximately twenty delegations were organized and

---

46 Ibid., 78–79.

47 Ibid., 86. In addition to model surveys, Li also pointed out two other crucial factors leading to the unreliability of agricultural data by 1958: the continuous upward revision of goals and disruptions in the people's commune movement (85).

48 At least one archival document (BMA 133-001-00120: 23) indicates that the meeting actually began a day earlier, on 25 June.

49 DSJ, 59–60. This likely coincided with the fourth and fifth plenary sessions of the Eighth Central Committee, which took place in May 1958. Xue Muqiao notes in his memoirs, Xue Muqiao: Huiyilu, 191, that it was here that people's subjective will (人的主观性; ren de zhuguanxing) was privileged over objective economic constraints (客观经济规律; keguan jingji guilu).

50 For summaries of the meeting, see SSB, "Ganqing chongtian, xinxin baibei," and Anon., "Jiehe xianshi douzheng gao tongji." Hebei's statistical reforms were first reported in the 29 April 1958 issue of RMRB, which carried an article and an editorial describing and praising the reforms. See Anon., "Yikao qunzhong jianli xianxiangshe tongjiwang," and Anon., "Tongji gongzuo de liangzhong zuofa."

51 Wang Huaxin, "Yige fengfu duocai de tongji gongzuo zhanlan." The exhibition contained approximately 1,000 items displayed in over 90 rooms spread across 3 floors.

dispatched throughout Hebei to various factories, construction sites, schools, streets and districts, townships, communes, and production and brigade teams for week-long site visits. In all, they visited nearly 100 grassroots statistical units. The conference reconvened in Baoding on 6 July, and during the following three days statements by twenty-four delegates, representing a range of provinces, districts, and municipalities, were heard.[52] In their speeches, the delegates promised to not only catch up with Hebei's reforms but also to completely reform statistical work within three months. They hoped to report their success as a gift to the nation at the upcoming National Day on 1 October.[53] The delegates brought the meeting to a close on 8 July by sending a telegram to Mao Zedong in which they publicly guaranteed to meet their three-month goal.[54]

In his speech at the Baoding Conference, Xue Muqiao claimed that Hebei's statistical reforms heralded the start of a new era in the history of Chinese statistics.[55] "What is a Great Leap Forward in statistics?" he asked, rhetorically. He proceeded to explain that a great leap in statistics was marked by two key characteristics.[56] The first, in line with Mao's general critique, was based on a rejection of dogmatism. Statistics was not an end unto itself but rather a tool to promote political and productive struggles. In a dramatic turnaround, Xue rejected the primacy of the very principles that he and the SSB had been espousing over the past several years. Xue now explained that a great leap in statistics could not be achieved merely by ensuring the correctness, timeliness, and completeness of periodic reports. And yet he warned that there were many who continued to hold on to such antiquated beliefs. What they practiced was nothing but "statistics for statistics sake" (为统计而统计; wei tongji er tongji). Another common mistake was to regard vigor and dynamism (轰轰烈烈;

52 The statements delivered by the representatives from Shanxi, Henan, Shaanxi, Guizhou, and Heilongjiang were republished as individual articles in *TJGZ*, no. 14 (1958). In an article in the same issue, an additional twenty-one statements by the representatives were summarized and reproduced in the order that they were delivered. Anon., "Ge sheng qu shi tongji gongzuo yuejin zhibiao."

53 "Learn from Hebei, Catch Up with Hebei, on National Day Report Our Success as a Gift" (学河北, 赶河北, 国庆节报捷, 献礼; *Xue Hebei, gan Hebei, guoqingjie baojie, xianli*), SSB, "Ganqing chongtian, xinxin baibei," 25.

54 Baoding Conference Delegates, "Xiang Mao zhuxi baozheng," 1. The promise was reiterated in *RMRB* on 1 August (Anon., "Jiehe xianshi douzheng gao tongji," 6).

55 Xue Muqiao, "Kuzhan san yue, gaibian quanguo tongji gongzuo mianmao."

56 Ibid., 3–5.

honghonglielie) in statistical work as a great leap forward in statistics.[57] Unless one also reflected on how statistical work contributed to a high tide in production, such vigor was of little use. Xue proceeded to equate both these types of errors, observing that statistical workers of the "vigor and dynamism" brand often ridiculed their "statistics-for-statistics" comrades. In truth, there was no difference between the two groups; much like Mencius' allegory about the soldiers who retreated fifty steps mocking those who retreated one hundred steps; both groups were guilty of misunderstanding the true nature of a great leap in statistics.

The second characteristic of a great leap in statistics was the mobilization of the masses to actively participate in statistical work. In this way, Xue explained, the theory of the mass line could be implemented in statistics: "Statistics is much like other kinds of work: It cannot rely solely on dedicated organizations or dedicated personnel, nor can it rely on a vertically integrated leadership structure." Instead, each level of administration was required to participate in statistical work, and even more significantly, the masses themselves could not remain indifferent to statistical work.[58] The idea of mass participation in statistical work was subsequently promoted by use of the phrase "the entire party and all the people [must] attend to statistics."[59]

It is likely that the Baoding meeting's rejection of statistics as it had been developed and practiced over the past several years came as a surprise to the statistical establishment. The experience of the Beijing Statistics Bureau serves as a case in point. The bureau had taken advantage of its proximity to Hebei to send its own investigative team well in advance of the Baoding meeting.[60] This eighteen-member team spent five days during the second half (21–25) of April in Luanping county (滦平县), studying statistical work in Luanhe (滦河) and other nearby districts. An initial report of their findings was filed as

---

57 This was likely a reference to a 29 April 1958 editorial in *RMRB* (Anon., "Tongji gongzuo de liangzhong zuofa"), which identified two methods of doing statistical work: "cold and cheerless" (冷冷清清; *lengleng qingqing*) and "on a grand and spectacular scale" (轰轰烈烈; *honghong lielie*). The editorial claimed that its purpose was to criticize the former and to promote the latter, and it did so based on evidence from the statistical reform in Hebei, pointing to a corresponding article on p. 3 of that day's issue of *RMRB* (Anon., "Yikao qunzhong jianli xianxiangshe tongjiwang").

58 Xue Muqiao, "Kuzhan san yue, gaibian quanguo tongji gongzuo mianmao," 4.

59 See, for instance, Editorial, "Quandang quanmin ban tongji," 3.

60 Hebei's experience was already reported in the national press by April 1958, if not earlier (Anon., "Yikao qunzhong jianli xianxiangshe tongjiwang").

early as 13 May.[61] A subsequent report was drafted on 23 June and filed on 1 July.[62] Tellingly, both reports made no mention of dogmatism. Instead, well in line with concerns from 1957 and earlier, they focused on the need to unify rural statistical work (统一农村中心工作; *tongyi nongcun zhongxin gongzuo*) across districts, villages, communes, and production brigades and teams. The claimed benefits of such unification included the resolution of familiar problems: excess and chaotic issuing of reports (乱发报表部; *luanfa baobiao*), chaotic demands for data (乱要数字; *luanyao shuzi*), and the propensity to supply false estimates (估计谎报; *guji huangbao*). At the same time, it was expected that such measures would increase the timeliness and correctness of the periodic reports, save labor time, and increase the efficiency of production-related activities.[63] All of these concerns took for granted the centrality of the periodic reporting system, which is precisely what came under attack at Baoding and in the months that followed.

The Beijing Statistics Bureau was caught off guard and moved quickly to remedy the situation. Within two weeks after the conclusion of the Baoding meeting, the bureau's party group (党组; *dangzu*) drafted a preliminary reform plan for consultation, which was to be submitted to the Beijing Municipal Party Committee and its secretary Wan Li (1916–2015).[64] The tone and tenor of this document was far more in line with Xue's concluding remarks at Baoding, which the report referenced in its opening paragraph. It reaffirmed the Baoding pledge to achieve a great leap in statistics by 1 October, and then proceeded to outline the two key features of Hebei's statistical success: politics in command and implementation of the mass line. In concrete terms, this translated into an attack on the efficacy of the periodic reporting system and the expertise of statistical workers; the periodic system had become overly rigid (呆板; *daiban*) and was mechanically implemented (生搬硬套; *shengban yingtao*). The result was that it could not keep pace with actual changes and was more like an almanac, only good for historical analysis and not real-time decision making. Statistical workers had also successfully mysticized (神秘化; *shenmi hua*) their work, isolating themselves from the masses, the party, and ergo from the key drivers of production.[65]

---

61 BMA 133-001-00120: 7–13.

62 BMA 133-001-00120: 14–22.

63 BMA 133-001-00120: 15.

64 "Guanyu gaige tongji gongzuo de qingshi (chugao)" 关于改革统计工作的请示(初稿(Request for Reform of Statistical Work (Preliminary Draft)). BMA 133-001-00120: 23–29. At this time, Wan Li was also one of Beijing's deputy mayors.

65 BMA 133-001-00120: 25.

The Beijing Statistics Bureau's recalibration act was likely repeated in other provinces and municipal bureaus. A succession of conferences was organized in Beijing in 1958 and 1959 to reinforce the message articulated at Baoding.[66] Restricted to provincial, municipal, and regional bureau chiefs, these conferences sought to discuss the progress made and problems encountered during the post-Baoding rectification of statistics. The first of these meetings took place four months after the Baoding meeting, and lasted from 25 October to 3 November. Summary materials produced in its wake, and meant for internal circulation, reiterated that the Baoding Conference had created a "new situation" in statistical work.[67] And yet the report clearly demonstrates the predicament the SSB found itself in. On the one hand, it acknowledged that statistical work had to adapt to the fast-changing situation during the GLF. On the other, it struggled to abandon its claims to expertise and the periodic reporting system, which it felt were fundamental to statistical work.[68] Even so, the shift toward typical sampling was undeniable. The report acknowledged that "[it is only] through practical work, especially by using typical sampling, [that] statistical cadres who are both red and expert can be cultivated."[69] This was especially true when it came to work in people's communes, where typical sampling had to be vigorously carried out.[70]

To aid the readjustment, publications and reports that sought to distill and spread the message of the conference or to claim a leap in the effectiveness of statistical work began to make an appearance shortly after the Baoding Conference. In August 1958, Hebei Statistical Press published *Tongji gongzuo yuejin jingyan* (The experience of a leap forward in statistical work), a collection of essays describing various aspects of statistical work in Hebei.[71] That volume was followed by *Baodingshi tongji gongzuo de gaochao* (The high tide of statistical work in Baoding).[72] Early in the following year, the Fujian Planning Committee published a volume on how to carry out statistical surveys, explaining that the establishment of the people's communes required new procedures for

---

66 Kong Dechao, "'Dayuejin' qijian de tongji gongzuo," provides a detailed listing of conferences and meetings related to the GLF.

67 BMA 133-001-00105: 1–10 (reproduced in SSB, *Tongji gongzuo zhongyao wenjian huibian*, Vol. 3, 218–30).

68 BMA 133-001-00105: 4, 6, 7.

69 BMA 133-001-00105: 4.

70 BMA 133-001-00105: 6.

71 Hebei Province Statistics Bureau, *Tongji gongzuo yuejin jingyan*.

72 Hebei Province Statistics Bureau, *Baodingshi tongji gongzuo de gaochao*.

statistical work.[73] Internal reports began to offer descriptions of these changes, frequently stressing the need for widespread implementation of typical sampling. The benefits of such a shift had been extolled in the summary of the November 1958 meeting:

> Since the Baoding Conference, statistical work has emerged on a new scale. Now, statistical authorities at all levels, from the central to the local, can within a few hours, or within a few tens-of-hours, master statistical data on all aspects of industrial and agricultural production and construction at the national, provincial, and regional levels. [And] within three or five days, they can conduct one-time comprehensive surveys and basically fulfill in very quick time the party leadership's various data requirements.[74]

The SSB had long strived to achieve the goal of meeting the party leadership's requirements, but the GLF seemingly helped realize that goal in mere months.

Regardless of whether or not such claims were true, and they were most certainly "true" only if a host of issues including accuracy and representativeness were kept in abeyance, the ability to make them pointed to a fundamental shift in the terrain of statistical work. The downgrading of the periodic reporting system and the promotion of mass participation in statistics effectively bypassed the tussle between socialist statistics and its nonsocialist alternatives in favor of a decisive turn to Maoist mass science, expressed via the promotion of the Maoist method of social investigation. This form of typical sampling had been largely ignored by the SSB during much of the 1950s. After the Baoding Conference, it was elevated to the status of orthodox method.

The Rise and Rise of Typical Surveys

"Typical surveys are an innovative survey method utilized in our nation under particular historical and social conditions; they are the concrete manifestation in statistical survey methodology of Chairman Mao's synthesis of Marxist-Leninist theory with the practice of revolution in China."[75] Feng Guoxi and the other authors of the March 1960 article within which this statement was made went on to list the various qualities of a typical survey. They claimed it could solve those problems that exhaustive enumeration could not and it

73 Fujian Province Planning Committee, *Zenyang kaizhan tongji diaocha yanjiu gongzuo.*

74 BMA 133-001-00105: 2.

75 Feng Guoxi et al., "Mao zhuxi lun dianxing diaochafa." The quoted section is reproduced from Garnaut, "Hard Facts and Half-Truths," 239.

could also help address problems where it was not necessary to carry out exhaustive enumeration. In addition, typical sampling also served to supplement the results of exhaustive enumeration. It was flexible and lightweight (灵活 轻便; *linghuo qingbian*); it helped save on manpower and reduced problems, thereby doing more with less. Using it, new things could be discovered or developed in a timely fashion. It was thus like a scalpel that dissected problems in the national economy and across society and offered specific analyses of complex social phenomena. In so doing, it also promoted continuous improvements in the style and methods of leadership. Finally, such typical sampling was important for the cultivation of red and expert cadres and helped reduce the problems in earlier statistical work, which was divorced not only from reality but also from the masses.[76]

Distinctive in Feng et al.'s description was the direct connection they made between typical sampling and the Maoist method of social investigation. As has been noted in earlier chapters, different forms of typical sampling had been used fairly extensively by the CCP in their base areas during the 1930s and 1940s. They may have been inspired by Mao's own two social surveys, in Hunan in 1927 and in Xunwu in 1930, but the method itself appears not to have been described as some sort of new Maoist theoretical synthesis. In the years following 1949, the reformulation of statistics as a socialist social science pushed typical surveys to secondary status. But they did not disappear entirely. Table 8.2 shows the probability by year that an article in the journals *TJGZTX*, *TJGZ*, and *JHTJ* mentioned the term "typical survey" (典型调查; *dianxing diaocha*) during the 1950s. The decade appears to divide neatly into two halves: prior to 1956, when less than 10 percent of the articles referred to the method; the years thereafter, when such references increased dramatically.[77]

A closer look at these articles also tells a story of a change in emphasis across the 1956 divide. The earliest articles from 1953 and 1954 tended to exhibit a balanced approach to the usefulness of typical surveys. For instance, in an article published in 1954, Liang Zhitang noted that when conducting typical

76 Feng Guoxi et al., "Mao zhuxi lun dianxing diaochafa," 79; Garnaut, "Hard Facts and Half-Truths," 230–240, provides a brief discussion of this methodology in the context of the reliability of famine-related data.

77 The corresponding number of articles for each year is: 5 (1953); 14 (1954); 22 (1955); 37 (1956); 60 (1957); 68 (1958); and 30 (1959). A total of 2,000 articles were published during these years. The division of the decade into two halves appears even more obvious if we restrict our search to occurrences of the term only in the titles of articles: four articles in the four-year period from 1953 to 1956, and fourteen articles in the period from 1957 to 1959, and nine in 1959 alone.

TABLE 8.2. Probability that an Article in *TJGZTX*, *TJGZ*, and *JHTJ* Mentions *Dianxing diaocha*

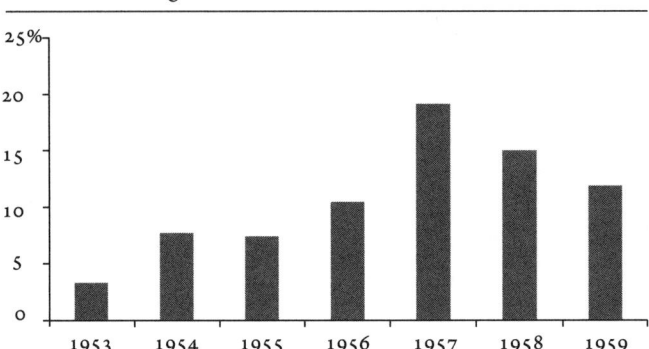

surveys in the agricultural sector, many statistical workers did not appreciate the importance of selecting a truly representative sample, a failure that could easily lead to skewed results. Accordingly, Liang devoted much space in the article to discussing different methods of selecting typical cases.[78]

The subject of typical surveys was addressed in significantly greater detail in two 1956 articles that sought to grapple more directly with their place within overall statistical work. The first of these was written by Gong Jianyao, one year before he was selected to visit the ISI in Calcutta. The title of the article is an indication of the struggles to come: "Typical Surveys and their Scientific Nature."[79] Gong began by noting that even though there had been tremendous interest and discussion in recent times regarding how best to use typical surveys, there was little clarity on what or how scientific they were. Gong undertook to clear the mist. He pointed out there were principally four views on the subject.[80] The first, exemplified in Zhang Zhihong's edited volume, *Nongye tongjixue* (农业统计学; Agricultural statistics), held that typical surveys were the same as typical sample surveys (典型抽样调查; *dianxing chouyang diaocha*). Zhang held that Mao's 1927 Hunan report was an "exceptional example" (卓越的范例; *zhouyue de fanli*) of such survey work. The second view regarded both typical surveys and typical sample surveys as forms of noncom-

78 Liang Zhitang, "Nongye tongji dianxing diaocha de jiben renshi he jiben zuofa." In an article published the previous year, Tianjin Municipal Statistics Bureau, "Siying xiaoxing gongchang shougongye dianxing diaocha gongzuo," reported on experience with typical surveys in the private small-scale handicraft industry.

79 Gong Jianyao, "Dianxing diaocha he tade kexuexing."

80 Ibid., 12.

prehensive survey methods (非全面调查方法; *fei quanmian diaocha fangfa*).
What distinguished them was how the sample to be surveyed was selected:
The latter relied on random sampling (随机抽样; *suiji chouyang*) methods,
the former did not. A third view stated that typical surveys were of two kinds:
One kind selected the typical based on classification and categorization of the
entire data, and then used the results from a more intensive study of the typical
case to extrapolate comprehensive figures (推算全面数字; *tuisuan quanmian
shuzi*); the second kind studied one or a few typical cases in detail to make
claims only about the general (i.e., average) situation (概括一般; *gaikuo
yiban*). The final, fourth, view held that typical surveys and typical sample
surveys were fundamentally different entities. The former picked out typical
cases from the entirety to make claims about the normal or average (一般;
yiban), whereas the latter was based on probability theory.

Having outlined their key features, Gong Jianyao offered his assessment.
According to Gong, the first definition, which effectively equated typical surveys with typical sample surveys, was not at all tenable, thereby also calling
into question the efficacy of Mao's own early survey work.[81] He felt that the
second and fourth were an improvement on the first, but the conflation of
probabilistic and nonprobabilistic methods still generated confusion. It was
the third definition that approached some degree of perfection. Accordingly,
Gong's own definition combined elements of the third while completely eschewing probabilistic methods. For him, a typical survey was a type of noncomprehensive survey method that consisted of two forms, each based on its
ability to generate comprehensive projections. The first called for in-depth
study and analysis of the characteristics and patterns of specific typical cases,
through which generalized claims (一般; *yiban*; i.e., about the average/ordi-
nary/norm) could be made about the patterns and problems affecting such
classes of things. It permitted claims that were not about some comprehensive
whole, but about patterns among a certain class of things within the whole.[82]
The second approach first classified the entirety that needed to be studied into
several categories. These categories were then studied in depth, and from each
was drawn a typical representative unit for further detailed investigation. The
results of such investigations could then be extrapolated to make claims about

81 A similar criticism was mounted by Zhang Minru, "Dui 'nongye tongjixue' yishu de
jidian yijian," in the same issue of *TJGZTX*. In the following year Zhang Zhihong would
publish a response in *TJGZ*— "Guanyu 'Dui nongye tongjixue yishu de jidian yijian de
shangque.'"

82 Gong Jianyao, "Dianxing diaocha he tade kexuexing," 13.

the entirety.[83] According to Gong, it was precisely the second method's ability to make projections about the whole that made it scientific. The first method was inferior because it only offered the ability to make claims about the norm. The remainder of Gong's article focused on how best to carry out the second type of typical surveying.

Gong's measured response and his sharply drawn distinction between typical surveys that were representative of a larger whole versus those that merely captured the general situation attracted the attention and ire of Shi Jiazhen. In an article published shortly thereafter, Shi used the distinction between inference (推论; *tuilun*) and projection (推算; *tuisuan*) to point out Gong's errors.[84] Shi countered Gong by claiming that to use as the sole criterion a method's ability to deliver comprehensive projections was incorrect. As an example, he cited the case of ascertaining how much time was needed by members of communes to attend to housework (家务劳动; *jiawu laodong*) and sideline work (经营副业; *jingying fuye*). The answer of two months was by itself sufficient to understand the situation and did not need any extrapolation or projection to a larger whole—it was representative of the norm, the general situation, without making claims for some larger totality. Even as Shi and Gong disagreed on the finer points, they found common ground in excluding probabilistic methods from their definition of typical sampling. Essential for both was the issue of representativeness, which they believed could be achieved through careful survey design and attention to local conditions.

Such dialogue continued into 1957, which, as we have seen in Table 8.2, witnessed a significant increase in interest in typical surveys—nearly twice as many articles as in the previous year. And so, Huang Jiantuo devoted his attention to developing a fourfold taxonomy for types of typical surveys.[85] Elsewhere, the Anhui Provincial Statistical Bureau acknowledged the strengths and weaknesses of the method, but also pointed out that the implementation of typical surveys was cheaper and more efficient than the periodic reporting system. Such articles did not call for the replacement of the periodic reporting system; instead, they promoted fruitfully combining each of the methods. The logic was straightforward: The incapacities generated by the periodic reporting system could be addressed, even overcome, through the use of typical surveys.[86]

83 Ibid.
84 Shi Jiazhen, "Guanyu dianxing diaocha de jige wenti."
85 Huang Jiantuo, "Wo duiyu dianxing diaocha fangfa de jidian cuqian renshi."
86 Anhui Statistics Bureau, "Women jinxing dianxing diaocha gongzuo de yidian tihui."

The semantic and substantive influence of the GLF becomes clearly discernible in articles from 1958. Keeping pace with some of the archival documents cited above, these articles complained that statistics and statistical workers had divorced themselves from reality. They had emphasized professionalization to the detriment of political awareness and commitment. Typical surveys provided a way out: They offered an escape from dogmatism, a direct route to the masses, and thereby the opportunity to combine the red and expert elements of statistical work.[87] The year also witnessed published examples of typical surveys, presumably an attempt by the SSB to spread best practices. For example, Luanping County Statistics Office—the same county visited by the Beijing Statistics Bureau team in April 1958—published a report on how Pioneer People's Commune had divided corn fields into three categories and used typical surveys to assess corn yields.[88]

The change in tone already discernible in 1958 took on a much starker and unequivocal quality beginning in 1959. In the year's first issue of *Jihua yu tongji* (Planning and Statistics; the rebadged version of *TJGZ*), Yang Bo[89] (1920–2016) published the first of a five-part article, each of which was entitled "How to Carry Out Typical Surveys," and appeared in successive issues of the journal.[90] Collectively, the five parts sought to correctly recognize the place of typical surveys within socialist statistical work, identify the kinds of problems typical surveys were particularly suited to solving, and explain how to carry out typical surveys. Yang began by noting that,

> In recent times, there has emerged an extremely valuable new phenomenon among statistical leaders at all levels: [they] often dispatch large numbers

---

87 Mo Rida, "Dianxing diaocha shi tongji gongzuo ganbu zouxiang hongzhuan de yitiao zhongyao daolu."

88 Luanping County Statistics Office, "Xianfeng renmin gongshe datun sanlei dikuai yumi chanliang dianxing diaocha."

89 Yang Bo was an official who had joined the party in 1940 and had been involved in financial and economic administration in Shandong. During the Civil War (1945–1949), along with Xue Muqiao he participated in the North China Finance and Economics Meeting in 1947. After 1949 he served as secretary of the trade office within the Government Administration Council's Finance and Economics Committee. In 1952 he was transferred to the SSB, where he continued to work in the research office on trade statistics and comprehensive statistics. He left the SSB in 1962. For more, see Zhonggong zhongyang zuzhibu yu Zhonggong zhongyang dangshi yanjiushi, *Zhongguo gongchandang lijie zhongyang weiyuan da cidian, 1921–2003*, 92.

90 Yang Bo, "Zenyang zuo dianxing diaocha" (no. 1, 1959); for the other four articles, see no. 2 (1959) through no. 5 (1959).

of cadres deep into the countryside, factories, mines, shops, basic construction sites, and other grassroots units to conduct highly specified, on-the-ground research work and thereby grasp the new circumstances and problems of the development of the national economy. . . . [Such activity] is the distinctive marker of changes in the style of statistics leadership and in work methods.[91]

Yang explained that prior to such changes, statistical bureaus at all levels primarily used one method: the periodic reporting system. They believed that this was the only useful way of completing statistical tasks. Typical surveys had little value within such an approach. Yang conceded that this was likely an outcome of the longstanding debate within the world of statistical research and work on the very nature of typical sampling and whether it was a scientific method. But he emphatically pointed out that those who dismissed typical sampling were in the wrong. The only measure of a method's scientific nature was whether it could help people correctly recognize the objective world. Typical surveys passed this test. Yang berated the SSB's dogmatism in ignoring typical surveys. He concluded his first article by noting that all three methods—periodic reports, censuses, and typical surveys—were scientific and had to be used based on specific circumstances and requirements.[92]

In the second article in the series, Yang explored the circumstances under which typical surveys were especially beneficial. The communes had a host of local problems that the periodic reporting system or the census method could not tackle. For instance, questions about the allocation of commune income or the response of the masses to specific demands could only be solved using typical surveys. Even more tellingly, and perhaps foreshadowing the crisis to come, Yang declared that typical surveys were the only means to assess the success of new phenomena. Therefore, the efficacy of the GLF's "deep plowing" (深耕细作; *shenggeng xizuo*) could only be ascertained via typical surveys. In similar fashion, in the new movement to build backyard blast furnaces, typical surveys were necessary to help resolve five key issues: selecting a site, selecting the ideal model furnace, fixing rent to account for production and depreciation, allocating suitable personnel, and, finally, selecting appropriate leadership. The periodic reporting system was not appropriate for any of these activities.

Yang devoted the third, fourth, and fifth articles in the series to providing a detailed exegesis on method, which he divided into eight parts. First and fore-

91 Yang Bo, "Zenyang zuo dianxing diaocha," no. 1 (1959), 36.

92 Ibid., 36–37.

most was the need to place politics in command and implement the mass line. A key element in this was entrusting leadership of typical surveys to local cadres. A second element called for the adoption of a common survey outline with regard to goals, questions, scale, and coordination with other teams. Third, each survey team was to be led as an independent combat unit. Fourth, it was essential to correctly select typical cases (典型对象; *dianxing duixiang*) and ensure their representativeness. Fifth, it was vital for teams in the same area or working on similar questions to coordinate among one another. Next, at the same time, the sorting and analysis of data should occur in situ and not later. Seventh, in terms of the individual surveyors, their key tasks were fourfold: to listen, to ask, to look, and to think. Finally, it was important to constantly summarize and share experiences in order to raise everyone's awareness.[93] Yang faithfully channeled Mao's "On Practice" to drive home many of his points and concluded with Mao's exhortation that cadres ought to be red and expert, and the redness should pierce deep into the expertise (又红又专, 红透专深; *you hong you zhuan, hong tou zhuan shen*).[94]

Articles published during the remainder of the year made increasingly strident calls for the aggressive adoption of typical surveys. These articles also began to consciously describe typical surveys as a Maoist method. In an article entitled "Expand the Use of Typical Surveys," Chen Yingzhong explained that 1959 was a year in which the GLF had taken on larger, better, and more complete proportions. Changes were legion and rapid, none more so than the emergence of people's communes.[95] Under such circumstances, "our statistical work must be greatly improved. We must improve survey methods, raise work levels, and improve the quality of our service. Therefore, in the present situation, to further and better use typical survey methods has taken on a real significance."[96] Chen's article was followed several issues later by two articles that discussed the use of typical surveys in the industrial and agricultural sectors. The latter, authored by the SSB, again called for the vigorous implementation of typical surveys.[97] Articles later in the year expanded on these themes. For instance, Lu Shutian published an article calling for the necessity of combining typical survey work with comprehensive

93 Yang Bo, "Zenyang zuo dianxing diaocha" (no. 3, 1959); Yang Bo, "Zenyang zuo dianxing diaocha" (no. 4, 1959); Yang Bo, "Zenyang zuo dianxing diaocha" (no. 5, 1959).

94 Yang Bo, "Zenyang zuo dianxing diaocha," no. 5 (1959), 37.

95 Chen Yingzhong, "Geng guangfan de yunyong dianxing diaocha fangfa."

96 Ibid., 36.

97 Wang Xinming, "Zai gongye qiye zhong kaizhan dianxing diaocha de tihui"; and SSB Agricultural Statistics Office, "Jianli nongcun jingji diaocha gang, dagao dianxing diaocha."

statistics.[98] By the time of publication of the article by Feng et al. in early 1960, the centrality of typical surveys as a key Maoist theoretical intervention and therefore as a method of determining influence had well and truly been established.[99]

Meetings in 1959: Mao as Method

Through the course of 1959, the SSB continued to struggle with the implications of the aggressive push to adopt typical surveys. Much as in the case of the Beijing Statistics Bureau in 1958, the SSB took time to adapt. This transition is captured in the reports that emanated from two important meetings that year. The first meeting was held in April and brought together the heads of various statistics bureaus at the province, city, and district levels. While acknowledging the massive changes wrought by the GLF, the meeting continued to stress that statistics was a specialized branch of knowledge that required expertise for effective execution. By the time of the November meeting, the attempt to cling on to the relevance of expertise gave way to an acceptance of the overarching primacy of politics and the interests of the party. The year also marked important institutional changes at the SSB. Xue Muqiao, who had led the SSB since its founding in 1952, was replaced by one of his deputies, Jia Qiyun (1914–2004).[100] The SSB's journal *TJGZ* also ceased to exist as an independent publication. At the start of the year, it was amalgamated with the journal *Jihua jingji* (计划经济; Economic planning) and reissued as *Jihua yu tongji* (计划与统计; Planning and statistics), which had a short-lived run of only fifteen issues before ceasing publication ca. mid-August.[101]

The 11–21 April session was the first countrywide meeting since the October–November 1958 meeting. Although sector-specific meetings had been

---

98 Lu Shutian, "Quanmian tongji bixu he dianxing diaocha xiang jiehe."

99 Feng Guoxi et al., "Mao zhuxi lun dianxing diaochafa," 81.

100 It is not clear if Xue Muqiao stepped down or was removed from his post. Li, *The Statistical System of Communist China*, 14, claimed that he was dismissed. But in his memoirs, Xue explained that he was asked by Chen Yun in May 1958 itself to divest himself of his duties at both the SSB and the SPC and instead to become deputy director of the Central Finance and Economics Group (中央财经小组), which had been set up that month under Chen Yun's leadership. Among the group's tasks was to produce *Jingji xiaoxi* (Economic Information Digest) for the top leadership (Xue Muqiao, *Xue Muqiao: Huiyilu*, 194). For more on Jia Qiyun, see *DSJ*, 6, 472.

101 Editorial Department, "Gao duzhe," 1.

organized in the preceding months, this meeting was the first to again bring together the heads of provincial-, city-, and district-level bureaus. Also present were other important participants, such as Jia Tuofu (1912–1967), then deputy director of the SPC. The summary report of the meeting is remarkable for highlighting unmistakably the SSB's attempt to hold on to its claims of expertise in a vastly changed political climate:

> Statistical work must operate under the leadership of the party and follow the mass line; it must be tightly integrated with both the party's central work and with the mass movement. This has a decisive significance for the improvement of the accuracy of statistical materials. But taking the mass line does not mean that we must abandon actual investigative statistical work and blindly believe the reports that the masses make, [i.e., we cannot simply] record the numbers that the masses give [to us]. *In the midst of the mass movement, the statistical bureau must exert its professional authority, it must use scientific statistical methods [to] reflect the objective reality and thereby uphold the truth.* Only in this way can we provide reliable statistical data so that party and government leaders have a good idea [of the state of affairs] (Emphasis in original, where the sentence is in italics).[102]

The summary then listed a host of proposed measures to achieve such a goal. The only clear concession in this list was the call to combine the use of comprehensive statistics with typical surveys and to use the latter to verify the former.

In his own remarks at the April meeting, which were circulated by the SSB in May 1959, Jia Qiyun noted that since 1958 several new measures had been adopted to overcome problems caused by an overt reliance on the Soviet Union. But these new measures had generated problems of their own, chief among which was that the pendulum had swung too much in the other direction: If earlier statistical work was riddled with dogmatism, then the reforms had led departments and districts to completely neglect the usefulness of specialized statistical institutions, going so far as to even reduce the number of statistical cadres.[103] Statistics, Jia asserted, remained a science and retained professional and technical elements.[104] It was this balance between professionalization and politics that remained unresolved:

102 BMA 133-001-00121: 4.
103 BMA 133-001-00121: 11–12.
104 BMA 133-001-00121: 14.

In the past few years, when carrying out statistical work [we] stressed the responsibilities of the statistical bureau but did not emphasize enough the leadership of the party and the masses. These days, if we only have the "mass line," and no dedicated statistical agency to take responsibility for applying scientific statistical methods to our work, this too cannot help the party offer effective leadership to statistical work; [like before] it is also [a] one-sided [solution]. Experience has proven that under the overall leadership of the party, the professional statistical authorities must work in conjunction with the masses to carry out statistical work.[105]

Jia's summary again reiterated the wider commitment to combine the use of comprehensive statistics with typical surveys.[106]

By the time of the next meeting in November 1959, the Lushan Conference had already taken place, completely politicizing the GLF and making principled objections of any kind effectively impossible.[107] The effect on the SSB was noticeable. In remarks delivered on 3 November, Jia Qiyun adopted an altogether different tone from his utterances months earlier. Now his principal objective was to stress the need to raise the "party spirit" of statistical work. Gone were references to expertise or professional training as independent and necessary qualities. Instead, they were replaced by the acknowledgment and exhortation to make statistics a tractable and handy tool of the party. Jia went so far as to declare that it was incumbent upon statistical workers to recognize that scientific statistical methods and the needs of the party were one and the same.[108] Accuracy remained an aspirational goal, but only the kind of accuracy that guaranteed the party's interests.[109] Jia reminded everyone that shortcomings in statistical work were but a single finger; the other nine fingers represented achievements and successes. It was the duty of statistical cadres to promote information about the nine fingers. As for the leftover solitary finger,

---

105 BMA 133-001-00121: 15.

106 BMA 133-001-00121: 17.

107 The Lushan Conference was a series of meetings held in July and August 1959, which included an expanded session of the Politburo (2 July to 1 August) and the Eighth Plenum of the Eighth Central Committee of the Communist Party (2–16 August). It was during this conference that Peng Dehuai, China's defense minister at the time, took a principled stand against the GLF and, by extension, against Mao's leadership. His downfall signaled the continuation of the GLF policies. See Kong Dechao, "'Dayuejin' qijian de tongji gongzuo," 32.

108 BMA 133-001-00121: 23.

109 BMA 133-001-00121: 24, 25.

it was to be reported to the superiors for review, but one had to exercise extreme caution lest such putatively damaging information be wantonly shared with all and sundry.[110]

The transformation of statistical work, as signaled in Jia's November 1959 speech and codified through both publications and meetings over the course of the year, received its final and decisive imprimatur with the publication and circulation of an intraparty document in April 1961. Published by the Confidential Office of the General Office of the Central Committee and entitled *Mao Zedong lun diaocha yanjiu* (Mao Zedong on investigation and research), the document was an attempt to produce a carefully selected compilation of Mao's writings over the years and to recast them as a coherent theory of social investigation.[111] It thus took the claims in Feng et al.'s 1960 article and gave them official validation. In a paragraph-long introduction, Mao explained that the first article in the collection, entitled "On Investigation Work," was a rebadged version of his 1930 essay "Oppose Book Worship." Mao claimed he had forgotten all about the essay but a chance encounter with a copy preserved in the Central Museum of the Revolution had led to a realization that "it still has some use, so I am having a good many copies printed as reference for comrades."[112]

Unsurprisingly, the emphasis in this essay and in the publication as a whole was on an ethnographic mode of analysis that placed the surveyor's personal on-the-ground experience at the heart of any kind of social research. Part seven of the essay, on the technique of investigation, offered seven guidelines: hold fact-finding meetings and undertake investigation through discussions; pick meeting attendees with care; decide on the size of the meeting; prepare a detailed outline for the investigation; ensure personal participation; probe deeply; and take your own notes. In a 1961 appendix to the piece, the CCP Center offered further instructions:

> The Centre asks that from now on, the leading personnel of Party committees above the county level, and first of all their first Party secretaries, conscientiously study comrade *Mao Zedong's ideological method and work method*, go down to the lowest levels (including in the rural and urban areas) and squat there, engaging in person in systematic *investigation of typical*

110 BMA 133-001-00121: 25.

111 Confidential Office of the General Office of the Central Committee, *Zhongfa* [1961] 261: *Mao Zedong on Investigation and Research*.

112 Ibid., 5.

cases, at least a couple of times each year, and make this the main task of leadership work (emphasis added).[113]

Mao's original rationale for survey work, "to determine correct tactics for [revolutionary] struggle, to determine which class is the mainstay of the revolutionary struggle, which class we ought to make an alliance with, which class we must strike down," had won the day.[114] Any residual concerns with typicality (or representativeness) were suppressed. Indeed, it would not be incorrect to label the surveys of 1959 "paradigmatic" rather than typical. Further work is needed to explore in detail variations in the places that were selected for surveying during the GLF. For starters, if they were places that proved relatively capable of weathering the transition to communes, rather than places that suffered greatly from famine, then such a "selection bias" would presumably have skewed the inferences that were being made about the general effects of the GLF campaign.

Aftermath

By the end of the summer of 1960, China's leaders, including Mao, had already come to realize that the GLF was a failure on most fronts.[115] However, a sense of how bad the situation was continued to elude them. It was an application of the Maoist investigation method in April 1961 that finally brought home the gravity of the situation. Tellingly, it also demonstrated how deeply reliant the method was on the quality of the investigator and on the duration of the survey. At the beginning of April, even as "Mao Zedong on Investigation and Research" was being finalized, Liu Shaoqi took a trip to his home county of Ningxiang in Hunan province. The trip has since become part of historical lore for the effect it had on him. Left speechless by the devastation he witnessed as he traveled from the provincial capital of Changsha to his native village of Huaminglou, Liu immediately decided to investigate further. Over the course of eighteen days, he and his team carried out a detailed investigation of the conditions in Tianhua Brigade (a model brigade within Changsha County's Guangfu People's Commune).[116] On 11 May, Liu wrote to Mao to describe

113 Ibid., 13–14.
114 Thompson, tr., *Report from Xunwu*, 27.
115 Yang, "Surviving the Great Leap Famine."
116 Zhou, ed., *The Great Famine in China*, 162. On Liu's approach to survey work, see Xing Hao, "Jingxin mouhua."

what he had seen. With the support of other leaders, including Zhou Enlai and Deng Xiaoping, Liu slowly began to reassert control over national policy. He reeled in the worst excesses of the GLF, disbanding the backyard furnaces and downsizing the communes.

In the world of statistics, these reforms included the appointment of Wang Sihua as new director of the SSB on 12 June.[117] Among Wang's chief challenges was the need to redress the statistical system's overwhelming reliance on typical surveys. This was far from a straightforward task. In a speech delivered in Guangzhou on 13 March, just one month before Liu Shaoqi's investigation of Tianhua brigade, Mao had reiterated his disdain for periodic reporting and the regular work of the SSB:

> To have a few [periodic] reports is okay; the SSB's job is [after all] statistical work. But in order to understand the situation [on the ground], we cannot rely primarily on [periodic] reports, nor can we rely entirely on step-by-step upward reporting [of data]. Rather, we must rely on personally engaging in investigation and research.[118]

The number of personnel within the SSB system was symptomatic of this unevenness and contempt for regular work. From a high of 200,000 personnel in 1956, the number would fall to 20,547 by 1962.[119] In March 1963, a national meeting was held to take stock of statistical work over the past thirteen years.[120] The GLF—again referred to as the period of the Second Five-year Plan—was discussed in a largely positive light, focusing on the advances made in the use of various survey methods and the increased alignment of statistical work with politics. But the meeting also acknowledged that problems of exaggerated numbers and too much decentralization had hamstrung these advances.[121] It was left to Xue Muqiao, now at the SPC, to demand a more thorough analysis of why such exaggeration had occurred and why it had stood uncorrected for two years.[122]

As the SSB's third director, Wang Sihua oversaw a range of new activities, including the restoration of various periodic reports and an expansion in the number of statistical personnel. Among the most visible achievements of the

---

117 *DSJ*, 78.

118 Cited in *DSJ*, 77.

119 Huang, "The Statistical Agency in China's Bureaucratic System," 61.

120 BMA 133-001-00262: 1–22.

121 BMA 133-001-00262: 4–5.

122 BMA 088-001-00761: 117.

period was the successful execution of China's second population census (1964). But these attempts at a technocratic restoration, themselves a part of Liu Shaoqi's wider reforms, ground to a halt in 1966. In the summer of that year, Mao launched the Great Proletarian Cultural Revolution. By the time Wang stepped down from the directorship of the SSB in 1969, statistical work had all but ceased.[123]

The influence of typical sampling survived the tumult of these years and found expression in the genre of official documentation (公文; *gongwen*) known as the investigation report (调查报告; *diaocha baogao*). Handbooks for writing official documents published in the 1950s do not contain any entries for *diaocha baogao*.[124] By the 1980s, however, *diaocha baogao* had become a standard feature of such handbooks, which routinely offered detailed expositions on how to produce them.[125] Handbooks from the 1980s typically list *diaocha baogao* with two other types of *baogao* (reports): responses to superiors (上复性报告; *shangfuxing baogao*) and clarification reports (知照性报告; *zhizhaoxing baogao*).[126] *Diaocha baogao* were themselves typically divided into two types. One type took the form of a thorough investigation of a specific case; the goal being to get to the truth of a particular matter or situation.[127] The other type was policy-related and broader in scope, typically investigating and responding to questions about methods, guiding principles, or policy measures. The handbooks identified Mao's 1927 *Report on an Investigation of the Peasant Movement in Hunan* as an example of the second kind.[128] Such reports were typically compiled by work teams composed of local people and

123 Statistical work during the early phase of the Cultural Revolution (1966–1969) was especially hard hit. If we take the annual number of events listed in the *DSJ* as a proxy, the years 1967–1969 stand out for almost no notable statistical activity. Those three years are the only years (barring 1949) that have single-digit entries (4, 5, and 1, respectively). The numbers for the early 1970s are marginally higher, averaging about 12 per year. The average for the period from 1949 to 1990 is 32 per year. See *DSJ*, 1–16.

124 I consulted the following four handbooks, none of which had an entry for *diaocha baogao*: Ling Xianfang, ed., *Renmin zhengfu gongwen chengshi*; Meng Ping and Zhu Li, eds., *Gongwen de jiben zhishi*; Zhongyang renmin zhengfu zhengwuyuan, *Gongwen chuli zanxing banfa*; and Su Fan, ed., *Xin gongwen shouce*.

125 In some instances, verification reports (核查报告; *hecha baogao*) and inspection reports (考察报告; *kaocha baogao*) were also used (Miao Fenglin, *Zhongguo gongwen xue*, 128).

126 Zhang Qingming, *Wenshuxue ji shiyong gongwen (xiuding ben)*, 196.

127 Miao Fenglin, *Zhongguo gongwen xue*, 128.

128 Zhang Qingming, *Wenshuxue ji shiyong gongwen (xiuding ben)*, 198; Miao Fenglin, *Zhongguo gongwen xue*, 128.

outsiders who had been sent to a locality. The team would spend some time obtaining a general understanding of local conditions, identify informants, gain their trust, and then use a snowball methodology to get to know more people. Only after having collected many viewpoints—a process that could take weeks—would the team synthesize the information into a report.

Conclusion: On Methods and Manipulation

The methodological debates traced here tell us that for much of the 1950s there was legitimate discontent with routine reporting and that typical sampling provided an enticing alternative. That it appeared more "red" and less "expert" than routine reporting only added to its attractiveness. At the same time, there should have been strong arguments against typical sampling, even in a Maoist framework—arguments based on the evident fact that the method was cheap, fast, and practical only if you trusted people who you knew from experience you could not trust. By 1959, such opposition was no longer tenable, creating the impression that a bona fide debate between proponents of different methodologies, whose merits were as yet unproven, was overtaken by a decision to suppress a methodology that could have bred doubts and uncertainties and to promote instead a methodology that produced only positive feedback.

The success of typical sampling raises questions about the relationship of these methods to mass science. Although there was a moment in 1958 when indeed everyone was asked to do statistical work, the evidence presented here indicates a different agenda. Before long, it was not the masses that were being told to alter their statistical work but rather the "leading personnel of party committees above the county level, and first of all their first party secretaries."[129] The dichotomy that then emerges at the height of the Great Leap, and shortly thereafter, is not between the scientist and the native informant (or even the scientist and the local political activist) but that between the scientist and the cadre. Fundamentally political in nature, statistics could not, in spite of protestations by the SSB, be left to people whose main training and outlook valued professionalization as much as, if not more than, political correctness (政治性; zhengzhixing). Even so, the rhetorical influence—if not the practice—of mass science, with its emphasis on antiexpertise and antiprofessionalization, persisted within the execution of socialist statistics into the 1970s. It

129 Confidential Office of the General Office of the Central Committee, *Zhongfa* [1961] 261: *Mao Zedong on Investigation and Research*, 13–14.

is this notion that has come to inform our understanding of statistics during the Mao era.[130]

The GLF continues to be understood as a multicausal event, and new explanations continue to be proposed. In a recent *American Political Science Review* article, James Kung and Shuo Chen argue that excess grain-procurement variations were patterned systematically on the political career incentives of Communist Party officials rather than on the conventionally assumed ideology or personal idiosyncrasies.[131] Numerous other explanations have been cited earlier in this chapter. To these causes we can add the central role played by the refashioning of statistical work. The promotion of typical surveys reduced the SSB's capacity to know and verify the numbers that were being reported. The extreme decentralization of the bureau down to units and individuals, who frequently had no statistical training, occurred in an environment that incentivized exaggeration. This proved to be a particularly volatile combination. Even as the SSB bemoaned exaggeration, it was essentially denied the capacity, whether through the periodic reporting system or large-scale random sampling, to assess and repudiate any of those exaggerations. As Li Choh-Ming had noted in his own study, "had the bureau been in a position to insist on complete enumeration, a large margin of error might yet have been present in the reports, but such fantastic claims as those of 1958 would have been checked with relative ease."[132] We can thus understand GLF statistics as part of the ongoing tussles over appropriate statistical methodology that spanned the entire decade of the 1950s.

130 See, for instance, Liu, *The Mirage of China*.

131 Kung and Chen, "The Tragedy of the *Nomenklatura*." Kung and Chen were subsequently criticized, in an ironic turn, for fabricating data to support their argument. See Yang, Xu, and Tao, "A Tragedy of the *Nomenklatura*?"

132 Li, *The Statistical System of Communist China*, 87.

9

Conclusion

IN 1974, LEO ORLEANS, then the China Research Specialist at the United States Library of Congress, wrote a short article pithily entitled "Chinese Statistics: The Impossible Dream."[1] Whether he was inspired by *Man of La Mancha* and Peter O'Toole's haunted rendition of the famous song cannot be known.[2] But his description of Chinese statistics as "notorious" and the Chinese as a "nation that does not value accuracy" suggest he may very well have. As he remarked in the article: "The contention here is that this is not the case, but that in fact, the Chinese themselves lack much of the statistical data we so desperately seek and that their traditional nonchalance toward accuracy in statistics persists among many of the people presently responsible for record-keeping."[3]

The study of modern China has long presented a paradox to the China specialist, to whom data have rarely been made available.[4] As the economist Thomas Rawski noted in a recent review of scholarship on the Chinese economy: "The common method underlying these studies, particularly those focused on the People's Republic, was the use of fragmentary data to reconstruct

1 Orleans, "Chinese Statistics: The Impossible Dream."

2 The 1972 movie was based on the eponymous 1965 Broadway musical, which in turn was adapted from the nonmusical 1959 teleplay *I, Don Quixote*. The original inspiration for all of these works was, of course, Miguel de Cervantes's seventeenth-century *The Ingenious Gentleman Don Quixote of La Mancha (El ingenioso hidalgo don Quijote de la Mancha)*. Both musical and movie are famous for the song "The Impossible Dream," in which Don Quixote expounds on his quest and his knightly duties.

3 Orleans, "Chinese Statistics: The Impossible Dream," 47.

4 Throughout the 1950s and the ensuing decades, statistics in the PRC remained a state secret.

the broader economic landscape. This approach arose from the paucity of systematic data prior to 1949 and from the limited publication of economic statistics under the People's Republic until the appearance of the *Zhongguo tongji nianjian* (国统计年鉴; China statistics yearbook) in 1981."[5] As a result, each occasion when the Chinese released data (e.g., census numbers in 1954, *Weida de shinian* in 1959, or production figures in 1970) generated tremendous excitement.[6]

And so, one gets the impossible dream of the China scholar tilting at the proverbial windmill of Chinese data. To abuse another metaphor, for many decades, statistics from the PRC was treated like gold dust: nearly impossible to obtain, but with the additional (and often justifiable) fear that they were the product of alchemy. Any moment one might discover what one held in one's hands was dust rather than gold. Orleans explained the frustrations of China watchers by noting: "Never having known the advantages (and, yes, the frustrations) of a good statistical base, they [the Chinese] don't miss it nearly as much as do many of us who look at China from the outside."[7]

The mainstream response in the West to the problem of Chinese data has been characterized by a tendency to not take Chinese statistics from the Mao years too seriously. The two extant book-length studies on PRC statistics exhibit this tendency, but in contrasting fashion. The first study was written by the economist Li Choh-Ming in the early 1960s, and the second was written by the anthropologist Xin Liu earlier this century. Li was primarily invested in gauging the accuracy of statistical data produced in the early PRC and in explaining the statistical failure associated with the 1959–1961 famine.[8] Liu's more recent anthropological study focused on the transition from what he characterizes as Maoist subjective mass science to objective science after 1978.[9] If the latter paints the entire Mao period from 1949 to 1978 with the broad brush of subjective mass science, then the former tends to neglect the

---

5 Rawski, "Studies of China's Economy," 175. See also the discussion in Perkins, "Research on the Economy of the People's Republic of China."

6 Excitement was often accompanied by incredulity. For instance, for a sense of the surprise and skepticism generated by the 1953 census numbers, see Krader and Aird, "Sources of Demographic Data on Mainland China"; Aird, "China's Population, Census and Vital Statistics"; Aird, "Estimating China's Population"; Coale, *Rapid Population Change in China*; and Cressey, "The 1953 Census of China." On 1970 production figures, see Orleans, "Chinese Statistics: The Impossible Dream," 51.

7 Orleans, "Chinese Statistics: The Impossible Dream," 47.

8 Li, *The Statistical System of Communist China*.

9 Liu, *The Mirage of China*.

scientific basis for the design of statistical work during the 1950s, blaming its collapse largely on the GLF of 1958. Aside from these book-length studies, journal articles by political scientists have offered institutional and structural analyses of statistical work, focusing in particular on the reform era. In similar fashion, comparisons with the Soviet reforms and the role of official statistics have also been carried out.[10]

But what if we take statistics from the first decade of the Mao era seriously? As the chapters in this book demonstrate, what emerges is a new understanding of the Chinese Communist project and its links to the wider world. A shift in focus away from questions of accuracy (the principal object of earlier scholarship) creates space to interrogate instead how statistics, a putatively neutral field, became the site for a fundamental theoretical battle about the nature of social reality. Chinese statisticians took great pains to draw distinctions between the natural and social worlds, distinctions that they believed reflected the most advanced and scientific principles. Statistics' proper domain, they resolved, was the social world, a place ruled by class analysis rather than uncertainty or chance.[11] The choice of statistical methods followed with elegant path dependency: Out went the "bourgeois" and "formalist" conceits of mathematical statistics and probability theory, embraced were the exhaustive methods of periodic reports and censuses.

The Soviet Union, whose expertise on definitional matters was eagerly embraced, also provided the model for a national statistical system. This system was a new product of the PRC, built largely from scratch, and at its apogee claimed a cadre of 200,000 statistical workers. Given China's geographic and demographic scale and the largely agrarian nature of its economy, the production and filling out of forms represented a mammoth effort for the SSB. Largely successful in the industrial sector, complete enumeration proved impracticable in the face of China's hundreds of thousands of villages (and later, thousands of communes). By 1956, the system's internal contradictions—delays, overproduction of reports, inadequately trained and insufficient personnel, reliability, and morale—all came to a head.

A potential solution was located in the unlikeliest of places. By the 1950s, the ISI in Calcutta had become one of the leading centers for statistical research and application and had extensively demonstrated the efficacy of

10 See, inter alia, Huang, "Information, Bureaucracy, and Economic Reforms"; Huang, "The Statistical Agency in China's Bureaucratic System"; Perkins, "Research on the Economy of the People's Republic of China."

11 A play on Gigerenzer et al., *The Empire of Chance*.

large-scale random sampling. A series of encounters with Indian planners and statisticians during 1956 gave Chinese statisticians the opportunity to discover the power of probabilistic methods in generating large amounts of accurate data with many fewer personnel. Zhou Enlai's visit to the ISI added an official seal of approval. Thus began a short-lived but lively exchange of personnel and expertise between the SSB and the ISI. But even as leading statistical officials prepared the theoretical and practical grounds for accommodating random sampling within Chinese statistical work, the Anti-Rightist movement exploded, rendering these efforts fraught. And within a year, the GLF left these attempts to adapt moot. Belying its status as an ancillary method, the ethnographic method of typical sampling was dramatically elevated to prime status, undermining not only the incipient dalliance with random sampling but also the existing network of periodic reports and censuses.

The story of midcentury China's struggles with generating statistical knowledge tells us much about the nature of the Chinese socialist state. It helps us better understand the constraints under which planning had to operate. Throughout the decade, Chinese statisticians and leading officials were very much the agents of the choices they made. They aggressively adopted the Soviet model in the early 1950s, energetically explored alternatives when that model began to fail them, and struggled with the pressures they were put under during the GLF. The system that they built was excellent at producing data; it actually incentivized the production of numbers. And numbers allowed for decisions to be made, setting in motion a vicious circle of data production and overproduction. At the same time, the system lacked any significant technology to check its numbers. A balanced use of ethnographic, exhaustive, and stochastic methods, which recognized the strengths and weakness of each, and used the results from one to check and strengthen those from another, remained beyond their grasp. The outcome was a Chinese state that, in spite of generating copious amounts of facts, remained poorly informed.

The tension between data generation and its use has broader implications for how we think about facts-based governance or the links between ignorance and decision making.[12] At an almost elemental level, even before we address issues of unintended consequences, it alerts us to the importance of initial assumptions about what exactly needs to be known: the object of collection and analysis. For China's socialist statisticians, statistics took as its object the

---

12 See Jerven, *Poor Numbers*, for an exploration of this tension in contemporary sub-Saharan African economies.

study of society and society alone. But based on the discussion presented here, we could go further and claim that statisticians' activities were additionally circumscribed by an interest in information that was "actionable." The state collected information in areas for which it had a transformative agenda. While these areas were no doubt large, as evidenced by the struggles described in this book, they also tended to exclude other domains that were potentially amenable to statistical analysis. Typically, these were areas in which the state had no broad or consistent policy to pursue: consumer preferences, hobbies, personal beliefs, sexual preferences, women's domestic work, and so on. Even the acknowledgment in 1957 that the state was unable to assess changes in the people's livelihood confined itself to outcomes of state action: family income and expenditures, purchasing power, wages, population, prices, and so forth. In other words, the primary objects of knowledge of the Chinese socialist state were its own actions and their consequences. While potentially stymieing effective policy making, at a broader level this focus also suggests a certain logical circularity. The state was unable to see and measure society as anything other than a response to its own policies.[13]

There were other facts also ignored by China's socialist statisticians. In actively dismissing the usefulness of pre-1949 social science research, they effectively denied themselves much useful information, especially about the Chinese countryside. Even though this knowledge was not a product of the complete enumeration valorized in the 1950s, it nonetheless was the outcome of serious efforts on the part of state and nonstate actors who carried out sustained local studies of various agrarian regions. Furthermore, given Yan'an's atypical nature, a reliance on the experience gained there meant that the CCP entered the 1950s with limited knowledge and experience about most of China. This lack of knowledge was compounded by a reliance on a system that generated copious amounts of data but not the capacity to make sense of them.

Contemporary comparisons with other countries are also instructive. It is important to recognize that the actions of Chinese statisticians and officials cannot be understood without acknowledging the contrasting Soviet and Indian influences. Doing so not only revises existing models of Cold War science but also globalizes the history of statistics and data, demonstrating wide-ranging developments in what has often been narrowly construed as a universal (if European) history. One way to expand our lens and look at the actions of states during the era without a priori privileging U.S.-centric and

13 My thanks to Jacob Eyferth for this penetrating insight.

Soviet-centric networks is to investigate the role of data and expertise in all their varied contexts. This would permit drawing comparisons and making connections not only with India, which is discussed here, but also with other countries in East Asia—Taiwan, the two Koreas, and Japan—that all emerged out of shared historical experiences and possess ties that extend back hundreds of years.

Such a wide-angle perspective may also identify historical patterns that at first glance appear distinct. The turn to local ethnography in the GLF period, for instance, has interesting echoes outside of China and reminds us that critiques of total enumeration possessed a rationale that cut across ideological and political boundaries. By the 1960s, people elsewhere had also begun to question the efficacy of centralized, state-led, top-down modes of development. Some, like the development economist Albert Hirschman, became disenchanted with large-scale statistical data, highlighting instead the importance of local initiatives, small-scale solutions, and "model regions."[14] Others, inspired by the possibilities of community development sought to present alternative paths to economic growth and development.[15] It was apparent to many that large, bureaucratically generated systems of data had limitations. These battles, between large aggregations of data and in-depth contextual knowledge, remain fundamental to debates within the social sciences today. Much as in the case of statistics in China in the 1950s, they frequently generate partisan loyalties, often with telling consequences.

By the early 1980s, the balance between the ethnographic and the exhaustive in China appeared to have returned to that of the mid-1950s. In an article published in 1982, S. Lee Travers observed that complete enumeration continued to supply the bulk of statistics on China's economy.[16] As for typical surveys, he explained:

> There are, of course, areas of interest to Chinese planners and social scientists, the data for which are not routinely compiled at local levels, e.g. data on family consumption patterns. In China, as elsewhere, sample surveys are used to fill such information gaps. Such surveys are commonly conducted using the technique of "typical example investigation" (*dianxing diaocha*).[17]

14 Adelman, *Worldly Philosopher.*
15 Immerwahr, *Thinking Small.*
16 Travers, "Bias in Chinese Economic Statistics."
17 Ibid., 478.

Travers went on to discuss the problems of bias that were unavoidable when using typical surveys. To demonstrate his point, he performed a test of data generated by a typical survey, conducted in 1979, of income and consumption patterns among rural residents. He found that "the probability of obtaining these sample results under a random sampling procedure approaches zero," and concluded that typical surveying could generate a substantial bias in the statistics.[18]

Travers's article was published at a time when Chinese statistics and statistical work were undergoing significant changes. A general restoration of a statistical infrastructure had already begun by 1972. In that year, the provincial bureaus in Inner Mongolia and Liaoning were restored. In the following year, they were joined by bureaus in Hebei, Jilin, Heilongjiang, and Sichuan. Shanxi, Zhejiang, Jiangsu, Henan, and Hunan followed in a slow trickle over the next several years. The floodgates were opened in 1978 and 1979, when sixteen provincial and city bureaus were opened.[19] In similar fashion, a whole range of periodic reports and surveys were reinstituted, and a China Statistical Society was set up in November 1979.[20] An altogether new development was the gradual opening up to statistical exchanges with countries in the West.[21] The PRC was admitted to the United Nations in 1971, but it was not until 1978 that Simon Goldberg, the Canadian director of the United Nations Statistics Division, led a team to China.[22] Statistical links with the United States (1979) and with Japan (1980) quickly followed suit.[23] In May

18 Ibid., 482–484.

19 See *DSJ* (pages in parentheses):

1972: Inner Mongolia, Liaoning (118)

1973: Hebei, Jilin, Heilongjiang, Sichuan (118, 119)

1975: Shanxi (122)

1976: Zhejiang (125)

1977: Jiangsu, Henan, Hunan (127, 128)

1978: Hubei, Guizhou, Tianjin, Shanghai, Gansu, Anhui, Ningxia, Guangdong, Yunnan (128–131, 133)

1979: Shaanxi, Jiangxi, Fujian, Guangxi, Shandong, Qinghai, Xizang (134–137, 140)

1980: Xinjiang (140)

20 *DSJ*, 138.

21 For contemporaneous exchanges involving Western and Chinese economists, see Gewirtz, *Unlikely Partners*.

22 *DSJ*, 131.

23 *DSJ*, 136, 141.

and July 1981, American and Chinese teams of statisticians exchanged visits.[24]

The most significant changes, however, took place in the world of statistical theory within China. On the surface, the dichotomy between statistics and mathematical statistics had endured undisturbed into the early 1980s. Some years ago, the Harvard statistician Xiao-Li Meng related to me that the first time he encountered a histogram with economic data was in graduate school in the United States in 1986. Prior to that, as a student of mathematics and mathematical statistics at Fudan University in Shanghai, his curriculum had been purely theoretical and avoided engagement with social or economic data.[25] But the seeds of change had already been planted by the late 1970s. In 1979, the statistician Dai Shiguang, the Renmin University professor who had accompanied Wang Sihua to the ISI in December 1956, rekindled the debate about the true nature and purpose of statistics. In an article published in the journal *Jingji yanjiu* (Economic Research) he set about demolishing what he called the theoretical cloak of Soviet socialist statistics. Using the rhetoric of "reform and opening up" and its renewed emphasis on science and technology—one of the four modernizations—Dai called for wholesale reform of statistics in China, including broader engagement with mathematical methods. His article was republished in the journal *Tongji* (Statistics), the recently reestablished descendant of *TJGZ*, and ignited major changes within the world of Chinese statistics.[26]

24 *DSJ*, 148.

25 Discussion with Xiao-Li Meng, Professor of Statistics and Dean of the Graduate School of Arts and Sciences, Harvard University, Cambridge, MA, 12 December 2013.

26 Dai Shiguang, "Jiji fazhan kexue de tongjixue."

Individuals, Institutions, and Statistical and Other Relevant Terms

ability　　才

accountant　　会计

activist (also enthusiast)　　积极分子

advanced experience　　先进经验

aggregate index　　总量指标

Agriculture Statistics Department　　农业司

annual bulletin　　年报

Anti-Imperialist League　　反帝大同盟

Anti-Japanese Society　　抗日救国会

Anti-Rightist Campaign　　反右派运动

arithmetic mean　　算术平均数

average index　　平均指标

averages　　平均数

Ba Jin (1904–2005)　　巴金

Bao Shichen (1775–1855)　　包世臣

Beijing Business and Vocational High School　　北京高级商业职业学校

Beijing Finance and Economics School　　北京财经学校

Beijing Public Enterprise Company　　北京市公营企业公司

Beijing School of Law and Politics　　北京法政专门学校

Beiping Second War Zone Mobilization Committee　　北平第二战区动员委员会

bias　　偏误

Bo Yibo(1908–2007)　　薄一波

bureaucratism　　官僚主义

cadre　　干部

Cai Yuanpei (1868–1940)　　蔡元培

capital construction statistics　　基本建设统计

Central Party School   中央党校
Central Politics School   中央政治学校
Central Printing House   中央印务馆
Central Research Institute   中央研究院
Central Statistical Association   中央统计协会
chain indices   环比指标
chaotic issuing of reports   乱发报表
Chen Da (Ta Ch'en, 1892–1975)   陈达
Chen Hansheng (1897–2004)   陈翰笙
Chen Yingzhong (1904–??)   陈应中
Chen Yun (1905–1995)   陈云
Chen Zhongsheng   陈钟声
China Statistical Society (1979)   中国统计学会
China Statistical Society (1929)   中国统计学社
comparative index   比较指标
complete enumeration periodic reporting system   全面定期统计报表制度
composite index   综合指标
Confidential Office of the General Office of the Central Committee   中央办公厅机要室
constant prices   不变价格
Construction and Engineering Bureau   建筑工程局
continuous sampling   连续抽样
control (also supervision)   监督

Dai Shiguang (1908–1999)   戴世光
Di Chaobai (1910–1977)   狄超白
distribution of goods   物资分配
Dong Biwu (1886–1975)   董必武
double track   双轨制
dynamic index   动态指标

Economic Research Institute of the Manchu Puppet Regime   伪满经济研究所
Education Bureau   教育局
"eight big formulas"   八大公式
eight immortals   八大元老
enthusiast (also activist)   积极分子
Expert   专

fault   毛病
Feng Guifen (1809–1874)   冯桂芬
Finance and Economics Committee   财政经济委员会
five-stage teaching method   五段教学方法
formal mathematical doctrine   形式数学
founder   奠基人
Full Wind Publishing Co.   长风出版社
future prospects   前途

Gao Gang (1905–1954)　高岗
general probability theory　概率论 or 几率论
geometric mean　几何平均数
Gong Jianyao (1927–)　龚鉴尧
group average　组平均数
grouping　分组
Great Leap Forward　大跃进
Gu Cheng (1882–1947?)　顾澄
Guo Gengji (1921–??)　过庚吉
Guo Ling (??–??)　郭凌
Guo Moruo (1892–1978)　郭沫若

He Lian (Franklin L. Ho, 1895–1975)　何廉
He Long (1896–1969)　贺龙
high tide　高潮
Hong Liangji (1746–1809)　洪亮吉
Hong Xiuquan (1814–1864)　洪秀全
Hou Wailu (1903–1987)　侯外庐
Housing Administration　房管局
Hu Daiguang (1919–2012)　胡代光
Hu Huanyong (1901–1998)　胡焕庸
Huang Jiantuo (??–??)　黄建拓
Hundred Flowers Movement　百花运动

ideal index formula　理想公式
Imperial Maritime Customs Service　大清皇家海关总税务司
independence of events　几个事件各各独立
index number　指标数
indicators　指数
industrial management　工业管理
industrial statistics　工业统计
inference　推论
infrastructure statistics　基建统计
International Society　国际学社
interpenetrating　交叉
interpenetrating sampling　交叉抽样
investigation meetings　调查会
investigator　调查员

Jia Qiyun (1914–2004)　贾启允
Jia Tuofu (1912–1967)　贾拓夫
Jin Guobao (Kuo-Pao King, 1893–1963; pseudonym, Jin Lüqin 金侣琴)　金国宝

large-scale observation　大量观察法
law of large numbers　大数法则

Lei Feng (1940–1962)　雷锋
Li Dazhao (1888–1927)　李大钊
Li Deyin (1902–1976)　厉德寅
Li Fuchun (1900–1975)　李富春
Li Zhenzhou (1920–2009)　李振周
Liang Qichao (1873–1929)　梁启超
lijia　里甲
Lin Biao (1907–1971)　林彪
Lin Fude (1925–2017)　林富德
Liu Dajun (D. K. Lieu, 1891–1962)　刘大钧
Liu Hongwan (??–??)　刘鸿万
Lin Lifu (1909–2001)　林里夫
Liu Pengnian (Bangnee Alfred Liu, 1901–88)　刘彭年
Liu Xin (??–??)　刘新
Local Industry Bureau　地方工业局

Ma Yinchu (1882–1982)　马寅初
Mao Zedong (1893–1976)　毛泽东
measure of central tendency (average numbers)　平均数
median　中位
mid-level vocational cadre　中等专业干部
mode　众数
model worker　模范工作者

Nanjing Higher Normal School　南京高等师范学校
National Chongqing University　国立重庆大学
National School of Commerce　国立商学院
National Shanghai School of Business　国立上海商学院
National Southwestern Associated University　国立西南联合大学
New Schools of Law and Administration　法政学堂
Ni Jiaxun (1931–)　倪加勋
noncomprehensive survey methods　非全面调查方法
North China Finance and Economics Meeting　华北财经会议
Northeast People's University　东北人民大学
numbers' game　数字游戏

observers　列席代表
original records　原始记录

party group　党组
"paradigmatic" survey　典型调查
Peng Dehuai (1898–1974)　彭德怀
people's subjective will　人的主观性
People's University　人民大学

period of recovery   恢复时期
period of transition   过渡时期
Pioneer People's Commune   先锋人民公社
probability sampling   机率抽样
projection   推算
Public Health Bureau   公共卫生局
Public Services Bureau   公用局
Public Works Bureau   建设局
Purge of Hidden Counter-revolutionaries   肃反运动
purposive method   典型研究

Qi Luming (??–??)   祁鹿鸣
qualitative   质
quantitative   量

random or stochastic sampling   随机抽样
rectification of original records   整顿原始记录
Red   红
relative index   相对指标
reorganization of educational institutions   （高校）院系调整
report   报告

sample   样本
sample household   调查户
sample survey   抽样调查
Sanitation Engineering Bureau   卫生工程局
sent down   下方
Shanghai Chinese Public School   上海中国公学
Shanghai Institute of Finance and Economics (SIFE)   上海财政经济学院
Shenyang Wholesale Price Index   沈阳市私商批发物价指数
side-line work   经营副业
single-track   单轨制
Sino-French University   中法大学
social investigation   社会调查
South Seas Tobacco Company   南洋烟草公司
State-Owned Whole Sale Price Index   国营批发物价指数
State Statistics Bureau (SSB)   国家统计局
Statistical Department of the Ministry of Education   教育部统计科
Statistics Office of the Finance and Economics Committee of the Government Administration
    Council of the Central People's Government   中央人民政府政务院财政经济委员会
    统计处
Statistical Office of the National Finance Committee   中财委统计总处
Statistical Survey Unit of the Northeast Finance and Economics Committee   东北财经
    委员会调查统计处

stratification 分组
strength index 强度指标
structural index 结构指标
subjective 主观
subjectivism 主观主义
Sun Shizheng (1919–??) 孙世铮
Sun Yefang (1908–1993) 孙冶方
supervision (also control) 监督

Tang Peijing (Pei-Ching Tang, 1903–1988) 唐培经
Third-Degree Model Worker 三等工作模范
Three- and Five-Antis 三反五反
time series 时间数列
total average 总平均数
total output value 总产值
training classes/courses 训练班
trend 动态数
Tsinghua University 清华大学
typical sampling (also typical survey) 典型调查

University of Commerce 商科大学

vertical leadership 垂直领导
virtue 德
vocational school 专科学校
volatility 流动幅度

weight 权数
Wan Li (1916–2015) 万里
Wang Anshi (1021–1086) 王安石
Wang Feixian (??–??) 王飞宪
Wang Jianzhen (??–??) 王建真
Wang Sihua (1904–1978) 王思华
Wang Shenming (pseudonym) 王慎铭
Wu Cangping (1922–) 乌沧萍
Wu Han (1908–1969) 吴晗
Wu Hui (1927–2018) 吴辉
Wu Zaoxi (1904–1979) 吴澡溪

Xu Baolu (Hsu Pao-Lu, 1910–1970) 许宝騄
Xu Qian (??–??) 徐前
Xue Muqiao (1904–2005) 薛暮桥

Yan Fu (1854–1921) 严复

Yang Bo (1920–2016)   杨波
Yang Jianbai (1911–2004)   杨坚白

Zhang Youyu (1898–1992)   张友渔
Zhou Peiyuan (1902–93)   周培源
Zhu De (1886–1976)   朱德
Zhu Junyi (Jennings P. Chu, 1892–1963)   朱君毅
Zou Yiren (1908–94)   邹依仁

BIBLIOGRAPHY

Archives and Libraries

In China

Beijing Municipal Archives (BMA)
Foreign Ministry Archives of the People's Republic of China, Beijing (WJB)
Institute of Economics, Chinese Academy of Social Sciences, Beijing
Institute of Modern History, Chinese Academy of Social Sciences, Beijing
National Library of China, Beijing
Peking University Library, Beijing
Resource Room, Department of Statistics, Renmin University
Tsinghua University Archives, Beijing
Tsinghua University Library, Beijing

In India

Indian Council for World Affairs, New Delhi
Institute for Economic Growth, New Delhi
National Archives of India (NAI), New Delhi
National Library, Kolkata
Nehru Memorial Museum and Library (NMML), New Delhi
P. C. Mahalanobis Memorial Museum and Archives (PCMMMA), Kolkata
Tata Institute of Fundamental Research (Archives), Mumbai

In the United States

Chinese Text Project (online resource)
Columbia University Library
Harvard University Library
Hoover Institution—Library and Archives, Stanford University
Rare Book and Manuscript Library, Columbia University
Sripati Chandrasekhar Papers, University of Toledo
Wolfram MathWorld (online resource)

(Oral History) Interviews

Debkumar Bose	Kolkata, 14 August 2009
V.P. Dutt	New Delhi, 30 August 2009
Gong Jianyao 龚鉴尧	Guangzhou, 14 September 2011
Meera Kosambi	Pune, 20 August 2012
Lin Fude 林富德	Beijing, 22 April 2011
Ni Jiaxun 倪加勋	Beijing, 29 March 2011; 7 September 2011
Narayan Sen	Singapore, 23 July 2012
Wu Cangping 邬沧萍	Beijing, 11 March 2011
Wu Hui 吴辉	Beijing, 8 September 2011
Yuan Wei 袁卫	Beijing, 25 March 2011

Journals and Newspapers

China

Caijing kexue (财经科学)
Caijing yanjiu (财经研究)
Guangming ribao (光明日报)
Jihua yu tongji (计划与统计)
Jingji zhoubao (经济周报)
Renmin ribao (人民日报) (*RMRB*)
Tongji (统计) (*TJ*)
Tongji gongzuo (Northeast Statistics Bureau) (统计工作) (*DBTJGZ*)
Tongji gongzuo (State Statistics Bureau) (统计工作) (*TJGZ*)
Tongji gongzuo tongxun (State Statistics Bureau) (统计工作通讯) (*TJGZTX*)
Xin jianshe (新建设)
Zhongguo gongye (中国工业)

India

Ananda Bazaar Patrika
The Hindu
Indian Statistical Institute—Annual Report(s)
Samvadadhvam
Sankhya
The Statesman
The Times of India

USSR

Izvestiia
Pravda
Vestnik Akademii Nauk SSSR

Vestnik Statistiki
Bulletin of the Academy of Sciences of the Armenian SSR: Social Science
Current Digest of the Russian Press

Books and Articles

Abraham, Itty. "The Contradictory Spaces of Postcolonial Techno-Science." *Economic and Political Weekly* 41, no. 3 (21 January 2006): 210–217.

Adelman, Jeremy. *Worldly Philosopher: The Odyssey of Albert O. Hirschman.* Princeton: Princeton University Press, 2013.

Aird, John S. "China's Population, Census and Vital Statistics." *Journal of the American Statistical Association* 56, no. 293 (March 1961): 44–51.

———. "Estimating China's Population." *Annals of the American Academy of Political and Social Science* 369, no. 1 (1967): 61–72.

Alonso, William, and Paul Starr, eds. *The Politics of Numbers.* New York: Russell Sage Foundation, 1987.

Altehenger, Jennifer E. *Legal Lessons: Popularizing Laws in the People's Republic of China, 1949–1989.* Cambridge, MA: Harvard University Asia Center, 2018.

Anderson, T. W., K. L. Chung, and E. L. Lehmann. "Pao-Lu Hsu 1909–1970." *Annals of Statistics* 7, no. 3 (1979): 467–470.

Anderson, Warwick. "Postcolonial Specters of STS." *East Asian Science, Technology and Society: An International Journal* 11, no. 2 (2017): 229–233.

Andreas, Joel. *Rise of the Red Engineers: The Cultural Revolution and the Origins of China's New Class.* Stanford: Stanford University Press, 2009.

Angang Steel Planning Office Statistics Section 鞍钢计划处统计科. "Ganxie Sulian zhuanjia dui women de bangzhu" 感谢苏联专家对我们的帮助 (Thanking the Soviet experts for their help), *TJGZ*, no. 21 (1957): 31–32.

Anhui Statistics Bureau 安徽省统计局. "Women jinxing dianxing diaocha gongzuo de yidian tihui" 我们进行典型调查工作的一点体会 (Our experience conducting typical surveys). *TJGZ*, no. 19 (1957): 21–22.

Anon. "Banianlai woguo fanyi chuban le daliang Sulian tongjixue zhuzuo: Woguo tongji chubanshe kaishi yu Sulian tongji xuejie jianli youyi de lianxi" 八年来我国翻译出版了大量苏联统计学著作: 我国统计出版社开始与苏联统计学界建立友谊的联系 (We have translated a large number of Soviet statistical works in the past eight years: Our statistical publishing house has begun to establish friendly relations with the Soviet statistical community). *TJGZ*, no. 20 (1957): 33.

———. *Communist China 1955–1959: Policy Documents and Analysis*, foreword by Robert R. Bowie and John King Fairbank. Cambridge, MA: Harvard University Press, 1963.

———. *Cong di yige wunian dao di erge wunian* 从第一个五年到第二个五年 (From the First Five Years to the Second Five Years). Hong Kong: Chang Feng chubanshe, Co., 1956.

———. "Ge sheng qu shi tongji gongzuo yuejin zhibiao" 各省区市统计工作跃进指标 (Indicators of a Great Leap Forward in statistical work at the province, regional, and municipal levels). *TJGZ*, no. 14 (1958): 26–28.

———. "Heluxiaofu fuhan shijie kexue gongzuozhe xiehui daibiaotuan Sulian yuan liji qianding tingzhi he shiyan xieding geguo renmin zheng yinqie qiwang guoji jushi yousuo huanhe, Meiguo yixie renwu queda jiao da rang yao huifu he shiyan" 赫鲁晓夫复函世界科学工作者协会代表团 苏联愿立即签订停止核试验协定 各国人民正殷切期望国际局势有所缓和   美国一些人物却大叫大嚷要恢复核试验, (Khrushchev's reply to the Delegation of the Association of International Scientists: The Soviet Union is willing to immediately sign an agreement to cease nuclear testing; People of all countries are eagerly awaiting an easing of the international situation; some American figures are screaming for a resumption of nuclear testing). *RMRB*, 13 August 1959, 4.

Anon. "Jiehe xianshi douzheng gao tongji: Tongji gongzuozhe juexin yikao qunzhong sangeyue gaibian gongzuo mianmao" 结合现实斗争搞统计: 统计工作者决心依靠群众三个月改变工作面貌 (Combine reality with struggle to execute statistical work: statistical workers are determined to rely on the masses and change the face of statistical work within three months). *RMRB*, 1 August 1958, 6.

———. *Jisuan jishu jiangyi* 计算技术讲义 (Lectures on the technology of calculation). Beijing: Renmin daxue chubanshe, 1955.

———. "Professor Mahalanobis Welcomes Chou En-Lai." *Samvadadhvam* 1, no. 3 (March 1957): 30–31.

———. "Tongji gongzuo de liangzhong zuofa" 统计工作的两种做法 (Two approaches to statistical work). *RMRB*, 29 April 1958, 1.

———. "Tongji, kuaiji he yewu jishu jisuan de xianghu lianxi he xianghu guanxi: Sulian tongji zhuanjia baogao jilu" 统计, 会计和业务技术计算的相互联系和相互关系: 苏联统计专家报告记录 (Relations and connections among statistics, accounting, and business calculations: A record of reports by Soviet statistical experts). *TJGZTX*, no. 2 (1956): 7–12.

———. "Yikao qunzhong jianli xian xiang she tongjiwang: Hebei yunyong tongji gongzuo cujin shengchan" 依靠群众建立县乡社统计网: 河北运用统计工作促进生产 (Relying on the masses to establish county-, village-, and commune-level statistical networks: Hebei uses statistical work to promote production). *RMRB*, 29 April 1958, 3

———. "Zhou zongli jiejian Yin tongji xuejia Mahalanuobisi Nan Hanchen huanyan Riben shizhang, difang yiyuan fangHua tuan" 周总理接见印统计学家马哈拉诺比斯南汉宸欢宴日本市长, 地方议员访华团 (Premier Zhou meets with Indian statistician Mahalanobis; Nan Hanchen hosts a banquet in honor of the Mayor of Japan and the local councillors visiting China). *RMRB*, 10 July 1957, 1.

Asen, Daniel S. *Death in Beijing: Murder and Forensic Science in Republican China*. Cambridge: Cambridge University Press, 2016.

Asimov, Isaac. *Foundation, Foundation and Empire, Second Foundation*. New York: Everyman's Library, 2010.

Ba Jin 巴金. *Ba Jin quanji: Di shiwu juan* 巴金全集: 第十五卷 (Collected works of Ba Jin. Vol. 15). Beijing: Renmin wenxue chubanshe, 1986.

———. "'Shuzi de shi,' xinfu de baozheng" '数字的诗,' 幸福的保证 ("The Song of numbers," the guarantee of happiness). *RMRB*, 16 July 1955, 3.

Bai Sheng 柏生. "Sulian zhuanjia zai Qinghua daxue" 苏联专家在清华大学 (Soviet Experts at Qinghua University). *RMRB*, 5 November 1953, 3.

Banister, Judith. *China's Changing Population*. Stanford: Stanford University Press, 1987.

Baoding Conference Delegates. "Xiang Mao zhuxi baozheng: Kuzhan sange yue, shixian quanguo tongji gongzuo de da gaige, da yuejin! 向毛主席保证: 苦战三个月, 实现全国统计工作的大改革, 大跃进! (A guarantee to Chairman Mao: Three months of hard work and [we shall] achieve great reforms and a Great Leap Forward in national statistical work!)." *TJGZ*, no. 14 (1958): 1.

Becker, Jasper. *Hungry Ghosts: Mao's Secret Famine*. New York: The Free Press, 1996.

Bernstein, Peter L. *Against the Gods: The Remarkable Story of Risk*. New York: John Wiley, 1998.

Bernstein, Thomas P. "Mao Zedong and the Famine of 1959–1960: A Study in Wilfulness." *The China Quarterly*, no. 186 (2006): 421–445.

Bernstein, Thomas P., and Hua-yu Li, eds. *China Learns from the Soviet Union, 1949–Present*. Lanham, MD: Lexington Books, 2010.

Berze, D., and S. Mehta, eds. "ISI Honorary Member Interviews: Professor C.R. Rao." *International Statistical Institute Newsletter* 29, no. 2 (2005). Accessed at http://isi.cbs.nl/Nlet/NLet052.htm#09HonoraryMemberInterviews

Bethlehem, J. G. "The Rise of Survey Sampling." CBS Discussion Paper: 01572–0314. The Hague: Statistics Netherland, 2009.

Bian, Morris L. *The Making of the State Enterprise System in Modern China: The Dynamics of Institutional Change*. Cambridge, MA: Harvard University Press, 2005.

Bianco, Lucien. *La récidive: Révolution russe, Révolution chinoise*. Paris: Gallimard, 2014.

Bideluofu 彼得罗夫 (А.И.Петров), ed. *Jingji tongjixue jiaocheng* 经济统计学教程 (Course on economic statistics). Beijing: Tongji chubanshe, 1957.

Blum, Alain, and Martine Mespoulet. *L'anarchie bureaucratique: Pouvoir et statistique sous Staline*. Paris: Éditions La Découverte, 2003.

Boecking, Felix. *No Great Wall: Trade, Tariffs and Nationalism in Republican China, 1927–1945*. Cambridge, MA: Harvard University Asia Center, 2017.

Boorman, Howard L., ed. *Biographical Dictionary of Republican China*, 5 vols. New York: Columbia University Press, 1967–1979.

Bouk, Dan. *How Our Days Became Numbered: Risk and the Rise of the Statistical Individual*. Chicago: University of Chicago Press, 2015.

Brazinsky, Gregg A. *Winning the Third World: Sino-American Rivalry during the Cold War*. Chapel Hill: University of North Carolina Press, 2017.

Bréard, Andrea. "Reform, Bureaucratic Expansion and Production of Numbers: Statistics in Early 20th Century China." Habilitationsschrift, Technische Universität Berlin, 2008.

Breiman, Leo. "Statistical Modeling: The Two Cultures." *Statistical Science* 16, no. 3 (2001): 199–231.

Brown, Jeremy. *City Versus Countryside in Mao's China: Negotiating the Divide*. Cambridge: Cambridge University Press, 2012.

Brown, Jeremy, and Matthew D. Johnson, eds. *Maoism at the Grassroots: Everyday Life in China's Era of High Socialism*. Cambridge, MA: Harvard University Press, 2015.

Brown, Jeremy, and Paul G. Pickowicz, eds. *Dilemmas of Victory: The Early Years of the People's Republic of China*. Cambridge, MA: Harvard University Press, 2007.

Buck, John Lossing. *Land Utilization in China: A Study of 16,786 Farms in 168 Localities and 38,256 Farm Families in Twenty-Two Provinces in China, 1929–1933*. Chicago: University of Chicago Press, 1937.

Buck, Peter. *American Science and Modern China, 1876–1936*. Cambridge: Cambridge University Press, 1980.

Campbell, Robert W. *A Bibliographical Dictionary of Russian and Soviet Economics*. London: Routledge, 2012.

Cao Shuji 曹树基. *Da jihuang: 1959–1961 nian de Zhongguo renkou* 大饥荒: 1959–1961年的中國人口 (Great famine: China's population 1959–1961). Hong Kong: Shidai guoji chuban youxian gongsi, 2005.

Cao, Shuji and Bin Yang. "Grain, Local Politics, and the Making of Mao's Famine in Wuwei, 1958–1961." *Modern Asian Studies* 49, no. 6 (2015): 1675–1703.

Chatterjee, Partha. "Development Planning and the Indian State." In *The State, Development Planning and Liberalisation in India*, edited by T. J. Byres, 82–103. New Delhi: Oxford University Press, 1997.

Cheek, Timothy, and Tony Saich, eds. *New Perspectives on State Socialism in China*. Armonk, NY: M.E. Sharpe, 1997.

Chen, Da. *Population in Modern China*. Chicago: University of Chicago Press, 1946.

Chen, Jiading, and Zheng Zhongguo. "Academic Achievements of Professor P.L. Hsu." Review essay on the website of the International Chinese Statistical Association, 2010. https://icsaimage.files.wordpress.com/2016/08/plhsu2.pdf.

Chen, Jian, and Yang Kuisong, "Chinese Politics and the Collapse of the Sino-Soviet Alliance." In *Brothers in Arms: The Rise and Fall of the Sino-Soviet Alliance, 1945–1963*, edited by Odd Arne Westad, 246–294. Stanford: Stanford University Press, 1998.

Chen Jun 陈俊. "Sulian zhuanjia zai Beijing daxue" 苏联专家在北京大学 (Soviet experts at Peking University). *RMRB*, 16 September 1953, 3

Chen, Theodore Hsi-en. "The Thought Reform of Intellectuals." In *Government of Communist China*, edited by George P. Jan, 477–485. San Francisco: Chandler Publishing Company, 1966.

Chen, X. R., K. T. Fang, and C. C. Yang, eds. *The Development of Statistics: Recent Contributions from China*. Harlow, Essex: Longman Scientific and Technical, 1992.

Chen, Yinghong. *Creating the "New Man": From Enlightenment Ideals to Socialist Realities*. Honolulu: University of Hawai'i Press, 2009.

Chen Yingzhong 陈应中. "Geng guangfan de yunyong dianxing diaocha fangfa" 更广泛地运用典型调查方法 (Expand the use of typical surveys). *Jihua yu tongji* 计划与统计 (Planning and statistics), no. 6 (1959): 34–36.

Chiang, Yung-chen. *Social Engineering and the Social Sciences in China, 1919–1949*. Cambridge: Cambridge University Press, 2001.

China. *First Five-year Plan for Development of the National Economy of the People's Republic of China in 1953–1957*. Beijing: Foreign Languages Press, 1956.

Chinese Maritime Customs Project. *List of Chinese Customs Publications, 1940*. Compiled from *Documents Illustrative of the Origin, Development, and Activities of the Chinese Customs Service*, Vol. VII (Shanghai: Statistical Department, Inspectorate General of Customs, 1940). Bristol: University of Bristol. Accessed at http://www.bristol.ac.uk/history/customs/customs bibliographies/cmcbibliography.pdf.

Chu, Jennings P. 朱君毅 (Zhu Junyi). "The Independently Controlled Statistical System of the

Chinese Government." In *Proceedings of the International Statistical Conferences, Vol. II: September 6–18, 1947*, Washington, DC, edited by William J. Bruce, 92–97. Calcutta: Alka Press, 1951. See also Zhu Junyi 朱君毅.

Chung Lin and Hsiao Lu. "How We Made the Census." *China Reconstructs* 4, no. 12 (December 1955): 9–11.

Clark, Hunter, Maxim Pinkovskiy, and Xavier Sala-i-Martin. "China's GDP Growth May Be Understated." Working Paper No. 23323 (April 2017), National Bureau of Economic Research, Cambridge, MA.

Coale, Ansley J. *Rapid Population Change in China, 1952–1982*. Washington, DC: National Academy Press, 1984.

Cohn, Bernard S. *An Anthropologist among the Historians and Other Essays*. Oxford: Oxford University Press, 1990.

Comte, Auguste. *A General View of Positivism*, tr. J.H. Bridges. Stanford: Academic Reprint, 1953. Original: *Discours sur l'ensemble du positivisme*. London, 1856.

Confidential Office of the General Office of the Central Committee 中央办公厅机要室. *Zhongfa* [1961] 261: *Mao Zedong on Investigation and Research* 毛泽东论调查研究. Reprinted by the CCP Hebei Provincial Committee General Office. Translated into English by Michael Schoenhals. Accessed at http://prchistory.org/wp-content/uploads/2015/02.

Connelly, Matthew. *Fatal Misconception: The Struggle to Control World Population*. Cambridge, MA: Belknap Press of Harvard University Press, 2008.

Coyle, Diane. *GDP: A Brief but Affectionate History*. Princeton: Princeton University Press, 2014.

Cressey, George B. "The 1953 Census of China." *The Far Eastern Quarterly* 14, no. 3 (1955): 387–388.

Cullather, Nick. *The Hungry World: America's Cold War Battle Against Poverty in Asia*. Cambridge, MA: Harvard University Press, 2010.

Dai Shiguang 戴世光. "Jiji fazhan kexue de tongjixue: Wei woguo zaori shixian sige xiandaihua fuwu" 积极发展科学的统计学: 为我国早日实现四个现代化服务 (Actively develop a scientific statistics: In service to the early realization of the Four Modernizations in our country). *Jingji yanjiu* 经济研究 (Economic research), no. 2 (1979): 63–67. Also published in *Tongji* 统计 (Statistics), no. 1 (1979): 23–27.

———. *Dai Shiguang wenji* 戴世光文集 (Collected works of Dai Shiguang). Beijing: Renmin daxue chubanshe, 2008.

———. "Yindu de quanguo chouyang diaocha: Shehui jingji diaocha" 印度的全国抽样调查: 社会经济调查 (India's National Sample Survey: A socio-economic survey). *TJGZ*, no. 9 (1957): 2–6.

Daston, Lorraine. *Classical Probability in the Enlightenment*. Princeton: Princeton University Press, 1988.

Daston, Lorraine, and Peter Galison. *Objectivity*. New York: Zone Books, 2007.

de Bary, William Theodore, and Richard Lufrano, comps. *Sources of Chinese Tradition, Vol. 2*, 2nd ed. New York: Columbia University Press, 2000.

De Grief, A. Alexis, and Mauricio Neito Olarte. "What We Still Do Not Know About South-North Technoscientific Exchange: North-centrism, Scientific Diffusion, and Social Studies of Science." In *The Historiography of Contemporary Science, Technology, and Medicine: Writing*

Recent Science, edited by Ronald E. Doel and Thomas Söderqvist, 239–259. London: Routledge, 2006.

DeMare, Brian James. *Mao's Cultural Army: Drama Troupes in China's Rural Revolution*. Cambridge: Cambridge University Press, 2015.

Desai, Meghnad. "Development Perspectives: Was There an Alternative to Mahalanobis?" In *India's Economic Reforms and Development: Essays for Manmohan Singh*, ed. Isher Judge Ahluwalia and I.M.D. Little, 40–48. New Delhi: Oxford University Press, 1998.

Desrosières, Alain. "How Real Are Statistics? Four Possible Attitudes." *Social Research* 68, no. 2 (July 2001): 339–355.

———. *The Politics of Large Numbers: A History of Statistical Reasoning*, tr. Camille Naish. Cambridge, MA: Harvard University Press, 1998.

———, ed. *Sampling Humans: The First Sample Surveys in Norway, Russia and the United States*. Berlin: Max-Planck Institute für Wissenschaftsgeschichte, 2001.

Diamant, Neil J. "Policy Blending, Fuzzy Chronology, and Local Understandings of National Initiatives in Early 1950s China." *Frontiers of History in China* 9, no. 1 (2014): 83–101.

———. *Revolutionizing the Family: Politics, Love, and Divorce in Urban and Rural China, 1949–1968*. Berkeley: University of California Press, 2000.

Dikötter, Frank. *Mao's Great Famine: The History of China's Most Devastating Catastrophe, 1958–1962*. New York: Walter & Co., 2010.

Dodge, Yadolah, ed. *The Oxford Dictionary of Statistical Terms*, 6th ed. Oxford: Oxford University Press, 2003.

Doel, Ronald E., and Thomas Söderqvist, eds. *The Historiography of Contemporary Science, Technology, and Medicine: Writing Recent Science*. London: Routledge, 2006.

Donnithorne, Audrey. *China's Economic System*. New York: Praeger, 1967.

Donoho, David. "50 Years of Data Science." Paper prepared for a presentation at the Tukey Centennial Workshop, Princeton NJ, 18 September 2015. Accessed at http://courses.csail .mit.edu/18.337/2015/docs/50YearsDataScience.pdf

Duanmu Mei 端木美. "Cong qingong jianxue dao Zhong-Fa daxue: 20 shiji Zhong-Fa wenhua jiaoliu xuemai" 从勤工俭学到中法大学: 20世纪中法文化交流的血脉 (From a work-study program to the Sino-French University: The lineage of Sino-French cultural exchanges in the twentieth century) *Zhongwai wenhua jiaoliu* 中外文化交流 (China & the World Cultural Exchange), no. 6 (2012): 44–51.

Eberhard-Bréard, Andrea. "Robert Hart and China's Statistical Revolution." *Modern Asian Studies* 40, no. 3 (2006): 605–629.

Editorial. "Quandang quanmin ban tongji" 全党全民办统计 (All the Party and all the people engage in statistics). *RMRB*, 13 August 1958, 3.

Editorial Department. "Gao duzhe" 告读者 (To the readers). *Jihua yu tongji* 计划与统计 (Planning and statistics), no. 1 (1959): 1.

Editorial Department of *Tongji* 本刊编辑部. "Jinian Wang Sihua tongzhi shishi wuzhou nian" 纪念王思华同志逝世五周年 (Commemorating the fifth anniversary of the passing of Comrade Wang Sihua). *Tongji* 统计 (Statistics), no. 7 (1983): 5–6, 14.

Editorial Note. "Sulian tongji kexue huiyi jueyi" 苏联统计科学会议决议 (Resolution of the Soviet Conference on Statistical Sciences). *TJGZTX*, no. 2 (1955): 1–4, 45.

Ekbladh, David. *The Great American Mission: Modernization and the Construction of an American World Order*. Princeton: Princeton University Press, 2010.

Elman, Benjamin A. *On Their Own Terms: Science in China, 1550–1900*. Cambridge, MA: Harvard University Press, 2005.

Engerman, David C. *The Price of Aid: The Economic Cold War in India*. Cambridge, MA: Harvard University Press, 2018.

Erickson, Paul, Judy L. Klein, Lorraine Daston, Rebecca Lemov, Thomas Sturm, and Michael D. Gordin. *How Reason Almost Lost its Mind: The Strange Career of Cold War Rationality*. Chicago: University of Chicago Press, 2013.

Eyferth, Jacob. *Eating Rice from Bamboo Roots: The Social History of a Community of Handicraft Papermakers in Rural Sichuan, 1920–2000*. Cambridge, MA: Harvard University Asia Center, 2009.

Ezhov, A. I. *Industrial Statistics* [English translation of *Statistika promyslennosti*]. Calcutta: Statistical Publishing House, 1958.

———. *Organisation of Statistics in the U.S.S.R.* Moscow: Progress Publishers, 1967.

———. *Soviet Statistics*, tr. V. Shneerson. Moscow: Foreign Languages Publishing House, 1957.

Fan, Fa-ti. "Collective Monitoring, Collective Defense: Science, Earthquakes, and Politics in Communist China." *Science in Context* 25, no. 1 (2012): 127–154.

Fang Bingzhu 方秉铸. "Cong zhishushang kan zichan jieji tongji sixiang de xuwei benzhi" 从指数上看资产阶级统计思想的虚伪本质 (A look at the hypocrisy of bourgeois statistics from the perspective of index numbers). In Northeast Statistics Bureau. *Tongji gongzuo: Di si xuanji* 统计工作：第四选集 (Statistical work: Fourth anthology), 32–33. Shenyang: Dongbei caijing chubanshe, 1952.

Fei, Xiaotong. *Earthbound China: A Study of Rural Economy in Yunnan*, rev. Eng. ed. Chicago: University of Chicago Press, 1945.

———. *Peasant Life in China: A Field Study of Country Life in the Yangtze Valley*. New York: E. P. Dutton, 1939.

Feng Guoxi 冯国熙 et al. "Mao zhuxi lun dianxing diaochafa" 毛主席论典型调查法 (Chairman Mao on typical investigation method). *Caijing kexue* 财经科学 (Finance and economics), no. 3 (1960): 76–81.

Feng Jiqing 冯杞靖. "Xijing tongji sixiang zhong de zichan jieji yidu" 洗净统计思想中的资产阶级遗毒 (Wash clean the bourgeois legacy in statistical thought). In Northeast Statistics Bureau, *Tongji gongzuo: Di si xuanji* 统计工作：第四选集 (Statistical work: Fourth anthology), 34–36. Shenyang: Dongbei caijing chubanshe, 1952.

Feng Zhidan 冯之丹. "Yindu renmin de jingji jianshe chengjiu" 印度人民的经济建设成就 (The achievements in economic construction of the Indian people). *RMRB*, 26 January 1956, 4.

Feyerabend, Paul. *Against Method*. London: Verso Press, 1993.

Fisher, Irving. "The Best Form of Index Number." *Quarterly Publications of the American Statistical Association* 17, no. 133 (March 1921): 533–551.

———. *The Making of Index Numbers: A Study of Their Varieties, Tests, and Reliability*. Boston: Houghton Mifflin Company, 1922.

Fisher, Margaret W., and Joan V. Bondurant. "Review Article: The Impact of Communist China on Visitors from India." *The Far Eastern Quarterly* 15, no. 2 (February 1956): 249–265.

"Focus: Science and Modern China." *Isis* 98, no. 3 (2007): 517–596.

Fong, H. D. "Recollections of Early Research on Chinese Economy." *The Tsing Hua Journal of Chinese Studies* 清华学报 3/4, no. 1 (1963): 69–86.

Foucault, Michel. "Governmentality." In *The Foucault Effect: Studies in Governmentality*, edited by Graham Burchell, Colin Gordon, and Peter Miller, 87–104. London: Harvester Wheatsheaf, 1991.

Friedman, Edward, Paul G. Pickowicz, and Mark Selden. *Revolution, Resistance, and Reform in Village China*. New Haven: Yale University Press, 2005.

Friedman, Jeremy Scott. *Shadow Cold War: The Sino-Soviet Competition for the Third World*. Chapel Hill: University of North Carolina Press, 2015.

Friedman, Milton. *Money Mischief: Episodes in Monetary History*. New York: Harcourt, Brace & Company, 1992.

Froyen, Richard. "GDP: One of the Great Inventions of the 20th Century." *Survey of Current Business* 80, no. 1 (January 2000): 6–8.

Fujian Province Planning Committee 福建省计划委员会. *Zenyang kaizhan tongji diaocha yanjiu gongzuo* 怎样开展统计调查研究工作 (How to commence statistical survey research). Fuzhou: Fujian renmin chubanshe, 1959.

Fukuyama, Francis. "What Is Governance?" *Governance* 26, no. 3 (2013): 347–368.

Gamble, Sydney D. *North China Villages: Social, Political, and Economic Activities Before 1933*. Berkeley: University of California Press, 1963.

———. *Ting Hsien: A North China Rural Community*. New York: International Secretariat of the Institute of Pacific Relations, 1954.

Gao, James Z. *The Communist Takeover of Hangzhou: The Transformation of City and Cadre, 1949–1954*. Honolulu: University of Hawai'i Press, 2004.

Garnaut, Anthony. "Hard Facts and Half-Truths: The New Archival History of China's Great Famine." *China Information* 27, no. 2 (2013): 223–246.

Ge Jiali 葛家理. "Zai Sulian zhuanjia de jiaodao xia" 在苏联专家的教导下 (Under the guidance of Soviet Experts). *RMRB*, 6 November 1954, 2.

Gelman, Andrew. "Little Data: How Traditional Statistical Ideas Remain Relevant in a Big-Data World." Minutes of a talk delivered at the University Seminar on Big Data and Digital Scholarship (no. 753), Columbia University, 17 September 2012.

———. "Statistics is the *Least* Important Part of Data Science." Blog post, 14 November 2013, 9:25 am. Accessed at http://andrewgelman.com/2013/11/14/statistics-least-important-part-data-science/

Gelman, Andrew, and Christian Hennig. "Beyond Subjective and Objective in Statistics." *Journal of the Royal Statistical Society: Series A (Statistics in Society)* 180, no. 4 (October 2017): 967–1033.

Gewirtz, Julian B. *Unlikely Partners: Chinese Reformers, Western Economists, and the Making of Global China*. Cambridge, MA: Harvard University Press, 2017.

Ghosh, Arunabh. "Accepting Difference, Seeking Common Ground: Sino-Indian Statistical Exchanges 1951–1959." *BJHS Themes* 1 (2016): 61–82.

———. "Before 1962: The Case for 1950s China-India History." *Journal of Asian Studies* 76, no. 3 (August 2017): 697–727

————. "India and Science and Technology in the Early PRC." *PRC History Review* 2, no. 3 (June 2017): 7–9.

————. "Lies, Damned Lies, and (Bourgeois) Statistics: Ascertaining Social Fact in Midcentury China and the Soviet Union." *Osiris* 33, no. 1 (January 2018): 149–168.

Ghosh, Arunabh, and Sören Urbansky, eds. "China from Without: Doing PRC History in Foreign Archives." *PRC History Review* 2, no. 3 (June 2017): 1–26.

Ghosh, J. K. "Mahalanobis and the Art and Science of Statistics: The Early Days." *Indian Journal of History of Science* 29, no. 1 (January 1994): 89–98.

Gieryn, Thomas F. "Boundary-Work and the Demarcation of Science from Non-Science: Strains and Interests in Professional Ideologies of Scientists." *American Sociological Review* 48, no. 6 (December 1983): 781–795.

————. *Cultural Boundaries of Science: Credibility on the Line.* Chicago: University of Chicago Press, 1999.

Gigerenzer, Gerd, Zeno Swijtink, Theodore Porter, Lorraine Daston, John Beatty, and Lorenz Krüger. *The Empire of Chance: How Probability Changed Science and Everyday Life.* Cambridge: Cambridge University Press, 1989.

Goikhman, Izabella. "Soviet-Chinese Academic Interactions in the 1950s: Questioning the 'Impact-Response' Approach." In *China Learns from the Soviet Union, 1949–Present,* edited by Thomas P. Bernstein and Hua-yu Li, 275–302. Lanham, MD: Lexington Books, 2010.

Gong Jianyao 龚鉴尧. "Dianxing diaocha he tade kexuexing" 典型调查和它的科学性 (Typical surveys and their scientific nature). *TJGZTX*, no. 19 (1956): 12–14.

————. "Shidao qinghuai" 师道情怀 (Sentiments from a career in teaching). *Zhongguo tongji* 中国统计 (China statistics), no. 5 (2001): 52–54.

————. *Shijie tongji mingren zhuanji* 世界统计名人传记 (Biographies of famous statisticians in the world). Beijing: Tongji chubanshe, 2000.

————. "Woguo chouyangfa yanjiu yu shijian de fazhan" 我国抽样法研究与实践的发展 (Development of the research and practice of sampling in our country). *Xi'an tongji xueyuan xuebao* 西安统计学院学报 (Bulletin of the Xi'an Institute of Statistics) 9, no. 2 (1994): 7–14.

————. "Woguo chouyangfa yanjiu yu shijian de fazhan (xu)" 我国抽样法研究与实践的发展(续) (Development of the research and practice of sampling in our country [continued]). *Xi'an tongji xueyuan xuebao* 西安统计学院学报 (Bulletin of the Xi'an Institute of Statistics) 10, no. 1 (1995): 12–19.

Gordin, Michael D. *The Pseudoscience Wars: Immanuel Velikovsky and the Birth of the Modern Fringe.* Chicago: University of Chicago Press, 2012.

————. "Statistique et révolution en Russie: Un compromis impossible (1880–1930), and L'anarchie bureaucratique: Statistique et pouvoir sous Staline." *Kritika: Explorations in Russian and Eurasian History* 5, no. 4 (Fall 2004) (New Series): 803–810.

Gore, A. P. "P. V. Sukhatme: A 'Social' Statistician." *Economic and Political Weekly* 32, no. 6 (8–14 February 1997): 257–258.

Gough, Ian. "Marx's Theory of Productive and Unproductive Labour." *New Left Review*, no. 76 (November–December 1972): 47–72.

Government of India, Ministry of Food and Agriculture. *Report of the Indian Delegation to China on Agricultural Planning & Techniques.* New Delhi: Government of India, 1956.

Government of India, Planning Commission. *Report of the Indian Delegation to China on Agrarian Cooperatives*. New Delhi: Government of India, May 1957.

Graminius, Carin. "Building a New China: Hukou Investigation Practices in Beijing and Tianjin, 1949–1950." *The PRC History Review* 2, no. 1 (February 2017): 1–10.

Graziosi, Andrea. "Political Famines in the USSR and China: A Comparative Analysis." *Journal of Cold War Studies* 19, no. 3 (Summer 2017): 42–103.

Greenhalgh, Susan. *Just One Child: Science and Policy in Deng's China*. Berkeley: University of California Press, 2008.

Grimmer, Justin. "We Are All Social Scientists Now: How Big Data, Machine Learning, and Causal Inference Work Together." *PS: Political Science & Politics* 48, 1 (2015): 80–83.

Gross, Miriam. *Farewell to the God of Plague: Chairman Mao's Campaign to Deworm China*. Berkeley: University of California Press, 2016.

Guha, Ramachandra. *An Anthropologist Among Marxists and Other Essays*. New Delhi: Permanent Black, 2001.

Gui Shizuo 桂世祚. "Ping Zou Yiren, Jin Guobao liangwei xiansheng de gongye tongjixue zhuzuo" 评邹依仁, 金国宝两位先生的工业统计学著作 (Review of Zou Yiren's and Jin Guobao's works on industrial statistics). *TJGZTX*, no. 23 (1956): 24–27.

Guo Gengji 过庚吉. "Fandui sanbu zichan jieji tongji sixiang dusu: Ping Jin Guobao zhu 'Gongye tongjixue yuanli' he Zou Yiren zhu 'Gongye tongji'" 反对散布资产阶级统计思想毒素: 评金国宝著 "工业统计学原理" 和邹依仁著 "工业统计" (Combat the spread of the toxin that is bourgeois statistical thought: A review of Jin Guobao's *Principles of Industrial Statistics* and Zou Yiren's *Industrial Statistics*). In Northeast Statistics Bureau. *Tongji gongzuo: Di si xuanji* 统计工作: 第四选集 (Statistical work: Fourth anthology), 30–31. Shenyang: Dongbei caijing chubanshe, 1952.

Haber, Maya. "Socialist Realist Science: Constructing Knowledge about Rural Life in the Soviet Union, 1943–1958." PhD diss., University of California, Los Angeles, 2013.

Hacking, Ian. "Biopower and the Avalanche of Printed Numbers." *Humanities in Society* 5, no. 3–4 (July 1982): 279–295.

———. "Making Up People." *London Review of Books* 28, no. 16 (17 August 2006): 23–26.

———. *The Taming of Chance*. Cambridge: Cambridge University Press, 1990.

Halberstam, David. *The Best and the Brightest*. New York: Random House, 1972.

Hall, B. E. Foster. "The Chinese Maritime Customs: An International Service, 1854–1950." Published by order of the Chinese Maritime Customs Project; Occasional Papers No. 5. Bristol: University of Bristol, 2015. Accessed at http://www.bristol.ac.uk/history/customs/papers/occasionalpaper5.pdf

Harris, Seymour E. "Appraisals of Russian Economic Statistics: Introduction." *The Review of Economics and Statistics* 29, no. 4 (November 1947): 213–214.

Hayhoe, Ruth. *China's Universities, 1895–1995: A Century of Cultural Conflict*. London: Routledge, 1996.

He Sheng 荷生. "Ping 'Tongjixue yuanli: Di yi juan'" 评 "统计学原理: 第一卷" (Review of *Principles of Statistics*, Vol. I). *Jingji zhoubao* 经济周报 (Economics weekly) 12, no. 3 (1950).

Hebei Province Statistics Bureau. *Baodingshi tongji gongzuo de gaochao* 保定市统计工作的高潮 (The high tide of statistical work in Baoding). Beijing: Tongjiju chubanshe, 1958.

————. *Tongji gongzuo yuejin jingyan* 统计工作跃进经验 (The experience of a Leap Forward in statistical work). Baoding: Hebei renmin chubanshe, 1958.

Heinzig, Dieter. *The Soviet Union and Communist China, 1945–1950: The Arduous Road to the Alliance*. Armonk, NY: M.E. Sharpe, 2004.

Hershatter, Gail. *The Gender of Memory: Rural Women and China's Collective Past*. Berkeley: University of California Press, 2011.

Ho, Denise Y. *Curating Revolution: Politics on Display in Mao's China*. Cambridge: Cambridge University Press, 2018.

Ho, Franklin Lien. *Index Numbers of the Quantities and Prices of Imports and Exports and of the Barter Terms of Trade in China, 1867–1928*. Tientsin: Chihli Press, 1930.

Holz, Carsten A. "China's Statistical System in Transition: Challenges, Data Problems, and Institutional Innovations." *Review of Income and Wealth* 50, no. 3 (September 2004): 381–409.

————. "The Quality of China's GDP Statistics." *China Economic Review* 30 (September 2014): 309–338.

Howell, Peter. "Once Again on Productive and Unproductive Labour." *Revolutionary Communist*, no. 3/4 (September 1975): 46–68.

Hsu, Kai-yu, ed. *Literature of the People's Republic of China*. Bloomington: Indiana University Press, 1980.

Hubei Province People's Committee 湖北省人民委员会. "Hubeisheng renmin weiyuanhui fachu tongzhi: Piping Baokangxian changqi tuoyan bubao tongji baobiao" 湖北省人民委员会发出通报: 批评保康县长期拖延不报统计报表 (Notification of the People's Committee of Hubei Province: Criticism of Baokang County for the long-term delay in submitting statistical reports). *TJGZTX*, no. 11 (1955): 48.

Hu Daiguang 胡代光. "Ganqing zichan jieji tongji sixiang de liudu" 肃清资产阶级统计思想的流毒 (Eliminate the pernicious influence of bourgeois statistical ideology). *TJGZTX*, no. 10 (1955): 23–27.

Hu, Danian. *China and Albert Einstein: The Reception of the Physicist and His Theory in China, 1917–1979*. Cambridge, MA: Harvard University Press, 2005.

————, ed. "Science, Technology, and Medicine in China's Cultural Revolution." *Endeavour* 41, no. 3 (September 2017): 67–150.

Huang Jiantuo 黄剑拓. "Wo duiyu dianxing diaocha fangfa de jidian cuqian renshi" 我对于典型调查方法的几点粗浅认识 (Superficial notes on methods of typical surveying). *TJGZ*, no. 6 (1957): 18–20.

Huang, Yasheng. "Information, Bureaucracy, and Economic Reforms in China and the Soviet Union." *World Politics* 47, no. 1 (October 1994): 102–134.

————. "The Statistical Agency in China's Bureaucratic System: A Comparison with the Former Soviet Union." *Communist and Post-Communist Studies* 29, no. 1 (1996): 59–75.

Immerwahr, Daniel. *Thinking Small: The United States and the Lure of Community Development*. Cambridge, MA: Harvard University Press, 2015.

Indian Statistical Institute. *Annual Report, April 1956–March 1957*. Kolkata.

Institut International de Statistique. *Bulletin de L'Institut International de Statistique: Compte Rendu De la XIXème Session de l'Institut International de Statistique*. Vol. 25. Tokyo, 1932.

Accessed at Bibliothèque nationale de France, http://gallica.bnf.fr/ark:/12148/bpt6
k61620q?rk=21459;2.

Institut International de Statistique. *Bulletin de L'Institut International de Statistique: Mémoires
divers portant sur l'Extrême-Orient. Présentés en Hommage à la XIXème Session de l'Institut
International de Statistique.* Tokyo, 1930. Accessed at Bibliothèque nationale de France, http://
gallica.bnf.fr/ark:/12148/bpt6k61611r?rk=21459;2.

International Statistical Institute. *Introduction: International Statistical Conferences, September
6–18, 1947, Washington, D.C.—Proceedings of the International Statistical Conference,* Vols. I
and II. Calcutta: Alka Press, 1951.

Iriye, Akira. "Internationalizing International History." In *Rethinking American History in the
Global Age,* ed. Thomas Bender, 48–60. Berkeley: University of California Press, 2002.

Jaffe, A.J. "A Review of the Censuses and Demographic Statistics of China." *Population Studies*
1, no. 3 (December 1947): 308–337.

Jasanoff, Sheila, ed. *Designs on Nature: Science and Democracy in Europe and the United States.*
Princeton: Princeton University Press, 2005.

———. *States of Knowledge: The Co-Production of Science and Social Order.* London: Routledge,
2006.

Jersild, Austin. *Sino-Soviet Alliance: An International History.* Chapel Hill: University of North
Carolina Press, 2014

Jerven, Morton. *Poor Numbers: How We are Misled by African Development Statistics and What
to Do About It.* Ithaca, NY: Cornell University Press, 2013.

Jiang Ling 江陵. "Zhonglei duo, xiangmu fan, biangeng duo, yaoqiu ji shi jianshao tongji baobiao de shihou le" 种类多，项目繁，变更多，要求急是减少统计报表的时候了 (Types
numerous, projects complex, changes many, and requirements urgent: It is time to reduce
the numbers of statistical reports). *RMRB,* 25 October 1956, 4.

Jiang Xinming 江新明. "Tongji yu kuaji de guanxi he qubie" 统计与会计的关系和区别 (The
relationship and differences between statistics and accounting). *Caikuai tongxun* 财会通
讯 (Communication of finance and accounting), no. S1 (1984): 25.

Jiang Xuepei 江学培. "Beijing daxue jinian: zhuming shuxuejia, tongjixuejia, Xu Baolu jiaoshou"
北京大学纪念：著名数学家，统计学家许宝騄教授 (Beijing University commemorates
the famous mathematician and statistician, Professor Xu Baolu). *Tongji* 统计 (Statistics),
no. 1 (1981): 39.

Jin Guobao 金国宝. "Dui 'Tongjixue dagang' de ziwo pipan" 对 "统计学大纲" 的自我批判
(Self-criticism of *An Outline of Statistics*). *TJGZ,* no. 3 (1957): 28–32.

———. *Gaoji tongjixue* 高级统计学 (Advanced statistics). Shanghai: Lixin Kuaiji tushu yongpinshu, 1951.

———. *Gongye tongjixue yuanli* 工业统计学原理 (Principles of industrial statistics). Shanghai:
Lixin kuaiji tushu yongpinshu, 1952 (republished in 1953).

———. *Shehui zhuyi zhidu xia de guomin shouru yanjiu* 社会主义制度下的国民收入研究
(Analysis of national income under the socialist system). Shanghai: Shanghai Finance Institute, 1956.

———. *Tongjixue* 统计学 (Statistics). Shanghai: Shangwu yinshuguan, 1935.

———. *Tongjixue dagang* 统计学大纲 (An outline of statistics). Shanghai: Shangwu yinshuguan, 1934.

Johnson, D. Gale. "China's Great Famine: Introductory Remarks." *China Economic Review* 9, no. 2 (1998): 103–109.

Kaple, Deborah A. "Soviet Advisors in China in the 1950s." In *Brothers in Arms: The Rise and Fall of the Sino-Soviet Alliance, 1945–1963*, edited by Odd Arne Westad, 117–140. Stanford: Stanford University Press, 1998.

Karabell, Zachary. *The Leading Indicators: A Short History of the Numbers That Rule Our World.* New York: Simon and Schuster, 2014.

Katzenstein, Peter J., and Stephen C. Nelson. "Worlds in Collision: Uncertainty and Risk in Hard Times." In *Politics in the New Hard Times: The Great Recession in Comparative Perspective,* edited by Miles Kahler and David A. Lake, 233–252. Ithaca, NY: Cornell University Press, 2013.

Kendall, D. G. "Andrei Nikolaevich Kolmogorov, 25 April 1903–20 October 1987." *Biographical Memoirs of Fellows of the Royal Society* 37 (November 1991): 300–319. Accessed at: https://doi.org/10.1098/rsbm.1991.0015

Kevles, Daniel J. *In the Name of Eugenics: Genetics and the Uses of Human Heredity.* Cambridge, MA: Harvard University Press, 1995.

Keynes, John Maynard. *Treatise on Probability.* London: Macmillan & Co., 1948 [1921].

Kirby, M. W. *Operational Research in War and Peace: The British Experience from the 1930s to 1970.* London: Imperial College Press, 2003.

Kirby, William C. "China's Internationalization in the Early People's Republic: Dreams of a Socialist World Economy." *The China Quarterly*, no. 188 (December 2006): 870–890.

———. "Continuity and Change in Modern China: Chinese Economic Planning on the Mainland and on Taiwan, 1943–1958." *Australian Journal of Chinese Affairs*, no. 24 (July 1990): 121–141.

———. *Germany and Republican China.* Stanford: Stanford University Press, 1984.

Knight, Frank H. *Risk, Uncertainty, and Profit.* Hart, Schaffner, and Marx Prize Essays, no. 31. Boston: Houghton Mifflin, 1921.

Knight, Nick. "Mao Zedong's On Contradiction and On Practice: Pre-Liberation Texts." *The China Quarterly*, no. 84 (December 1980): 641–668.

Knorr-Certina, Karin. *Epistemic Cultures: How the Sciences Make Knowledge.* Cambridge, MA: Harvard University Press, 1999.

Kong Dechao 孔德超. "'Dayuejin' qijian de tongji gongzuo" '大跃进' 期间的统计工作 (Statistical work during the "Great Leap Forward"). *Dangshi wenhui* 党史文汇 (Corpus of party history), no. 4 (2001): 31–33.

Kotz, Samuel, and Eugene Seneta. "Lenin as a Statistician: A Non-Soviet View." *Journal of the Royal Statistical Society. Series A (Statistics in Society)* 153, no. 1 (1990): 73–94.

Kozlov (?) 科兹洛夫, T. I. et al. *Tongji yiban lilun wenti* 统计一般理论问题 (Common theoretical problems in statistics). Beijing: Tongji chubanshe, 1957.

Krader, Lawrence, and John S. Aird. "Sources of Demographic Data on Mainland China." *American Sociological Review* 24, no. 5 (October 1959): 623–630.

Krige, John, and Helke Rausch, eds. *American Foundations and the Coproduction of World Order in the 20th Century.* Göttingen: Vandenhoeck & Ruprecht, 2012.

Krotevich, Sergei K. "Vsekitayskaya perepis' naseleniya 1953 g." (The all-China population census of 1953). *Vestnik Statistiki* (Journal of Statistics), no. 5 (September–October 1955): 31–50.

Krotevich, Sergei K. 克拉戴维寄, 谢康. *Zhongguo 1953 nian quanguo renkou diaocha* 中国1953 年全国人口调查 (The All-China population census of 1953). Beijing: Tongji chubanshe, 1956.

Krüger, Lorenz, Lorraine J. Daston, and Michael Heidelberger, eds. *The Probabilistic Revolution*, Vol. 1: *Ideas in History*. Cambridge, MA: MIT Press, 1987.

Krugman, Paul. "Asimov's Foundation Novels Grounded my Economics." *The Guardian*, 4 December 2012.

Kulp, Daniel Harrison. *Country Life in South China: The Sociology of Familism*, Vol. 1, *Phenix Village, Kwantung, China*. New York: Bureau of Publications, Teachers College, Columbia University, 1925.

Kung, James Kai-sing, and Shuo Chen. "The Tragedy of the *Nomenklatura*: Career Incentives and Political Radicalism during China's Great Leap Famine." *American Political Science Review* 105, no. 1 (February 2011): 27–45.

Lam, Tong. *A Passion for Facts: Social Surveys and the Construction of the Chinese Nation-State, 1900–1949*. Berkeley: University of California Press, 2011.

Lampland, Martha. "False Numbers as Formalizing Practices." *Social Studies of Science* 40, no. 3 (2010): 377–404.

Lampland, Martha, and Susan Leigh Star, eds. *Standards and Their Stories: How Quantifying, Classifying, and Formalizing Practices Shapes Everyday Life*. Ithaca, NY: Cornell University Press, 2009.

Landefeld, J. Steven. "GDP: One of the Great Inventions of the 20th Century." *Survey of Current Business* 80, no. 1 (January 2000): 6–14. Accessed at https://bea.gov/scb/pdf /BEAWIDE/2000/0100od.pdf

Lardy, Nicholas R., and Kenneth Lieberthal, eds. *Chen Yün's Strategy for China's Development: A Non-Maoist Alternative*. Armonk, NY: M.E. Sharpe, 1983.

Latour, Bruno. *Science in Action: How to Follow Scientists and Engineers through Society*. Cambridge, MA: Harvard University Press, 1987.

Lee, Christopher J., ed. *Making a World after Empire: The Bandung Moment and Its Political Afterlives*. Athens: Ohio University Press, 2010.

Lee, Hong Yung. *From Revolutionary Cadres to Party Technocrats in Socialist China*. Berkeley: University of California Press, 1991.

Leek, Jeff, Blakeley B. McShane, Andrew Gelman, David Colquhoun, Michèle B. Nuijten, and Steven N. Goodman. "Five Ways to Fix Statistics." *Nature* 551, no. 7682 (30 November 2017): 557–559.

Lei, Sean Hsiang-Lin. "How Did Chinese Medicine Become Experiential? The Political Epistemology of *Jingyan*." *Positions: East Asia Cultures Critique* 10, no. 2 (Fall 2002): 333–364.

———. *Neither Donkey Nor Horse: Medicine in the Struggle Over China's Modernity*. Chicago: Chicago University Press, 2014.

Lemov, Rebecca M. *Database of Dreams: The Lost Quest to Catalog Humanity*. New Haven: Yale University Press, 2015.

Lenin, Vladimir I. "How to Organise Competition?" In *Lenin Collected Works*, 3rd Printing, Vol. 26, 404–415. Moscow: Progress Publishers, 1977.

———. "Letters to the Central Statistical Board" (dated 16 August 1921). In *Lenin Collected Works*, 2nd Printing, Vol. 33, 30–32. Moscow: Progress Publishers. 1973.

———. "To G. M. Krzhizhanovsky" (dated 26 May 1921). In *Lenin Collected Works,* 2nd Printing, Vol. 35, 497–98. Moscow: Progress Publishers, 1973.

Levy, Jonathan. *Freaks of Fortune: The Emerging World of Capitalism and Risk in America.* Cambridge, MA: Harvard University Press, 2012.

Li, Choh-Ming. "Communist China's Statistical System: 1949–1957." *The American Economic Review* (Papers and Proceedings of the Seventy-Third Annual Meeting of the American Economic Association) 51, no. 2 (May 1961): 499–517.

———. *Economic Development of Communist China: An Appraisal of the First Five Years of Industrialization.* Berkeley: University of California Press, 1959.

———. *The Statistical System of Communist China.* Berkeley: University of California Press, 1962.

———. "Statistics and Planning at the Hsien Level in Communist China." *The China Quarterly,* no. 9 (January–March 1962): 112–123.

Li Deng 李澄. "Jinian Wang Sihua tongzhi" 纪念王思华同志 (Commemorating Comrade Wang Sihua). *Renwen zazhi* 人文杂志 (Journal of humanities), no. 6 (1984): 64–65, 68.

Li, Lillian M. *Fighting Famine in North China: State, Market, and Environmental Decline, 1690s–1990s.* Stanford: Stanford University Press, 2007.

Li Ming 厉明, ed. "Li Ming tongzhi weishenme buyuan zuo tongji gongzuo?" 厉明同志为什么不愿作统计工作? (Why is Comrade Li Ming unwilling to do statistical work?). *TJGZTX,* no. 7 (1955): 40.

Li Yuren 李育仁. "Wo dui tongji gongzuo de renshi" 我对统计工作的认识 (My understanding of statistical work). *TJGZTX,* no. 10 (1955): 53.

Li Zhenzhou 李振周. "Chedi pipan yu kefu women nongye tongji gongzuo zhong de youqing baoshou sixiang: Hubeisheng tongjiju Li Zhenzhou fujuzhang fayan zhaiyao" 彻底批判与客服我们农业统计工作中的右倾保守思想: 湖北省统计局李振周副局长发言摘要 (Thoroughly critique and overcome rightist conservative ideas in our agricultural statistics work: Abstract of the speech by Li Zhenzhou, deputy director of the Hubei Province Statistical Bureau). *TJGZTX,* no. 4 (1956): 10–11.

Liang Qichao 梁启超. "Lishi tongjixue" 历史统计学 (Historical statistics). In *Liang Qichao quanji: Di shisi juan* 梁启超全集: 第十四卷 (Collected works of Liang Qichao: Vol. 14), 4045–4050. Beijing: Beijing chubanshe, 1999.

Liang Zhitang 梁之汤. "Nongye tongji dianxing diaocha de jiben zhishi he jiben zuofa" 农业统计典型调查的基本知识和基本做法 (Basic knowledge and practice regarding typical surveys of agricultural statistics). *TJGZTX,* no. 7 (1954): 44–46.

Lie, Trygve. "Opening Address to the International Statistical Institute's 1947 Meeting." In International Statistical Institute. *Introduction—International Statistical Conferences, September 6–18, 1947, Washington, D.C.—Proceedings of the International Statistical Conference,* Vol. I, 151–152. Calcutta: Alka Press, 1951.

Lieu, D.K. 刘大钧 (Liu Dajun). "Collecting Statistics in China." *The American Statistician* 2, no. 6 (December 1948): 12–13.

———. "Statistical Work in China." Presented at the Nineteenth Session of the International Statistical Institute, Tokyo, 1930.

Lin, Yaohua. *The Golden Wing: A Sociological Study of Chinese Familism.* New York: Oxford University Press, 1947.

Lindsay, Michael. *Notes on Educational Problems in Communist China, 1941–47*. New York: International Secretariat, Institute of Pacific Relations, 1950.

Ling Xianfang 凌献方, ed. *Renmin zhengfu gongwen chengshi yanjiu (zengding ban)* 人民政府公文程式研究 (增订版) (Research on official documents of the People's Government [Updated Edition]). Shanghai: lixin kuaiji tushu yongpinshe, 1951.

Liu Chang 刘畅 and Zhang Yun 张云. *Ershi shiji Zhongguo de tongjixue* 二十世纪中国的统计学 (The discipline of statistics in twentieth-century China). Beijing: Dangjian duwu chubanshe, 2000.

Liu, Lydia H. *Translingual Practice: Literature, National Culture, and Translated Modernity: China, 1900–1937*. Stanford: Stanford University Press, 1995.

Liu, Xin. *The Mirage of China: Anti-Humanism, Narcissism, and Corporeality of the Contemporary World*. New York: Berghahn Books, 2009.

Lu Shutian 卢曙天. "Quanmian tongji bixu he dianxing diaocha xiang jiehe: Jinianlai wuzi tongji diaocha gongzuo de jiben tihui" 全面统计必须和典型调查想结合:几年来物资统计调查工作的基本体会 (Comprehensive statistics and typical surveys must be combined: The basic experience of survey work in goods and materials statistics in recent years). *Jihua yu tongji* 计划与统计 (Planning and statistics), no. 15 (1959): 19–22.

Luanping County Statistics Office 滦平县统计科. "Xianfeng renmin gongshe datun sanlei dikuai yumi chanliang dianxing diaocha" 先锋人民公社大屯三类地块玉米产量典型调查 (Three categories of typical survey of corn yields in Xianfeng People's Commune). *TJGZ*, no. 21 (1958): 9–10.

Luo Zhufeng 罗竹风. *Hanyu dacidian: Di jiu juan* 汉语大词典: 第九卷 (Grand dictionary of Chinese: Vol. 9). Shanghai: Hanyu dacidian chubanshe, 1994.

Lü, Xiaobo, and Elizabeth J. Perry, eds. *Danwei: The Changing Chinese Workplace in Historical and Comparative Perspective*. Armonk, NY: M.E. Sharpe, 1997.

Lüthi, Lorenz M. *The Sino-Soviet Split: Cold War in the Communist World*. Princeton: Princeton University Press, 2008.

Lynn, Jonathan, and Anthony Jay, eds. *The Complete Yes Minister: The Diaries of a Cabinet Minister by the Right Hon. James Hacker MP*. Topsfield, MA: Salem House, 1984.

Ma Min 马敏 and Lu Hanwen 陆汉文. "Minguo shiqi zhengfu tongji gongzuo yu tongji ziliao shulun" 明国时期政府统计工作与统计资料述论 (Government statistical work and statistical materials during the Republican period). *Huazhong shifan daxue xuebao (Renwen shehui kexue ban)* 华中师范大学学报 (人文社会科学版) (Journal of Central China Normal University [Humanities and social sciences issue]) 44, no. 6 (November 2005): 116–129, 160.

MacFarquhar, Roderick. *The Origins of the Cultural Revolution*, Vol. 2: *The Great Leap Forward 1958–1960*. New York: Columbia University Press, 1983.

———. *The Politics of China, 1949–1989*. Cambridge: Cambridge University Press, 1993.

MacFarquhar, Roderick and John K. Fairbank, eds. *The Cambridge History of China*, Vol. 14: *The People's Republic, Part 1: The Emergence of Revolutionary China, 1949–1965*. Cambridge: Cambridge University Press, 1987.

MacKenzie, Donald A. *An Engine, Not a Camera: How Financial Models Shape Markets*. Cambridge, MA: MIT Press, 2006.

Mahalanobis, P. C. "Some Impressions of a Visit to China 19 Jun–11 July 1957," 37 pp. Unpublished typed manuscript in PCMMMA, Kolkata.

———. "Statistics Must Have a Purpose." *Samvadadhvam* 1, no. 1 (July 1956): 3–10.

———. "Why Statistics?" *Sankhyā: The Indian Journal of Statistics* 10, no. 3 (September 1950): 195–228.

Mann, Michael. "The Autonomous Power of the State: Its Origins, Mechanisms and Results." *European Journal of Sociology* 25, no. 2 (1984): 185–213.

Manning, Kimberly Ens, and Felix Wemheuer, eds. *Eating Bitterness: New Perspectives on China's Great Leap Forward and Famine.* Vancouver: University of British Columbia Press, 2011.

Mao, Tse-Tung 毛泽东. *A Critique of Soviet Economics by Mao Tsetung,* tr. Moss Roberts; annotated by Richard Levy, with an introduction by James Peck. London: Monthly Review Press, 1977.

———. "Some Questions Concerning Methods of Leadership," 1 June 1943. In *Selected Works of Mao Tse-Tung,* Vol. 3, 117–129. Peking: Foreign Languages Press, 1965.

———. "Preface to *Rural Surveys,*" 17 March 1941. In *Selected Works of Mao Tse-Tung,* Vol. 3, 11–14. Peking: Foreign Languages Press, 1965.

———. "Red and Expert—31 January 1958," as appearing in the Red Guard Publication *Long Live Mao Zedong Thought* (1969). Accessed at http://www.marxists.org/reference/archive /mao/selected-works/volume-8/mswv8_04.htm

———. *Report on an Investigation of the Peasant Movement in Hunan.* Peking: Foreign Languages Press, 3rd ed., 1965.

———. *Selected Works of Mao Tse-Tung,* Vol. 1. Peking: Foreign Languages Press, 1965.

———. *Selected Works of Mao Tse-Tung,* Vol. 3. Peking: Foreign Languages Press, 1965.

———. *Selected Works of Mao Tse-Tung,* Vol. 4. Peking: Foreign Languages Press, 1961.

———. *Selected Works of Mao Tsetung,* Vol. 5. Peking: Foreign Languages Press, 1977.

Martin, Bernard. *Strange Vigour: A Biography of Sun Yatsen.* Port Washington, NY: Kinnikat Press, 1970.

Marx, Karl. *Capital,* Vol. 1: *A Critique of Political Economy,"* tr. from the German third ed. Samuel Moore and Edward Bibbins Aveling; ed. Frederick Engels; revised and amplified according to the fourth German edition by Ernest Untermann. New York: Modern Library, 1906.

———. *Economic and Philosophic Manuscripts of 1844,* tr. Martin Milligan. Mineola, NY: Dover Publications, 2007.

Maximov. G. (Максимов, Г.). "Naselenie SSSR (Kvykhodu v svet 16-tomnogo izdaniia itogov Vsesoiuznoi perepisi naseleniia 1959 g." (Population of the USSR [Towards a conclusion in light of the 16-volume publication on the results of the all-union population census in 1959]). *Vestnik Statistiki* (Journal of statistics), no. 3 (1963): 3–13.

Maximov, G., and A. Isupov (Максимов, Г., Исупов, А.). "Natsional'nosti SSSR po dannym Vsesoiuznoi perepisi naseleniia" (Nationalities of the USSR according to the data of the all-union population census). *Vestnik Statistiki* (Journal of statistics), no. 4 (1960): 65–75.

Mazur, Mary G. "The United Front Redefined for the Party-State: A Case Study of Transition and Legitimation." In *New Perspectives on State Socialism in China,* edited by Timothy Cheek and Tony Saich, 51–75. Armonk, NY: M.E. Sharpe, 1997.

McCormick, Ted. *William Petty and the Ambitions of Political Arithmetic.* Oxford: Oxford University Press, 2009.

McCullagh, Peter. "John Wilder Tukey, 16 June 1915–26 July 2000." *Biographical Memoirs of Fellows of the Royal Society* 49 (December 2003): 537–555.

Megill, Allan. "Introduction: Four Senses of Objectivity." In *Rethinking Objectivity*, edited by Allan Megill, 1–20. Durham, NC: Duke University Press, 1994.

Meisner, Maurice J. *Mao's China and After: A History of the People's Republic*. New York: Free Press, 1999.

Mencius 孟子. "Wan Zhang—II 万章下 (Wan Zhang, second part)," in *The Works of Mencius* 孟子. Available at Chinese Text Project, http://ctext.org/mengzi/wan-zhang-ii.

Meng Ping 梦萍 and Zhu Li 竹立, eds. *Gongwen de jiben zhishi* 公文的基本知识 (Basic knowledge of official documents). Shanghai: Huiwentang shuju, 1953.

Menon, Nikhil. "Planned Democracy: Development, Citizenship, and the Practices of Planning in Independent India, c. 1947–1966." PhD diss., Princeton University, 2017.

Mespoulet, Martine. *Statistique et révolution en Russie: Un compromis impossible, 1880–1930.* Rennes: Presses Universitaires de Rennes, 2001.

Meyskens, Covell Franklin. "Maoist China's Hinterland War Machine: The Cold War, Industrial Modernity, and Everyday Life in China's Third Front, 1964–1980." PhD diss., University of Chicago, 2015.

Miao Fenglin 苗枫林. *Zhongguo gongwen xue* 中国公文学 (Study of official documents in China). Ji'nan: Qilü shuju, 1988.

Mills, Harriet C. "Thought Reform: Ideological Remolding in China." In *Government of Communist China*, edited by George P. Jan, 486–498. San Francisco: Chandler Publishing Company, 1966.

Mitchell, Timothy. *Rule of Experts: Egypt, Techno-Politics, Modernity*. Berkeley: University of California Press, 2002.

Mo Rida 莫日达. "Dianxing diaocha shi tongji gongzuo ganbu zouxiang hongzhuan de yitiao zhongyao daolu" 典型调查是统计工作干部走向红专的一条重要道路 (Typical surveys provide an important path for cadres involved in statistical work to move toward red and expert). *TJGZ*, no. 22 (1958): 9–10.

Mohun, Arwen. *Risk: Negotiating Safety in American Society*. Baltimore: Johns Hopkins University Press, 2013.

Morgan, Mary S. *The World in the Model: How Economists Work and Think*. Cambridge: Cambridge University Press, 2012.

Mullaney, Thomas. *Coming to Terms with the Nation: Ethnic Classification in Modern China*. Berkeley: University of California Press, 2011.

N. 廖佐夫. "Xin Zhongguo tongji de gaikuang yu tongji gongzuozhe de renwu" 新中国统计的概况与统计工作者的任务 (Summary of statistics in new China and the tasks of statistical workers). In Northeast Statistics Bureau. *Tongji gongzuo: di san xuan ji* 统计工作：第三选辑 (Statistical work: Third anthology), 7–9, 10. Shenyang: Dongbei caijing chubanshe, January 1952.

National Bureau of Statistics 国家统计局. "Huiwang: Lishishang de dongbei tongjiju" 回望：历史上的东北统计局 (Looking back: The history of the Northeast Statistics Bureau). *Zhongguo tongji* 中国统计 (China statistics), no. 8 (2010): 22–23.

———. *Zhonghua renmin gongheguo tongji dashiji 1949–2009* 中华人民共和国统计大事记

1949–2009 (Chronicle of statistical events in the People's Republic of China 1949–2009). Beijing: Zhongguo tongji chubanshe, 2009.

Naughton, Barry. "The Pattern and Legacy of Economic Growth in the Mao Era." In *Perspectives on Modern China: Four Anniversaries*, edited by Kenneth Lieberthal et al., 226–254. Armonk, NY: M.E. Sharpe, 1991.

Nelson, Stephen C., and Peter J. Katzenstein. "Uncertainty, Risk, and the Financial Crisis of 2008." *International Organization* 68, no. 2 (Spring 2014): 361–392.

Nixon, J. W. *A History of the International Statistical Institute, 1885–1960*. The Hague: International Statistical Institute, 1960.

Northeast Statistics Bureau. 东北人民统计局 *Tongji gongzuo: Di san xuanji* 统计工作：第三选辑 (Statistical work: Third anthology). Shenyang: Dongbei caijing chubanshe, January 1952.

———. *Tongji gongzuo: Di si xuanji* 统计工作：第四选集 (Statistical work: Fourth anthology). Shenyang: Dongbei caijing chubanshe, 1952.

———. *Tongji gongzuo: Di wu xuanji* 统计工作：第五选集 (Statistical work: Fifth anthology). Shenyang: Dongbei caijing chubanshe, 1953.

———. *Tongji gongzuo: Di yi xuanji* 统计工作：第一选集 (Statistical work: First Anthology) , 4th ed. Shenyang: Dongbei caijing chubanshe 1951.

Ó Gráda, Cormac. "The Ripple That Drowns? Twentieth-century Famines in China and India as Economic History." *The Economic History Review* 61 (2008): 5–37.

———. *Famine: A Short History*. Princeton: Princeton University Press, 2010.

Onoprienko, G. K. *Bibliograficheskii Ukazatel Statei i Materialov po Statistike i Uchetu. Zhurnal Vestnik Statistiki za 50 let (1919–1968)* (Bibliography of Essays and Materials on Statistics and Accounting: Fifty Years of the Journal "Vestnik Statistiki" [1919–1968]). Moscow: Statistika, 1971.

Orleans, Leo A. "Chinese Statistics: The Impossible Dream." *The American Statistician* 28, no. 2 (May 1974): 47–52.

———. *Professional Manpower and Education in Communist China*. Washington, DC: U.S. Government Printing Office, 1961.

Ostrovitianov, K.V., and J. M. "The Discussion on Statistics Summed Up." *Soviet Studies* 6, no. 3 (January 1955): 321–331.

Oxford English Dictionary, online ed. Oxford: Oxford University Press.

Panikkar, K. M. *In Two Chinas: Memoirs of a Diplomat*. London: G. Allen & Unwin, 1955.

Parsons, Talcott. "The Professions and Social Structure." *Social Forces* 17, no. 4 (May 1939): 457–467.

Passin, Herbert. "Sino-Indian Cultural Relations." *The China Quarterly*, no. 7 (July–September 1961): 85–100.

Patriarca, Silvana. *Numbers and Nationhood: Writing Statistics in Nineteenth-Century Italy*. Cambridge: Cambridge University Press, 1996.

Perkins, Dwight H. "Appendix A: Reliability of Chinese Statistics." In *Market Control and Planning in Communist China*, 215–225. Cambridge, MA: Harvard University Press, 1966.

———. "Research on the Economy of the People's Republic of China: A Survey of the Field." *Journal of Asian Studies* 42, no. 2 (February 1983): 345–372.

Phalkey, Jahnavi, and Tong Lam, eds. *Science of Giants: China and India in the Twentieth Century.* *BJHS Themes* 1 (2016).

Pickering, W.S.F., and Geoffrey Walford, eds. *Durkheim's Suicide: A Century of Research and Debate.* London: Routledge, 2000.

Pierson, Paul. "Increasing Returns, Path Dependence, and the Study of Politics." *American Political Science Review* 94, no. 2 (June 2000): 251–267.

Poovey, Mary. *A History of the Modern Fact: Problems of Knowledge in the Sciences of Wealth and Society.* Chicago: University of Chicago Press, 1998.

Porter, Theodore M. "Chance Subdued by Science." *Poetics Today* 15, no. 3 (Fall 1994): 467–478.

———. *Karl Pearson: The Scientific Life in a Statistical Age.* Princeton: Princeton University Press, 2004.

———. *The Rise of Statistical Thinking, 1820–1900.* Princeton: Princeton University Press, 1986.

———. *Trust in Numbers: The Pursuit of Objectivity in Science and Public Life.* Princeton: Princeton University Press, 1995.

Rajan, Raghuram. "Professor Raghuram Rajan talks about his return to Chicago Booth." Media Relations and Communications, The University of Chicago Booth School of Business, 2 February 2017. Accessed at https://newschicagobooth-sites.uchicago.edu/newsroom/professor-raghuram-rajan-talks-about-his-return-chicago-booth

Rao, C. R. "Prasantha Chandra Mahalanobis, 1893–1972." *Biographical Memoirs of Fellows of the Royal Society* 19 (December 1973): 454–492.

Rawski, Thomas G. "Studies of China's Economy." In *A Scholarly Review of Chinese Studies in North America,* 174–191. Ann Arbor: Asia Past & Present: New Research from AAS, no. 11, 2013.

Ren Chengyou 任承佑 et al., eds. *Gongwen yu gongwen xiezuo (xiuding ban)* 公文与公文写作 (修订版) (Official documents and writing official documents [rev. ed.]). Chongqing: Xi'nan shifan daxue chubanshe, 1995.

Renmin University of China. 中国人民大学. *Tongjixue cankao ziliao—yi* 统计学参考资料—I (Reference Materials on Statistics—Vol. 1). Beijing: Renmin University, 1952.

Renmin University of China, Department of Statistics. 中国人民大学统计教研室. *Gongye tongjixue cankao ziliao* 工业统计学参考资料 (Reference materials on industrial statistics). Beijing: Renmin University Statistics Department, 1951.

———. *Maoyi tongjixue cankao ziliao—yi* 贸易统计学参考资料—I (Reference materials on trade statistics—Vol. 1). Beijing: Renmin University Statistics Department, 1951.

———. *Nongye tongjixue cankao ziliao* 农业统计学参考资料 (Reference materials on agricultural statistics). Beijing: Renmin University Statistics Department, 1952.

———. *Tongji lilun zhong de jige wenti* 统计理论中的几个问题 (Some questions in statistical theory). Beijing: Xinjianshe zazhi she, 1952.

———. *Tongjixue jiangyi (chugao), 1–6 fence* 统计学讲义(初稿), 1–6 分册 (Lectures on statistics [first draft], Vol. 1–6). Beijing: Zhongguo renmin daxue chubanshe, 1959.

———. *Tongjixue yuanli jiangyi: Shangxia ce* 统计学原理讲义: 上下册 (Lectures on statistical principles: Vols. 1 and 2). Beijing: Zhongguo renmin chubanshe, 1956.

Renmin University of China, Department of Trade Statistics. 中國人民大學, 貿易統計教研

室. *Jingji tongjixue cankao ziliao—yi* 经济统计学参考资料—I (Reference materials on economic statistics: Vol. 1). Beijing: Renmin University, 1953.

Renmin University of China, School of Statistics. 中国人民大学, 统计学院. *Xiaoyou jiniance* 校友纪念册 (Alumni Records). N.p.: Zhongguo renmin daxue chubanshe, 2007.

Riskin, Carl. *China's Political Economy: The Quest for Development since 1949*. Oxford: Oxford University Press, 1988.

———. "Maoism and Motivation: Work Incentives in China." In *China's Uninterrupted Revolution: From 1840 to the Present*, edited by Victor Nee and James Peck, 415–461. New York: Pantheon Books, 1975 [1973].

———. "Seven Questions about the Chinese Famine of 1959–1961." *China Economic Review* 9, no. 2 (1998): 111–124.

Rogaski, Ruth. *Hygienic Modernity: Meanings of Health and Disease in Treaty-Port China*. Berkeley: University of California Press, 2004.

Rosensweig, Phil. "Robert S. McNamara and the Evolution of Modern Management." *Harvard Business Review* 88, no. 12 (1 December 2010): 86–93.

Rowe, William T. "Bao Shichen (1775–1855), an Early Nineteenth-Century Agrarian Reformer." *Frontiers of History in China* 9, no. 1 (2014): 1–31.

———. *Speaking of Profit: Bao Shichen and Reform in Nineteenth-Century China*. Cambridge, MA: Harvard University Asia Center, 2018.

Rudra, Ashok. *Prasanta Chandra Mahalanobis: A Biography*. New Delhi: Oxford University Press, 1996.

Scarlett, Zachary. "China After the Sino-Soviet Split: Maoist Politics, Global Narrative and the Imagination of the World." PhD diss., Northeastern University, 2014.

Schmalzer, Sigrid. *The People's Peking Man: Popular Science and Human Identity in Twentieth-Century China*. Chicago: University of Chicago Press, 2008.

———. *Red Revolution, Green Revolution: Scientific Farming in Socialist China*. Chicago: Chicago University Press, 2016.

———. "Self-Reliant Science: The Impact of the Cold War on Science in Socialist China." In *Science and Technology in the Global Cold War*, edited by Naomi Oreskes and John Krige, 75–106. Cambridge, MA: MIT Press, 2014.

———. "Youth and the 'Great Revolutionary Movement' of Scientific Experiment in 1960s–1970s Rural China." In *Maoism at the Grassroots: Everyday Life in China's Era of High Socialism*, edited by Jeremy Brown and Matthew D. Johnson, 154–178. Cambridge, MA: Harvard University Press, 2015.

Schneider, Laurence A. *Biology and Revolution in Twentieth Century China*. Lanham, MD: Rowman & Littlefield, 2003.

Schrader, Charles R. *History of Operations Research in the United States Army*, Vol. 1: *1942–1962*. Washington, DC: Office of the Deputy Under Secretary of the Army for Operations Research, U.S. Army, 2006.

Schumpeter, Joseph A. *Ten Great Economists: From Marx to Keynes*. London: Allen and Unwin, 1952.

Schurmann, Franz. *Ideology and Organization in Communist China*. Berkeley: University of California Press, 1966.

Scott, James C. *The Art of Not Being Governed: An Anarchist History of Upland Southeast Asia*. New Haven: Yale University Press, 2009.

————. *Seeing Like a State: How Certain Schemes to Improve the Human Condition Have Failed*. New Haven: Yale University Press, 1998.

Semidevkin, N. P. 谢木杰夫金. "He qingnian tongzhi tantan tongji xuexi: Sulian tongji zhuanjia Xiemujiefujin tongzhi 1956 nian 5 yue 28 ri zai Xi'an tongji xuexiao de baogao 和青年同志谈谈统计学习: 苏联统计专家谢木杰夫金同志1956年5月28日在西安统计学校的报告 (Conversations with young comrades about studying statistics: A report on Soviet statistical expert Xiemujiefujin's visit to the Xi'an School of Statistics). *TJGZTX*, no. 14 (1956): 13–15.

Sen, Amartya Kumar. *Development as Freedom*. New York: Anchor Books, 1999.

————. *Poverty and Famines: An Essay on Entitlement and Deprivation*. Oxford: Clarendon Press, 1982.

Sen, Tansen. *India, China, and the World: A Connected History*. Lanham, MD: Rowman & Littlefield, 2017.

Seneta, E. "A Sketch of the History of Survey Sampling in Russia." *Journal of the Royal Statistical Society, Series A (General)* 148, no. 2 (1985): 118–125.

————. "Vestnik Statistiki" (Journal of Statistics). In *Encyclopedia of Statistical Sciences*. New York: John Wiley & Sons, Inc., 2004. Accessed at http://dx.doi.org/10.1002/0471667196.ess2866.pub2

Seow, Victor Kian Giap. "Carbon Technocracy: East Asian Energy Regimes and the Industrial Modern, 1900–1957." PhD diss., Harvard University, 2014.

Shanghai Institute of Finance and Economics (SIFE), Statistics Department. 上海财经学院统计系. "Pipan zichan jieji tongji zhishu de fan kexuexing he fandong benzhi" 批判资产阶级统计指数的反科学性和反动本质 (Criticizing the anti-scientific and reactionary nature of bourgeois statistical indicators). *TJGZTX*, no. 17 (1956): 30–33, 20.

Shen, Grace Yen. *Unearthing the Nation: Modern Geology and Nationalism in Republican China*. Chicago: University of Chicago Press, 2014.

Shen Zhihua 沈志华. *Sulian zhuanjia zai Zhongguo (1948–1960)* 苏联专家在中国 (1948–1960) (Soviet experts in China [1948–1960]). Beijing: Zhongguo guoji guangbo chubanshe, 2003.

Shi Jiazhen 施家珍. "Guanyu dianxing diaocha de jige wenti" 关于典型调查的几个问题 (Some questions regarding typical surveys). *TJGZTX*, no. 22 (1956): 4–6.

Shi Qing 石青. "Pipan jiben jianshe tongji gongzuo zhong tuoli shiji de xianxiang" 批判基本建设统计工作中脱离实际的现象 (Criticisms regarding statistical work in capital construction being divorced from reality). *TJGZTX*, no. 19 (1956): 1–3.

Shi Tai 施泰. "Tongji mingren lu: Jin Guobao" 统计名人录: 金国宝 (Who's who in statistics: Jin Guobao). *Tongji yu yuce* 统计与预测 (Statistics and forecasting), no. 6 (1994): 60–61.

————. "Zou Yiren" 邹依仁. *Tongji yu yuce* 统计与预测 (Statistics and forecasting), no. 6 (2001): 52–53.

Shwed, Uri, and Peter Bearman. "The Temporal Structure of Scientific Consensus Formation." *American Sociological Review* 75, no. 6 (2010): 817–840.

Si Gengsheng 司更生. "Fanlan chengzai de diaocha tongji biao" 泛滥成灾的调查统计表 (Swamped by statistical tables). *RMRB*, 12 September 1953, 3.

Singh, K. Natwar. *My China Diary 1956–88*. New Delhi: Rupa & Co., 2009.

Sismondo, Sergio. *An Introduction to Science and Technology Studies*. Chichester: Wiley-Blackwell, 2nd. ed., 2010.

Smith, Aminda M. *Thought Reform and China's Dangerous Classes: Reeducation, Resistance, and the People*. Lanham, MD: Rowman & Littlefield, 2013.

Smith-Peter, Susan. "Defining the Russian People: Konstantin Arsen'ev and Russian Statistics Before 1861." *History of Science* 45, no. 1 (2007): 47–64.

Song Tao 宋涛 ed. *20 shiji Zhongguo xueshu dadian: Jingjixue, xiace* 20世纪中国学术大点: 经济学下册 (Chinese academic canon in the twentieth century: Economics, Vol. 2). Fuzhou: Fujian jiaoyu chubanshe, 2005.

Song Xuguang 宋旭光. "Xin Zhongguo jingji tongji fazhan de lishi kaocha" 新中国经济统计发展的历史考察 (Historical investigation of the development of economic statistics in new China). *Caijing wenti yanjiu* 财经问题研究 (Research on financial and economic issues), no. 3 (2013): 21–26.

Soviet Central Statistical Administration 苏联统计局, ed. *Tongji lilun, di 1–10 zhang* 统计理论第1–10章 (Statistical theory, chapters 1–10). N.p.

———. 苏联统计局, ed. *Tongji lilun, di 11–26 zhang* 统计理论第11–26章 (Statistical theory, chapters 11–26). N.p.

Spence, Jonathan D. *The Search for Modern China*. New York: W.W. Norton 1990.

Stalin, Joseph. "Organisational Report of the Central Committee," May 24, 1924. In *Works*, Vol. 6, 1924, 199–230. Moscow: Foreign Languages Publishing House, 1953.

State Statistics Bureau (SSB). "Communiqué of the Major Statistics from the First National Population Census as Related by the State Statistics Bureau of the People's Republic of China (November 1, 1954)." *Chinese Sociology and Anthropology* 16, no. 3–4 (Spring–Summer 1984): 62–65.

———. "Ganqing chongtian, xinxin baibei: Quanguo shengshi tongji gongzuo Baoding xianchang huiyi shengli bimu" 干劲冲天, 信心百倍: 全国省市统计工作保定现场会议胜利闭幕 (Energetic and brimming with confidence: The national, provincial-, and city-level on-the-spot meeting in Baoding comes to a successful conclusion). *TJGZ*, no 14 (1958): 25.

———. *Guanyu fazhan guomin jingji de di yige wunian (1953 nian dao 1957 nian) jihua zhixing jieguo de gongbao* 关于发展国民经济的第一个五年 (1953年到1957年) 计划执行结果的公报 (Bulletin concerning the results of the first five-year [1953–1957] plan for the development of the national economy). Beijing: Zhongguo tongji chubanshe, 1958.

———. *Guanyu yijiu wusan niandu guomin jingji fazhan he guojia jihua zhixing jieguo de gongbao* 关于一九五三年度国民经济发展和国家计划执行结果的公报 (Bulletin concerning the results of national economic development and plan implementation for 1953). Beijing: Tongji chubanshe, 1953.

———. *Guanyu yijiu wusi niandu guomin jingji fazhan he guojia jihua zhixing jieguo de gongbao* 关于一九五四年度国民经济发展和国家计划执行结果的公报 (Bulletin concerning the results of national economic development and plan implementation for 1954). Beijing: Tongji chubanshe, 1955.

———. *Guanyu 1956 niandu guomin jingji jihua zhixing jieguo de gongbao* 关于1956年度国民经济计划执行结果的公报 (Bulletin concerning the results of national economic plan implementation for 1956). Beijing: Tongji chubanshe, 1957.

State Statistics Bureau (SSB). *Guojia tongjiju guanyu 1958 nian guomin jingji fazhan qing-kuang de gongbao* 国家统计局关于1958年国民经济发展情况的公报 (Bulletin concerning national economic development in 1958). Beijing: Zhongguo tongji chubanshe, 1958.

———. *Jiceng nongye tongji gongzuo jingyan huibian* 基层农业统计工作经验汇编 (Compilation of experiences carrying out grassroots agricultural statistics work). Beijing: Tongji chubanshe, 1956.

———. *Jisuanji de susuanfa* 计算机的速算法 (Methods of rapid calculation using calculating machines). Beijing: Tongji chubanshe, 1958.

———. *Re'ai women de tongji gongzuo* 热爱我们的统计工作 (Ardently love our statistical work). Beijing: Tongji chubanshe, 1956.

———. *Sulian tongji zhuanjia tanhua jilu xuanbian* 苏联统计专家谈话记录选编 (Conversations with Soviet statistical experts: Selections). Beijing: Tongji chubanshe, 1957.

———. *Tantan tongji ganbu de zixue* 谈谈统计干部的自学 (Discussions [concerning] self-study by statistical cadres). Beijing: Tongji chubanshe, 1957.

———. *Ten Great Years: Statistics of the Economic and Cultural Achievements of the People's Republic of China.* Peking: Foreign Languages Press, 1960. Translation of: *Weida de shinian: Zhonghua renmin gongheguo jingji wenhua jianshe chengjiu de tongji* 伟大的十年: 中华人民共和国经济文化建设成就的统计. Beijing: Renmin chubanshe, 1959.

———. *Tongji gongzuo zhongyao wenjian huibian* 统计工作重要文件汇编, 3 Vols. (Compilation of important documents relating to statistical work, 3 Vols.). Beijing: Tongji chubanshe, 1955, 1957, 1959.

———. *Tongji jiangzuo jiangyi: Di 1–11 jiang* 统计讲座讲义: 第1–11讲 (Statistical lectures: No.1 to 11). Beijing: Tongji chubanshe, 1957.

———. *Weida de shinian: Zhonghua renmin gongheguo jingji wenhua jianshe chengjiu de tongji* 伟大的十年: 中华人民共和国经济文化建设成就的统计 (Ten great years: Statistics of the economic and cultural achievements of the People's Republic of China). Beijing: Renmin chubanshe, 1959

———. *Woguo gongye tongji gongzuo de jingyan* 我国工业统计工作的经验 (The experience of industrial statistical work in our country). Beijing: Tongji chubanshe, 1958.

———. *Yindu Mahalanuobisi jiaoshou fangHua tongji baogaoji* 印度马哈拉诺比斯教授访华统计报告集 (Report on the visit to China of the Indian statistician Professor Mahalanobis). Beijing: Zhongguo tongji chubanshe, 1958.

———. *Zenyang zuo gongye tongji fenxi gongzuo* 怎样做工业统计分析工作 (How to carry out analysis of industrial statistics). Beijing: Zhongguo tongji chubanshe, 1956.

State Statistics Bureau, Agricultural Statistics Office 国家统计局农业统计司. "Jianli nongcun jingji diaocha gang, dagao dianxing diaocha" 建立农村经济调查纲, 大搞典型调查 (Build a program of rural economic surveys, vigorously implement typical surveys). *Jihua yu tongji* 计划与统计 (Planning and statistics), no. 9 (1959): 9–10.

Stiffler, Douglas Alden. "Building Socialism at Chinese People's University: Chinese Cadres and Soviet Experts in the People's Republic of China, 1949–1957." University of California, San Diego, PhD diss., 2002.

———. "Resistance to the Sovietization of Higher Education in China." In *Universities Under Dictatorship*, edited by John Connelly and Michael Grüttner, 213–243. University Park: Penn State University Press, 2005.

Stigler, Stephen M. *The History of Statistics: The Measurement of Uncertainty before 1900*. Cambridge, MA: Belknap Press of Harvard Univ. Press, 1986

Strauss, Julia C. *Strong Institutions in Weak Polities: State Building in Republican China, 1927–1940*. Oxford: Clarendon Press, 1998.

———, ed. "The History of the People's Republic of China, 1949–1976." *The China Quarterly*. Special Issue, no. 7. Cambridge: Cambridge University Press, 2007.

Strumilin, Stanoslav G. 斯特鲁米林. *Tongjixue* 统计学 (Statistics). Beijing: Tongji chubanshe, 1957.

Su Fan 苏翻, ed. *Xin gongwen shouce* 新公文手册 (New handbook for official documents). Shanghai: Beixin shuju, 1953.

Thakurdas, Purushotamdas et al. *Memorandum Outlining a Plan of Economic Development for India*. Harmondworth, Eng.: Penguin Books, 1945.

Tang Qinghong 唐庆洪. "Zichan jieji tongji zhishu lilun pipan" 资产阶级统计指数理论批判 (Criticism of the bourgeois theory of index numbers). *Caijing yanjiu* 财经研究 (Journal of finance and economics), no. 1 (1956): 13–19.

Thampi, Madhavi, ed. *India and China in the Colonial World*. New Delhi: Social Science Press, 2005.

Thaxton, Ralph. *Catastrophe and Contention in Rural China: Mao's Great Leap Forward Famine and the Origins of Righteous Resistance in Da Fo Village*. Cambridge: Cambridge University Press, 2008.

Thompson, Malcolm. "The Birth of the Chinese Population: A Study in the History of Governmental Logics." University of British Columbia, PhD diss., 2013.

Thompson, Roger R., tr. *Report from Xunwu*. Stanford: Stanford University Press, 1990.

Tianjin Municipal Statistics Bureau 天津市统计局. "Siying xiaoxing gongchang shougongye dianxing diaocha gongzuo" 私营小型工场手工业典型调查工作 (Typical survey of the private small-scale handicraft industry), *TJGZTX*, no. 9 (1953): 21–22, 27.

Tie Hua 铁华 and Liu Xin 刘新. "Lun tongjixue de duixiang ji qi kexue jichu: Ping Gou Shisheng zhu 'Tongjixue yuanli'" 论统计学的对象及其科学基础:评勾适生著"统计学原理" (Discussion of the object and theoretical basis of statistics: A review of Gou Shisheng's *Principles of Statistics*). *Xin jianshe* 新建设 (New construction) 4, no. 5 (1951): 60–63.

Tobin, James. *Essays in Economics: Macroeconomics*. New York: North-Holland, 1972; reprint, Cambridge, MA: MIT Press, 1987.

Tooze, J. Adam. *Statistics and the German State, 1900–1945: The Making of Modern Economic Knowledge*. Cambridge: Cambridge University Press, 2001.

Travers, S. Lee. "Bias in Chinese Economic Statistics: The Case of the Typical Sample Investigation." *The China Quarterly*, no. 91 (September 1982): 478–485.

Trescott, Paul B. *Jingji Xue: The History of the Introduction of Western Economic Ideas into China, 1850–1950*. Hong Kong: Chinese University Press, 2007.

U, Eddy. "The Making of *Zhishifenzi*: The Critical Impact of the Registration of Unemployed Intellectuals in the Early PRC." *The China Quarterly*, no. 173 (March 2003): 100–121.

United Nations Statistical Office. "The Preparation of Sampling Survey Reports." *Statistical Papers, Series C, No. 1 (Revised)*. Lake Success, NY: 1950.

———. "Recommendations for the Preparation of Sampling Survey Reports (Provisional Issue)." *Statistical Papers*, Series C, No. 1, Rev. 2. New York, 1964.

United States Bureau of Foreign and Domestic Commerce. *National Income, 1929–32.* Washington, DC: U.S. Government Printing Office, 1934.

Vaidyanathan, A., C. R. Rao, T. N. Srinivasan, and J. N. Bhagwati. "Pitambar Pant: An Appreciation." *Economic and Political Weekly* 8, no. 17 (28 April 1973): 774–777.

Van de Ven, Hans J. *Breaking with the Past: The Maritime Customs Service and the Global Origins of Modernity in China.* New York: Columbia University Press, 2014.

Vogel, Ezra F. *Canton Under Communism: Programs and Politics in a Provincial Capital, 1949–1968.* Cambridge, MA: Harvard University Press, 1969.

———. "From Revolutionary to Semi-Bureaucrat: The 'Regularisation' of Cadres." *The China Quarterly*, no. 29 (January–March 1967): 36–60.

Vukovich, Daniel F. *China and Orientalism: Western Knowledge Production and the P.R.C.* London: Routledge, 2012.

Wakelam, Randall Thomas. *The Science of Bombing: Operational Research in RAF Bomber Command.* Toronto: University of Toronto Press, 2009.

Walker, Kenneth R. *Food Grain Procurement and Consumption in China.* Cambridge: Cambridge University Press, 1984.

Wallace, Jeremy L. "Juking the Stats? Authoritarian Information Problems in China." *British Journal of Political Science* 46 (2016): 11–29.

Wang Huaxin 王华新. "Yige fengfu duocai de tongji gongzuo zhanlan" 一个丰富多彩的统计工作展览 (A richly colored exhibition of statistical work). *TJGZ*, no. 14 (1958): 28–29.

Wang Jianzhen 王健真. "Dui zichan jieji zhishu 'lilun' de pipan" 对资产阶级指数'理论'的批判 (Criticism of the bourgeois theory of index numbers). *TJGZTX*, no. 11 (1955): 40–43.

Wang Sihua 王思华. "Jieshao Yindu tongji gongzuo, jiji kaizhan quanguo chouyang diaocha" 介绍印度统计工作, 积极开展全国抽样调查 (Introducing India's statistical work, actively carrying out a national sample survey). *TJGZ*, no. 6 (1957): 8–10, 17.

———. "Kaocha Yindu tongji gongzuo de baogao" 考察印度统计工作的报告 (Report on Indian statistical work). Report delivered at the Sixth Annual Statistical Work Conference, Beijing, September 1957. In SSB, *Tongji gongzuo zhongyao wenjian huibian: Di sanji* 统计工作重要文件汇编: 第三辑 (Compilation of important documents relating to statistical work: Vol. III), 90–97. Beijing: Tongji chubanshe, November 1959.

———. "Sannianlai dongbei tongjiju gongzuo zongjie" 三年来东北统计局工作总结 (Summary of statistical work performed by the Northeast Statistical Bureau during the past three years). In Northeast Statistics Bureau. *Tongji gongzuo: Di wu xuanji* 统计工作: 第五选集 (Statistical work: Fifth anthology), 1–10. Shenyang: Dongbei caijing chubanshe, 1953.

———. "Tongji gongzuo de renwu" 统计工作的任务 (The tasks of statistical work). In Northeast Statistics Bureau. *Tongji gongzuo: Di yi xuanji* 统计工作: 第一选集 (Statistical work: First anthology), 7–14. Shenyang: Dongbei caijing chubanshe, 1951.

———. *Wang Sihua tongji lunwenji* 王思华统计论文集 (Collected statistical works of Wang Sihua). Beijing: Zhongguo tongji chubanshe, 1986.

———. *Zibenlun jieshuo* 资本论解说 (Theory of *Das Kapital*) Beijing: Shenghuo, dushu, xinzhi sanlian shudian, 1950.

Wang Sili 王思立. "Pipan wo zai tongjixue zhuzuo zhong de zichan jieji sixiang" 批判我在统

计学著作中的资产阶级思想 (Criticism of the presence of bourgeois ideology in my statistical works). *TJGZ*, no. 10 (1957): 27–30, 26.

Wang, Tong-eng. *Economic Policies and Price Stability in China*. Berkeley: Institute of East Asian Studies, University of California, 1980.

Wang Xiangyuan 王相源. "Wo yiding ba gongzuo zuode genghao" 我一定把工作做得更好 (I must do my job better). *TJGZTX*, no. 10 (1955): 52–53. Reproduced in RWTG, 38.

Wang Xinming汪新明. "Zai gongye qiye zhong kaizhan dianxing diaocha de tihui" 在工业企业中开展典型调查的体会 (The experience of implementing typical surveys in industrial enterprises). *Jihua yu tongji* 计划与统计 (Planning and statistics), no. 10 (1959): 34–36.

Wang Youjun 王幼军and Andrea Bréard. "Tongjixue zai jindai Zhongguo de yanbian: Cong shehui tongji dao shuli tongji" 统计学在近代中国的演变: 从社会统计到数理统计 (The evolution of statistics in modern China: From social statistics to mathematical statistics). *Shanghai jiaotong daxue xuebao (zhexue shehui kexue ban)* 上海交通大学学报 (哲学社会科学版) (Newsletter of Shanghai Communications University [Philosophy and social sciences issue] 18, no. 3 (2010): 61–67, 74.

Wang, Zhengxian 王正宪. "Guanyu jiben tongjixue yuanli jiaocai de yixie yijian" 关于几本统计学原理教材的一些意见 (A few recommendations for textbooks on basic statistics)." *TJGZTX*, no. 16 (1956): 31–33.

Wang, Zuoyue. "The Chinese Developmental State During the Cold War: The Making of the 1956 Twelve-Year Science and Technology Plan." *History and Technology*, no. 31, 3 (2015): 180-205.

———. "The Cold War and the Reshaping of Transnational Science in China." In *Science and Technology in the Global Cold War*, edited by Naomi Oreskes and John Krige, 343–370. Cambridge, MA: MIT Press, 2014.

Ward, Michael. *Quantifying the World: UN Ideas and Statistics*. Bloomington: Indiana University Press, 2004.

Watts, Duncan. *Everything Is Obvious: How Common Sense Fails*. New York: Crown Business, 2011.

Wemheuer, Felix. *Famine Politics in Maoist China and the Soviet Union*. New Haven: Yale University Press, 2014.

Westad, Odd Arne, ed. *Brothers in Arms: The Rise and Fall of the Sino-Soviet Alliance, 1945–1963*. Washington, DC: Woodrow Wilson Press, and Stanford: Stanford University Press, 1998.

Wu Cangping 邬沧萍. *Wu Cangping zixuanji* 邬沧萍自选集 (Personally selected writings of Wu Cangping). Beijing: Renmin daxue chubanshe, 2007.

Wu Hui 吴辉. "Zhongguo de tongji diaocha fangshi" 中国的统计调查方式 (China's statistical survey methods). In *Zhongguo shehui zhuyi tongji gongzuo de jianli yu fazhan* 中国社会主义统计工作的建立与发展 (The establishment and development of socialist statistical work in china), edited by State Statistics Bureau, Institute of Statistical Science, 31–41. Beijing: Zhongguo tongji chubanshe, 1985.

Xin Hua 辛华, ed. *Eyu xingming yiming shouce* 俄语姓名译名手册 (Handbook of translations of Russian names). Beijing: Shangwu yinshuguan, 1982.

Xing Hao 邢浩. "Jingxin mouhua: Liu Shaoqi de diaocha yanjiu zhi dao: Yi 1957 nian nanxia diaoyan wei li" 精心谋划: 刘少奇的调查研究之道: 以1957年南下调研为例 (Meticulous planning: A discussion of Liu Shaoqi's approach to conducting surveys, using the southern surveys of 1957 as a case study). *Dang de wenxian* 党的文献 (Party documents), No. 6 (2014): 116–118.

Xu Qian 徐前. "Tongji xueshuo fazhan bianhua de lunkuo" 统计学说发展变化的轮廓 (An outline of the developments and changes in statistical theory). In *Tongjixue yuanli ziliao huibian* 统计学原理资料汇编 (Compilation of materials on statistical theory), edited by Yang Zengwu, Fu Chensheng, and Xu Qian 扬曾武, 傅春生, 徐前, 1–35. Beijing: Zhong-yang guangbo dianshi daxue chubanshe, 1983.

———— et al., eds. *Jingji tongjixue jiangyi* 经济统计学讲义 (Lectures on statistical economics). Beijing: *Tongji chubanshe*, 1957.

Xu Qian, Jiang Zhao Lin Fude, Zheng Yao, Liu Xin, and Zhou Fugong 徐前, 江昭, 林富德, 郑尧, 刘新, 周复恭. *Tongji lilun yiban wenti jianghua* 统计理论一般问题讲话 (Common problems in statistical theory). Beijing: Tongji chubanshe, 1956.

Xu Qian 徐前 and Liu Xin 刘新. "Guanyu zichan jieji tongji lilun de pipan" 关于资产阶级统计理论的批判 (Criticisms regarding bourgeois statistical theory). *Xin jianshe* 新建设 (New construction), no. 1 (1953): 14–20, 2; and no. 2 (1953): 25–28.

————. "Guanyu zichan jieji tongji lilun de pipan" 关于资产阶级统计理论的批判 (Criticisms regarding bourgeois statistical theory). *TJGZTX*, no. 8 (1955): 28–34, 21.

Xu, Xianchun. "The Establishment, Reform, and Development of China's System of National Accounts." *Review of Income and Wealth* Series 55, Special Issue no. 1 (July 2009): 442–465.

Xu Youchun 徐友春. *Minguo renwu dacidian* 民国人物大辞典 (Biographical dictionary of Republican China). Shijiazhuang: Hebei renmin chubanshe, 1991.

Xue Muqiao 薛暮桥. "Di yige wunian jihua qijian woguo tongji gongzuo de chubu jingyan he jinhou renwu" 第一个五年计划期间我国统计工作的初步经验和今后任务 (The preliminary experience and recent tasks of statistical work during our country's first five-year plan period). *TJGZ*, no. 21 (1957): 1–21.

————. "Kuzhan san yue, gaibian quanguo tongji gongzuo mianmao: Shixian quanguo tongji gongzuo dayuejin!" 苦战三月, 改变全国统计工作面貌: 实现全国统计工作大跃进! (Three months of hard work, change the face of statistical work: realize a Great Leap Forward in national statistical work!). *TJGZ*, no. 14 (1958): 2–7.

————. "Woguo de jihua gongzuo he tongji gongzuo bixu xiang Sulian xuexi" 我国的计划工作和统计工作必须向苏联学习 (Our nation's planning and statistics work must learn from the Soviet Union). *RMRB*, 12 November 1957, 7.

————. *Xue Muqiao: Huiyilu* 薛暮桥: 回忆录 (Xue Muqiao: Memoirs). Tianjin: Tianjin ren-min chubanshe, 2nd. ed, 2006.

Yang Bo 杨波. "Zenyang zuo dianxing diaocha" 怎样做典型调查 (How to carry out typical surveys). *Jihua yu tongji* 计划与统计 (Planning and statistics), no. 1 (1959): 36–37.

————. "Zenyang zuo dianxing diaocha" 怎样做典型调查 (How to carry out typical surveys). *Jihua yu tongji* 计划与统计 (Planning and statistics), no. 2 (1959): 36–37.

————. "Zenyang zuo dianxing diaocha" 怎样做典型调查 (How to carry out typical surveys). *Jihua yu tongji* 计划与统计 (Planning and statistics), no. 3 (1959): 24–25.

————. "Zenyang zuo dianxing diaocha" 怎样做典型调查 (How to carry out typical surveys). *Jihua yu tongji* 计划与统计 (Planning and statistics), no. 4 (1959): 40–41.

———. "Zenyang zuo dianxing diaocha" 怎样做典型调查 (How to carry out typical surveys). *Jihua yu tongji* 计划与统计 (Planning and statistics), no. 5 (1959): 36–37.

Yang, Dali L. *Calamity and Reform: State, Rural Society, and Institutional Change since the Great Leap Famine*. Stanford: Stanford University Press, 1996.

———. "Surviving the Great Leap Famine: The Struggle Over Rural Policy, 1958–1962." In *New Perspectives on State Socialism in China*, edited by Timothy Cheek and Tony Saich, 262–302. Armonk, NY: M.E. Sharpe, 1997.

Yang, Dali L., Huayu Xu, and Ran Tao. "A Tragedy of the *Nomenklatura*? Career Incentives, Political Loyalty and Political Radicalism during China's Great Leap Forward." *Journal of Contemporary China* 23, no. 89 (2014): 864–883.

Yang, Dennis Tao. "China's Agricultural Crisis and Famine of 1959–1961: A Survey and Comparison to Soviet Famines." *Comparative Economic Studies* 50, no. 1 (March 2008): 1–29.

Yang Jianbai 杨坚白. "Sulian zhuanjia bangzhu women jianli le kexue de tongji gongzuo" 苏联专家帮助我们建立了科学的统计工作 (Soviet experts have helped us establish scientific statistical work). *RMRB*, 5 February 1953, 3.

Yang Jisheng 杨继绳. *Mubei: Zhongguo liushi niandai dajihuang jishi* 墓碑: 中国六十年代大饥荒纪实. Hong Kong: Cosmos Books, 2008. Published in English as *Tombstone: The Great Chinese Famine, 1958–1962*, tr. Stacy Mosher and Guo Jian. New York: Farrar, Straus and Giroux, 2012.

Yang Kuisong 杨奎松. "Mao Zedong shi ruhe faxian da jihuang de" 毛泽东是如何发现大饥荒的? (How did Mao Zedong discover the famine?). *Jianghuai wenshi* 江淮文史 (Jiangsu-Anhui literature and history), no. 3 (2014): 4–29.

———. "The Sino-Soviet Alliance and Nationalism: A Contradiction." Parallel History Project on NATO and the Warsaw Pact, The Cold War History of Sino-Soviet Relations, June 2005. Accessed at http://www.php.isn.ethz.ch/lory1.ethz.ch/publications/areastudies /documents/sinosov/Kuisong.pdf. Originally published in the English edition of *Social Sciences in China* 16, no. 2 (Summer 2005): 86–99.

Yang Zengwu, Fu Chensheng, and Xu Qian 扬曾武, 傅春生, 徐前, eds. *Tongjixue yuanli ziliao huibian* 统计学原理资料汇编 (Compilation of materials on statistical theory). Beijing: Zhongyang guangbo dianshi daxue chubanshe, 1983.

Yao Naiqiang, ed. *Hanying shuangjie Xinhua zidian* 汉英双解: 新华字典 (Xinhua dictionary of English and Chinese). Beijing: Shangwu yinshuguan, 2000.

Yates, Frank, and Kenneth Mather. "Ronald Aylmer Fisher, 1890–1962." *Biographical Memoirs of Fellows of the Royal Society* 9 (November 1963): 91–129.

Yu Shengqi 俞圣祺. "Zai Sulian zhuanjia bangzhu xia de Zhongguo renmin daxue" 在苏联专家帮助下的中国人民大学 (The People's University under the aid of the Soviet experts). *RMRB*, 8 November 1954, 2.

Yu Yue 于越. "Kuaiji dang eryi yi" 会计当而已矣. (If accounting is proper, then it is enough). *Tongji yanjiu* 统计研究 (Statistical research), no. 1 (1986): 74.

Yue, Meng. "Hybrid Science versus Modernity: The Practice of the Jiangnan Arsenal, 1864–1897." *East Asian Science, Technology, and Medicine*, no.16 (1999): 13–52.

Yugow, Alexander. "Economic Statistics in the U.S.S.R." *The Review of Economics and Statistics* 29, no. 4 (November 1947): 242–246.

Zarkovic, S. S. "Note on the History of Sampling Methods in Russia." *Journal of the Royal Statistical Society. Series A (General)* 119, no. 3 (1956): 336–338.

Zelin, Madeleine. *The Magistrate's Tael: Rationalizing Fiscal Reform in Eighteenth- Century Ch'ing China.* Berkeley: University of California Press, 1984.

Zhak, D., N. Koshliak, and I. Maximov, (Жак, Д., Н. Кошляк, И. Максимов). "Opyt makhanizatsii ucheta v kolkhozakh" (The experience of the mechanization of accounting in Kolkhoz)." *Vestnik Statistiki* (Journal of statistics), no. 2 (1954): 63–68.

Zhang Minru 张敏如. "Dui 'nongye tongjixue' yishu de jidian yijian" 对 "农业统计学" 一书的几点意见 (Some suggestions concerning the book *Agricultural Statistics*). *TJGZTX*, no. 12 (1956): 26–29.

Zhang Qingming 张清明. *Wenshuxue ji shiyong gongwen (xiuding ben)* 文书学及实用公文 (修订本) (Paperwork and practical official documents) rev. ed. Wuhan: Wuhan daxue chubanshe, 1984.

Zhang Zhihong 张志鸿, ed. "Guanyu 'Dui nongye tongjixue yishu de jidian yijian de shangque'" 关于 '对农业统计学' 一书的几点意见的商榷 (A discussion on the opinions about the book *Agricultural statistics*). *TJGZ*, no. 5 (1957): 27–30, 33.

———, ed. *Nongye tongjixue* 农业统计学 (Agricultural statistics). Shanghai: Lixin kuaiji tushu yongpinshe, 1953.

Zhang Zhiji 张知几, ed. *Dazhong tongji* 大众统计 (People's statistics). Shanghai: Xin zhishi chubanshe, 1956.

Zhang Zhiji 张知几 and Zou Yiren 邹依仁, eds. *Dazhong tongjixue* 大众统计学 (People's statistics). Shanghai: Lixin kuaiji tushu yongpinshe chuban, 1953.

Zhonggong zhongyang zuzhibu yu Zhonggong zhongyang dangshi yanjiushi 中共中央组织部与中共中央党史研究室 (Central Organization Department of the Chinese Communist Party and Central Party Office of the Chinese Communist Party). *Zhongguo gongchandang lijie zhongyang weiyuan da cidian, 1921–2003* 中国共产党历届中央委员大辞典, 1921–2003 (Dictionary of the successive Central Committees of the Chinese Communist Party, 1921–2003). Beijing: Zhonggong dangshi chubanshe, 2004.

Zhongyang Makesi Engesi Liening Sidalin zhuzuo bianyiju 中央马克思恩格斯列宁斯大林著作编译局 (Central Compilation Bureau of the Works of Marx, Engels, and Stalin), ed. *Zhengzhi jingjixue jiaokeshu: Yiming huibian* 政治经济学教科书: 译名汇编 (Political economy textbook: A compilation of translated names). Beijing: Shidai chubanshe, 1956.

Zhongyang renmin zhengfu zhengwuyuan 中央人民政府政务院 (State Council of the Central People's Government). *Gongwen chuli zanxing banfa* 公文处理暂行办法 (Interim measures for processing official documents). Beijing: Renmin chubanshe, 1951.

Zhou Enlai [Chou En-lai] 周恩来. "On the Question of Intellectuals." In *Communist China 1955–1959: Policy Documents and Analysis,* foreword by Robert R. Bowie and John K. Fairbank, 128–144. Cambridge, MA: Harvard University Press, 1963.

Zhou, Xun, ed. *The Great Famine in China, 1958–1962: A Documentary History.* New Haven: Yale University Press, 2012.

Zhu Hong'en 朱鸿恩. "Du 'Sulian tongji zhuanjia tanhua jilu xuanbian'" 读 "苏联统计专家谈话记录选编" (Reading "Selected lectures by Soviet Statistical Experts"). *TJGZ*, no. 21 (1957): 32–33.

Zhu Junyi 朱君毅. *Minguo shiqi de zhengfu tongji gongzuo* 民国时期的政府统计工作 (Government statistical work during the Republican Period). Beijing: Zhongguo tongji chubanshe, 1988. See also Chu, Jennings P.

Zou Yiren 邹依仁. "Guanyu zichan jieji tongji xueshu sixiang de ziwo pipan" 关于资产阶级统计学术思想的自我批判 (Self-criticism regarding bourgeois statistical thought). *TJG-ZTX*, no. 8 (1956): 25–28.

———. "Lun tongji shi yimen shehui kexue" 论统计是一门社会科学 (Discussion of statistics as a social science). *Caijing yanjiu* 财经研究 (Journal of finance and economics), no. 4 (1957): 45–49.

Zou Yiren 鄒依仁, Zheng Deru 郑德如, Yang Hui 杨惠, Xie Jia 谢嘉, Zhang Weiming 张维铭, Jiang Shiju 蒋士驹, Wang Sili 王思立, Chu Fengyi 褚鳳仪, and Jin Guobao 金国宝. "Guanyu shuli tongji zai shehui jingji tongji kexue zhong de diwei yu zuoyong wenti de taolun (Shanghai caijing xueyuan tongjixi xueshu zuotanhui jilu)" 关于数理统计在社会经济统计科学中的地位与作用问题的讨论(上海财经学院统计系学术座谈会记录) (Discussion regarding the status and use of mathematical statistics in statistical work in a socialist economy [Record of a forum in the Department of Statistics in the Shanghai Institute of Finance and Economics]), *Caijing yanjiu* 财经研究 (Journal of finance and economics), no. 3 (1958): 57–62, 65.

Zu Ting'an 祖廷安. "Siben gongye tongji shuji" 四本工业统计书籍 (Four books on industrial statistics). *TJGZ*, no. 1 (1958): 25.

A NOTE ON THE TYPE

This book has been composed in Arno, an Old-style serif typeface in the
classic Venetian tradition, designed by Robert Slimbach at Adobe.